Fracture of
Functionally Graded Materials

Elsevier Science Internet Homepage - http://www.elsevier.com

Consult the Elsevier homepage for full catalogue information on all books, journals and electronic products and services.

Elsevier Titles of Related Interest

CARPINTERI
Minimum Reinforcement in Concrete Members.
ISBN: 008-043022-8

FRANÇOIS & PINEAU
From Charpy to Present Impact Testing.
ISBN: 0-08-043970-5

FUENTES *ET AL.*
Fracture Mechanics: Applications and Challenges.
ISBN: 008-043699-4

JONES
Failure Analysis Case Studies II.
ISBN: 008-043959-4

MACHA *ET AL.*
Multiaxial Fatigue and Fracture.
ISBN: 008-043336-7

MARQUIS & SOLIN
Fatigue Design of Components.
ISBN: 008-043318-9

MARQUIS & SOLIN
Fatigue Design and Reliability.
ISBN: 008-043329-4

MOORE *ET AL.*
Fracture Mechanics Testing Methods for Polymers, Adhesives
and Composites.
ISBN: 008-043689-7

MURAKAMI
Metal Fatigue Effects of Small Defects and Nonmetallic
Inclusions
ISBN: 008-044064-9

RAVICHANDRAN *ET AL.*
Small Fatigue Cracks: Mechanics, Mechanisms & Applications.
ISBN: 008-043011-2

RÉMY and PETIT
Temperature-Fatigue Interaction.
ISBN: 008-043982-9

TANAKA & DULIKRAVICH
Inverse Problems in Engineering Mechanics II.
ISBN: 008-043693-5

VOYIADJIS *ET AL.*
Damage Mechanics in Engineering Materials.
ISBN: 008-043322-7

VOYIADJIS & KATTAN
Advances in Damage Mechanics: Metals and Metal Matrix Composites.
ISBN: 008-043601-3

WILLIAMS & PAVAN
Fracture of Polymers, Composites and Adhesives.
ISBN: 008-043710-9

Related Journals

Free specimen copy gladly sent on request. Elsevier Science Ltd, The Boulevard, Langford Lane, Kidlington, Oxford, OX5 1GB, UK

Acta Metallurgica et Materialia
Cement and Concrete Research
Composite Structures
Computers and Structures
Corrosion Science
Engineering Failure Analysis
Engineering Fracture Mechanics
European Journal of Mechanics A & B
International Journal of Fatigue
International Journal of Impact Engineering
International Journal of Mechanical Sciences
International Journal of Non-Linear Mechanics
International Journal of Plasticity
International Journal of Pressure Vessels & Piping
International Journal of Solids and Structures
Journal of Applied Mathematics and Mechanics
Journal of Construction Steel Research
Journal of the Mechanics and Physics of Solids
Materials Research Bulletin
Mechanics of Materials
Mechanics Research Communications
NDT&E International
Scripta Metallurgica et Materialia
Theoretical and Applied Fracture Mechanics
Tribology International
Wear

To Contact the Publisher

Elsevier Science welcomes enquiries concerning publishing proposals: books, journal special issues, conference proceedings, etc. All formats and media can be considered. Should you have a publishing proposal you wish to discuss, please contact, without obligation, the publisher responsible for Elsevier's mechanics and structural integrity publishing programme:

Dean Eastbury
Senior Publishing Editor, Materials Science & Engineering
Elsevier Science Ltd
The Boulevard, Langford Lane
Kidlington, Oxford
OX5 1GB, UK

Phone: +44 1865 843580
Fax: +44 1865 843920
E.mail: d.eastbury@elsevier.com

General enquiries, including placing orders, should be directed to Elsevier's Regional Sales Offices – please access the Elsevier homepage for full contact details (homepage details at the top of this page).

Fracture of Functionally Graded Materials

Editors

R.H. Dodds
University of Illinois, USA

K.-H. Schwalbe
GKSS, Germany

Guest Editor

G.H. Paulino
University of Illinois at Urbana-Champaign, USA

Reprinted from the journal
Engineering Fracture Mechanics, Volume 69 Issue 14-16, 2002.

Each paper appears in the same format as it was published in the
journal; citations should be made using the original publication details.

2002

PERGAMON
An Imprint of Elsevier Science
Amsterdam – Boston – London – New York – Oxford – Paris
San Diego – San Francisco – Singapore – Sydney – Tokyo

ELSEVIER SCIENCE Ltd
The Boulevard, Langford Lane
Kidlington, Oxford OX5 1GB, UK

Reprinted from Engineering Fracture Mechanics, Volume 69 Issue 14-16, 2002

First edition 2002

Library of Congress Cataloging in Publication Data
A catalog record from the Library of Congress has been applied for.

British Library Cataloguing in Publication Data
A catalogue record from the British Library has been applied for.

ISBN: 0-08-044160-2

⊗ The paper used in this publication meets the requirements of ANSI/NISO Z39.48-1992 (Permanence of Paper).
Printed in by Great Britain by Polestar Wheatons Ltd, Exeter

Engineering Fracture Mechanics

Volume 69 Numbers 14–16 2002

CONTENTS

FRACTURE OF FUNCTIONALLY GRADED MATERIALS

Guest Editor: G. H. Paulino

PERGAMON

Engineering Fracture Mechanics 69 (2002) 1519–1520

Engineering Fracture Mechanics

www.elsevier.com/locate/engfracmech

Editorial

Fracture of functionally graded materials

Scientific research on functionally graded materials (FGMs) considers, in a large sense, functions of gradients in materials comprising thermodynamic, mechanical, chemical, optical, electromagnetic, and/or biological aspects. In essence, FGMs are characterized by spatially varied microstructures created by non-uniform distributions of the reinforcement phase with different properties, sizes and shapes, as well as by interchanging the role of reinforcement and matrix materials in a continuous manner, as illustrated by Figs. 1 and 2. The second figure shows an example of a large-bulk ceramic/metal engineering FGM. This new concept of engineering the material microstructure marks the beginning of a paradigm shift in the way we think about materials and structures as it allows one, due to recent advances in material processing, to fully integrate material and structural design considerations.

In the present edition, a collection of technical papers is presented that represents current research interests with regard to the fracture behavior of FGMs. The papers include a balance amongst theoretical, computational, and experimental techniques. All the participants have contributed to advancing the state of knowledge in FGMs, and this special issue demonstrates that our understanding of fracture of FGMs is becoming increasingly clear. However, it also indicates areas for further development, such as constraint

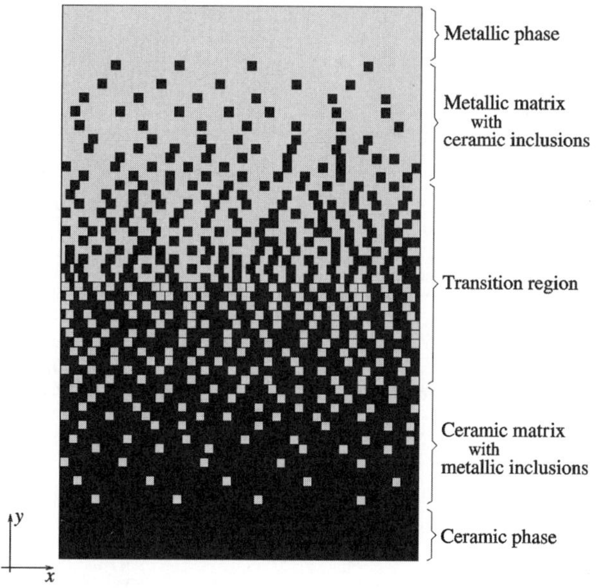

Fig. 1. Schematic illustration of an FGM with continuously graded microstructure.

0013-7944/02/$ - see front matter © 2002 Published by Elsevier Science Ltd.
PII: S0013-7944(02)00045-0

Fig. 2. Detail of an 11-layer $ZrO_2(3Y)$/stainless-steel FGM of 100 mm diameter and 15 mm thickness. The illustration above is a scanning electron micrograph (SEM) of a layer containing 10% zirconia and 90% stainless-steel volume fraction. Notice the submicron ceramic particles clustered around the metal microparticles. The material was synthesized using spark plasma sintering technique (SPS) by Sumitomo Co. (Japan). The SEM was obtained by graduate student Zhaoxu Dong at the Center for Microanalysis of Materials, University of Illinois, which is partially supported by the US Department of Energy under grant DEFG02-96-ER45439.

effects full (i.e. for the entire range of material composition) experimental characterization of engineering FGMs under static and dynamic loading, development of fracture criteria with predictive capability, multiphysics and multiscale (space and time) failure considerations, and connection of research with industrial applications. In fact, this latter aspect needs to be emphasized so that FGMs can find wide use in engineering applications.

The Editor would like to thank Prof. Dodds for his invitation to assemble this set of papers on Fracture of FGMs, the authors and reviewers for their scientific contributions and constructive criticism, and the staff of Elsevier for their timely publication of this special issue.

Glaucio H. Paulino
Guest Editor
Newmark Laboratory
Department of Civil and Environmental Engineering
University of Illinois at Urbana-Champaign
205 N. Mathews Ave.
Urbana, IL 61801, USA
Tel.: +1-217-333-3817
fax: +1-217-265-8041
E-mail address: paulino@uiuc.edu
URL: http://cee.ce.uiuc.edu/paulino

Engineering Fracture Mechanics 69 (2002) 1521–1555

PERGAMON

Engineering Fracture Mechanics

www.elsevier.com/locate/engfracmech

Statistical fracture modeling: crack path and fracture criteria with application to homogeneous and functionally graded materials

T.L. Becker Jr. [a], R.M. Cannon [b], R.O. Ritchie [b,*]

[a] *Department of Mechanical Engineering, University of California, Berkeley, CA 94720, USA*
[b] *Department of Materials Science and Engineering, University of California, Berkeley, and Materials Sciences Division, Lawrence Berkeley National Laboratory, Berkeley, CA 94720, USA*

Received 19 February 2001; received in revised form 22 January 2002; accepted 23 January 2002

Abstract

Analysis has been performed on fracture initiation near a crack in a brittle material with strength described by Weibull statistics. This nonlocal fracture model allows for a direct correlation between near crack-tip stresses and failure. Predictions are made for both the toughness and average fracture initiation angle of a crack under mixed-mode loading. This is pertinent for composites and is especially interesting for functionally graded materials (FGMs), where the stress and strength fields vary from the homogeneous form away from the crack tip. Both analytic and finite element analyses of FGMs reveal that gradients in Weibull scaling stress $\sigma_0(x, y)$ usually lead to a dramatic decrease of initiation fracture toughness; moreover, gradients normal to the crack result in a crack growing toward the weaker material. When comparing FGMs with gradients in Young's modulus in the direction of the crack path, $E(x)$, and the same stress-intensity factor K, the crack growing into the steeper negative gradient will be tougher, if m, the Weibull modulus, is low; with growth in the stiff direction, the effect is opposite. These effects offset the higher-stress intensity for cracks growing into more compliant material, and the crack-tip shielding when growing into a stiffer material based upon expectations for the applied load. Perpendicular gradients in modulus can cause a far-field mode I loading to produce mixed-mode loading of the crack tip and other asymmetric adjustments in the stress field; the gradient induces non-coplanar cracking that depends strongly on m. The distribution of damage near a crack tip will vary strongly with m. For high m materials, failure is dominated by the very near-tip parameters, and effects of gradients are minimized. With low m, distributed damage leading to toughening can be exaggerated in FGMs. Finally, consideration is given to the role of several higher-order terms in the stress field.
© 2002 Elsevier Science Ltd. All rights reserved.

Keywords: Functionally graded materials; Fracture toughness; Crack kinking; Mixed-mode fracture; Weibull statistics

* Corresponding author. Tel.: +1-510-486-5798; fax: +1-510-486-4881.
E-mail address: roritchie@lbl.gov (R.O. Ritchie).

Nomenclature

x, y, z	Cartesian coordinates, crack lies along $y = 0$ in x-direction
r, θ	polar coordinates
ξ, ς	coordinate variables
$\sigma_{xx}, \sigma_{xy}, \sigma_{yy}$	stress components referenced to $\{x, y\}$ coordinate system
σ_{yield}	yield strength
σ_1	maximum principal stress
$\sigma_{\theta\theta}$	Hoop stress
a	crack length
B	characteristic thickness in z direction
$K_{\text{I}}, K_{\text{II}}$	mode I and mode II stress-intensity factors
K_{Ic}	plane-strain fracture toughness value
G	strain energy release rate
G_{c}	critical strain energy release rate at fracture
ψ	phase angle of crack tip, $\psi = \tan^{-1}(K_{\text{II}}/K_{\text{I}})$
$f^{\text{I}}, f^{\text{II}}$	nondimensional mode I and II functions of θ for elastic crack stresses
f_1^{ψ}	nondimensional function of θ and ψ for elastic crack principal stress
Φ	global failure probability
K_{Φ}, G_{Φ}	probable fracture toughness, K or G, to achieve failure probability Φ
$p(\sigma)$	failure probability in limit of small volume
p^*	volume-weighted failure probability in crack-tip field
m	Weibull modulus
σ_0	Weibull scaling stress
σ_{u}	Weibull cutoff stress
g	function to describe dependence of property variation on location
b, c	gradient parameter, value at crack tip
E	Young's modulus
v	Poisson's ratio
$\bar{x}, \bar{y}$	coordinates of average location of crack initiation site
ϕ	angle of average crack initiation event, $\tan^{-1}(\bar{y}/\bar{x})$
f	percentage of fracture initiations occurring between $-\pi$ and θ_f
θ_f	angle defining the fth-percentile of fracture probability
$\bar{r}$	average distance of crack initiation
r_x, r_n	distance of maximum, minimum local failure probability
ρ	cutoff radius near crack tip
R	outer radius of integration for analysis of infinite bodies
F	nondimensional mixed-mode toughness function for homogeneous material
$\theta_{\sigma}^*, \theta_{G}^*$	optimal kink angles for max $\sigma_{\theta\theta}$ or G criterion
J	Jacobian of mapped quadratic finite element
w_i	Gauss–Legendre integration weights
q	Toughness dependence on radial band (p or R) exponent $= (M - 4)/2m$

1. Introduction

The earliest report on the relationship between the volume of a brittle material and its measured strength is credited to Leonardo Da Vinci, who observed that the strength of a wire in tension decreased as its length increased [1]. Since then, extreme-value Weibull statistics [2] have been developed to model the dependence of probabilistic failure on both applied stress and affected volume of a bulk material. This dependence can explain phenomena such as the enhanced measured strengths of brittle materials in bending vs. those in tension. Similarly, the fracture toughness of material that undergoes a stress-controlled fracture with extension initiated away from a pre-existent crack tip can be predicted by applying Weibull statistical analysis to bodies containing cracks or notches.

One area where Weibull statistics can play a descriptive role is in the fracture of composites or layered materials. Important developing applications in aerospace, power generation, microelectronics and bio-engineering demand characteristics that are often unobtainable in any single material. Traditional design of such a material seeks to invoke the desirable characteristics of each of the constituent phases in order to meet such requirements. However, the internal stresses caused by the elastic and thermal properties mismatch at an interface of two bulk materials can mitigate the successful implementation of such composites. To address this problem, FGMs have been developed to satisfy the needs for properties that are unavailable in any single material and for graded properties to offset adverse effects of discontinuities.

The introduction of gradual compositional changes removes large-scale interface-induced stress singularities and can even result in stress-free material joints. This gradual change in composition over length scales that are significant compared to the overall dimensions of the body is what distinguishes FGMs from other composites. The gradient shape is a design parameter determining the behavior of an FGM structure. Associated complications involving spatially varying elastic constants have demanded re-examination of the elastic crack problem. Although the fracture mechanics solutions for stress and displacement fields in homogeneous and continuously graded materials agree in the asymptotic near-tip limit [3], the scope of the material gradient effects on fracture behavior are not well comprehended.

These crack-induced stresses are affected by modulus gradients, however. For example, far-field tensile loading on elastic materials with gradients in modulus can cause mixed-mode (tensile and shear) crack loading. This will obviously influence the extension criteria for the simple situation of cracks in materials that grow continuously from the tip, which are known to kink or deflect when loaded in mixed mode. Alternatively in certain classes of FGMs (e.g., composites), crack extension will entail a nonlocal criteria involving reinitiation some distance away from the crack tip. In such an FGM, these reinitiations will depend on the nonlocal stress field that reflects both the stress-intensity factor and the gradient-induced effects, i.e., an FGM will fracture at a stress intensity, K, different from that in a homogeneous material. Weibull statistical analysis provides a tool for describing the relationship between the probability of fracture and the distributions of stresses and strengths near such a crack; the effects on the crack path of the gradient-induced stresses can also be assessed. Effects on toughness and crack trajectory from gradients in strength can similarly be explored though this framework.

To describe the mixed-mode crack extension criteria of materials governed by Weibull statistics near a prior crack, which is the purpose of this work, relationships for the average location of fracture initiation near an elastic crack tip and for the probable toughness are developed. These relationships are first applied to infinite, homogeneous bodies and results are compared with classical deterministic behavior for kinking and extension in response to stresses exactly at the tip. The analysis is then applied to infinite FGMs. Although the analysis depends on failure initiating at locations away from the crack tip, in both these applications, effects of higher-order terms in the stress field are not considered. In order to test the dependence on conventional geometry-specific stresses away from the crack tip or on gradient induced, higher-order terms, two fracture mechanics specimen geometries with various modulus or strength gradients are examined using the finite element method (FEM).

2. FGM fracture mechanics

A common, although not universal, feature of FGMs is that they are multiphase composites where the microstructure can often be described as a particle of one phase embedded in a matrix of another. In such an FGM, the gradient shape has little influence at small length scales, as the mechanics are dominated by the size, shape and interface conditions of the particle. Continuum mechanics analyses of an FGM usually emphasize the large-scale phenomenon and treat the FGM as a smoothly graded material, employing only "effective" properties of the isotropic homogenized composite at any given location. It is through the variation of these effective material properties (e.g., elastic modulus) that the nature of the spatial variations of the constituents is considered to affect the mechanical behavior of an FGM.

For FGMs with compositions including a brittle phase, e.g., ceramic, intermetallic or glass, fracture is an important design limitation. If an FGM is a brittle/brittle composite (e.g., Mo/SiO$_2$ [4], Al$_2$O$_3$/Si$_3$N$_4$ [5], SiC/TiC [6], ZrO$_2$/Al$_2$O$_3$ [7]), linear-elastic fracture mechanics can be used to characterize such failure, providing that effects of the gradient in elastic modulus are accounted along with the heterogeneous nature of the fracture resistance for the multiphase microstructure.

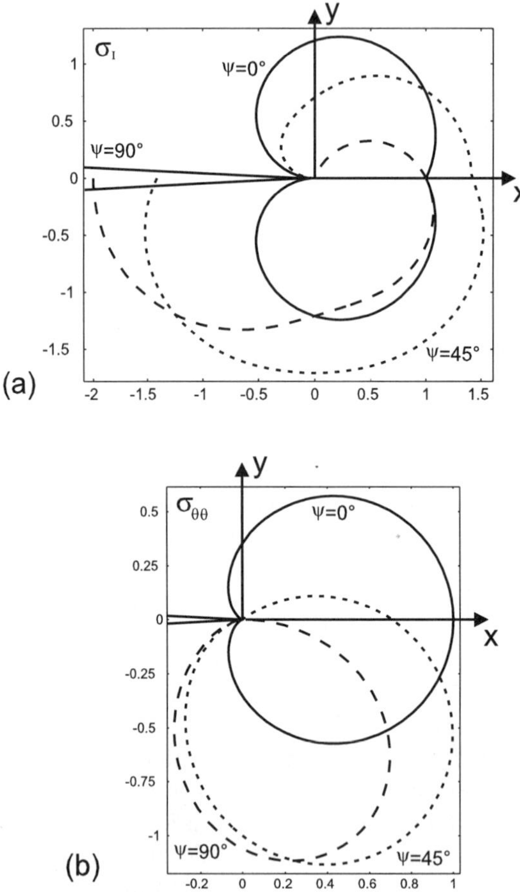

Fig. 1. Isostress contours for (a) maximum principal stress σ_1 and (b) hoop stress $\sigma_{\theta\theta}$ for loading phase angles $\psi = 0°, 45°, 90°$. Contour shapes indicate relative stress levels, with larger r corresponding to higher stress at that θ.

The study of fracture mechanics for FGMs has yielded linear-elastic crack-tip stress field solutions for various elastic gradients and boundary conditions. A fundamental result is that for locations asymptotically close to the crack tip, the stress fields for a homogeneous material and an FGM are identical [8,9]. In this limit, the stresses near the crack tip are described by the singular terms of the classical solution of Williams [10]

$$\sigma \approx \frac{K_{\mathrm{I}} f^{\mathrm{I}}(\theta)}{\sqrt{2\pi r}} + \frac{K_{\mathrm{II}} f^{\mathrm{II}}(\theta)}{\sqrt{2\pi r}} \tag{1}$$

with

$$f^{\mathrm{I}}(\theta) = \left[\cos\left(\frac{\theta}{2}\right)\left(1 + \sin\left(\frac{\theta}{2}\right)\right) \right] \quad \text{and} \quad f^{\mathrm{II}}(\theta) = \left[-\sin\left(\frac{\theta}{2}\right) + \sqrt{1 - \frac{3}{4}\sin^2\left(\frac{\theta}{2}\right)} \right],$$

for $\sigma = \sigma_1$, the maximum principal stress. Fig. 1 compares isostress contours for principal and hoop stresses for a sharp elastic crack with phase angles of $\psi = 0°$, $45°$ and $90°$ ($K_{\mathrm{II}} = 0$, $K_{\mathrm{I}} = K_{\mathrm{II}}$, $K_{\mathrm{I}} = 0$).

As in homogeneous materials, K_{I} and K_{II}, the modes I and II stress-intensity factors, characterize the symmetric and antisymmetric stress fields in the neighborhood of a crack tip; however, their values for a given geometry and load will differ vs. the homogeneous case in both magnitude, $|K| = \sqrt{K_{\mathrm{I}}^2 + K_{\mathrm{II}}^2}$, and phase angle, $\psi = \tan^{-1}(K_{\mathrm{II}}/K_{\mathrm{I}})$ [11].

With variable modulus $E(x,y)$, strains associated with the Williams crack-tip field do not satisfy the equations of compatibility. At finite distances away from the crack tip (where the elastic modulus differs from that at the tip), the solution field must take on a different character. Jin and Batra [12] estimated the size of the near-tip zone over which the stresses of the Williams singularity for homogeneous material will asymptotically govern (the K-dominant region); this is where $(|\nabla E|/E) \ll (1/r)$, $(|\nabla^2 E|/E) \ll (1/r^2)$. Therefore, the steeper the gradient in E, the smaller the region wherein Eq. (1) pertains. Outside the K-dominant regions, the stresses depend on the variation in E, among other factors. However, extant solutions have not been expressed directly in a series form for which the behavior of the next order terms can be readily appreciated.

The mixed-mode analysis of Erdogan and co-workers [3,11] justified describing the asymptotic stress field near a crack by pointwise multiplication of the conventional singular terms by the variation in Young's modulus, in the form $E(x,y)/E(0,0) = e^{(\beta x + \gamma y)}$. It is proposed that more generally: [1]

$$\sigma_{ij} \approx \frac{E(x,y)}{E(0,0)} \left[\frac{K_{\mathrm{I}}}{\sqrt{2\pi r}} f^{\mathrm{I}}_{ij}(\theta) + \frac{K_{\mathrm{II}}}{\sqrt{2\pi r}} f^{\mathrm{II}}_{ij}(\theta) \right], \tag{2}$$

for any $E(x,y)$ which is a continuous function of position (taking Poisson's ratio, v, as a constant). $E(0,0)$ is the value at the crack tip.

Eq. (2) approximates the effect of modulus variation, $E(x,y)$, on the stress over a wider region than that in which Eq. (1) pertains. Eq. (2) satisfies the equations of compatibility exactly, although it does not satisfy the conditions for equilibrium. It will therefore also be limited in its own region of dominance.

The issue of the direction of crack growth in homogeneous materials was first addressed by Erdogan and Sih [14], who predicted that a crack loaded in mixed mode would kink along an angle, θ_σ^*, which

---

[1] See Becker et al. [13] for a derivation of this type of approximations for FGMs with spatially varying Young's modulus.

corresponds to the maximum hoop stress, $\sigma_{\theta\theta}$, at the tip. Later analyses that explored the mode-mixity and strain energy release rate, G, of a kinked crack gave more rigorous results, e.g., the optimal θ_G^* gives maximum G; these converge with the hoop-stress criteria at small kink angles (see discussions by Cotterell and Rice [15] and Hutchinson and Suo [16]). The first analysis of crack kinking in FGMs [17] was limited to application of the first-order homogeneous analysis of Cotterell and Rice [15]; implicit here is the assumption that the gradient in Young's modulus only affects the crack tip through mode-mixity, ψ. This is true for crack kinks of infinitesimal length growing continuously from the tip, but these assumptions cannot be sustained if material inhomogeneity leads to longer kinks or for fracture events that are initiated ahead of the crack tip in materials with variable strength. Recently, an FEM analysis by Becker et al. [18] revealed information about adjustments in the stress field that arise from a sigmoidal gradient in elastic modulus aligned normal to a small crack. In this instance, ψ changes over a dimension $r \sim 1/b$, attaining its asymptotic value for $rb < 0.2$, where $b \approx |\nabla E|/E$ is the gradient coefficient. In contrast, a T-like stress term that adjusts σ_{xx}, only emerges at much smaller dimensions, i.e., at $r < 1/10b$, and does not approach its asymptotic limit until far smaller values of r; it influenced kink angles, θ_G^*, even for $rb \sim 0.01$. This gradient-induced T-stress depends strongly on externally applied shear loading [18], unlike the T-stress for homogeneous materials that describes an increment in σ_{xx} that is independent of position [10,19].

3. Ritchie–Knott–Rice fracture modeling

3.1. Weibull statistics

The effects of variability in strength of brittle materials can be described using extreme value statistics [2]. For a body experiencing spatially variable stress σ, the probability of failure Φ, is

$$\Phi = 1 - \exp\left[\int_{\mathrm{Vol}} -\frac{p(\sigma)}{V_0}\,\mathrm{d}V\right], \tag{3}$$

where V_0 is a reference volume and $p(\sigma)$ the strength distribution, i.e., the cumulative failure probability in the small-volume limit. The three-parameter Weibull strength distribution has the form:

$$p(\sigma) = \begin{cases} \left(\dfrac{\sigma_1 - \sigma_u}{\sigma_0}\right)^m & \text{for } \sigma - \sigma_u > 0, \\ 0 & \text{for } \sigma - \sigma_u \leqslant 0, \end{cases} \tag{4}$$

where σ_1 is maximum principal stress and σ_u, σ_0 and m are material constants to be determined by experiment. The lower bound cutoff strength, σ_u, is often set to zero for ceramics and glasses, thus defining the two-parameter Weibull distribution, where σ_0 is the scaling strength.

The Weibull function expresses the failure probability in terms of both the stress level and the volume over which stress is distributed. The severity of this volume dependence is largely determined by the Weibull modulus, m. [2] In the limit of a purely deterministic material, as $m \to \infty$, failure is dictated by the maximum stress at any point, with no scatter. For a high-quality engineering ceramic, m is in the range of 20–50. However, for brittle/brittle composites, the presence of internal interfaces, local residual stresses, and more microstructural variability often produce greater scatter in strength and so lower Weibull modulus. Indeed, bend tests of a SiO_2/Mo FGM have indicated a Weibull modulus of 4 pertains [20].

---

[2] The cutoff stress σ_u also can mitigate the effect of volume. When a two-parameter distribution is applied to an infinite body, any nonzero far-field stress leads to a failure probability of unity. Invoking a cutoff stress allows modeling of failure in an infinite body.

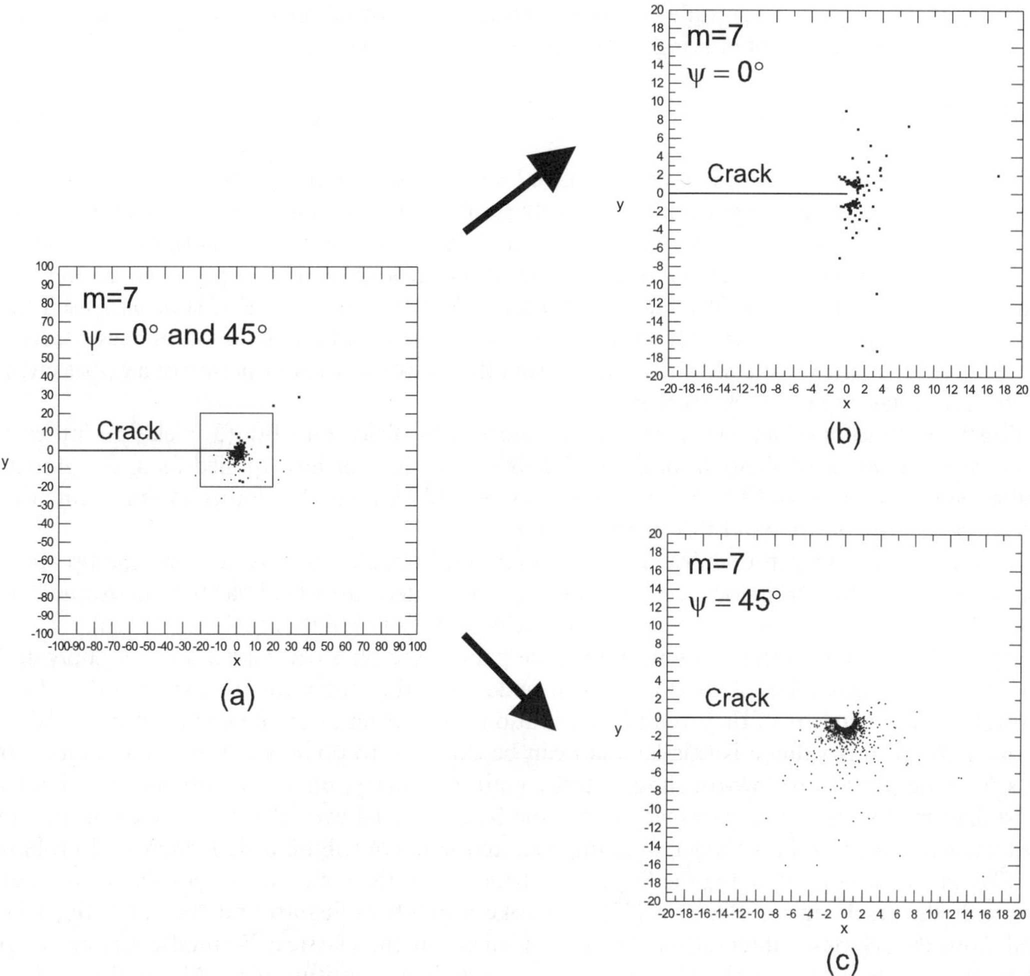

Fig. 2. Results of Monte Carlo simulations for a brittle material near a sharp elastic crack. Mode I, ($\psi = 0°$, square symbols) and mixed-mode ($\psi = 45°$, circles) cases are displayed for Weibull modulus, $m = 7$. Points around the crack tip indicate the location of a simulated fracture initiation. Strong skewing toward the negative y-direction is evident for the mixed-mode case, as is the greater number of fracture initiations indicating a decreased toughness.

To demonstrate characteristic fracture initiation behavior near a crack in a homogeneous material with strength described by Weibull statistics using the maximum principal stress, Fig. 2 displays the results of Monte Carlo simulations (details are in Appendix A). Each point indicates the location of a fracture initiation for mode I (square symbols) and mixed mode, $\psi = 45°$, (circles) loading. First, for the mode I case, these data are scattered away from the x-axis, but are roughly symmetric about $y = 0$, giving a bimodal distribution with an average angle of 4°. For the mixed-mode case, the data are strongly skewed toward the negative y-direction, at an average angle of $-81°$, and failures are found near the crack flank where $\sigma_{\theta\theta}$ vanishes. This indicates that, based on averaging several events, there are tendencies for straight-ahead crack extension under mode I loading and downward growth under positive phase-angle loading. In addition, for the same number of trials, the number of fracture initiations under mixed-mode loading exceeds that for mode I loading at the same $|K|$ (764 sites vs. 111). This indicates a lower resistance to fracture initiation pertains at a given magnitude of stress intensity. This would result in a lower measured fracture toughness. A framework for quantifying such expected trends is needed.

3.2. Statistical fracture modeling

Nonlocal, statistical crack extension criteria have been successfully applied for marginally brittle metals but can be problematic, even if appealing, for purely elastic materials. The near-tip region of a crack is an extreme example of the competition between stress and volume. At the tip of a sharp elastic crack, stress is infinite but is experienced over exactly zero volume. Away from the crack tip, an increasing volume is exposed to decreasing stress. The Ritchie–Knott–Rice (RKR) fracture model [21] in part used this competition to motivate the description of fracture toughness and its variability in low-toughness steels in terms of the stresses at a "characteristic distance" ahead from the crack tip, where fracture of a brittle inclusion or particle triggers catastrophic crack growth.

The direct substitution of singular fracture mechanics stress fields into Eq. (3) yields an integral that is not finite for most values of Weibull modulus. [3] This means that for any applied load, the global failure probability would be unity and the fracture toughness would be zero. A number of arguments or adjustments have been invoked to avoid this nonphysical result.

For materials with sufficient ductility, crack-tip blunting truncates the stresses near the tip and renders the integral (Eq. (3)) finite. Numerical calculation [23] of a blunted stress field has been performed as well as pseudo-analytic application [24] of a plastic notch field within the HRR singularity [25,26].

For elastic sharp cracks, special attention must be paid to the physical implications of applying Eq. (3) near the crack tip. Regardless of the mechanism underlying the statistical fracture process, the characterizing statistical parameters of the strength distribution will depend upon the volume of material tested to obtain them. When this volume is smaller than can be expected to possess any of the dominant fracture elements (carbide particle in low-toughness steels, grain boundary, etc.), the subsequent failure will no longer be described by the same parameters as is the bulk material even if it is statistical in nature. Such small volumes are likely to be stronger than implied, and may contribute little incremental probability of failure. The implications of applying the Weibull distribution to the region of small volume and high-stress gradient near the crack tip led Beremin [27] to invoke a cutoff radius around the crack tip, which was excluded from the region of integration. Evans [28] ruled out the classical Weibull equation for $p(\sigma)$ on related physical grounds and analyzed a sharp crack tip with a substitute $p(\sigma)$. Alternatively, the explicit modeling of a notch rather than a sharp crack allows for integration of the stress field over the entire material region, using Eqs. (3) and (4) with the results depending the notch size [23,29].

Previous authors (e.g., Refs. [22,24,28]) have investigated the incremental failure probability ahead of a crack tip, with the distance having the maximum failure probability deemed to be a characterizing parameter. These analyses utilized the first term of a stress field expansion (either Williams or HRR) within an infinite body, which also required the use of a far-field cutoff criteria.

3.3. Crack trajectory

For mixed-mode fracture, considering the single distance of maximum failure probability is insufficient compared to analyzing the full planar stress field. Although analyses of continuous crack kinking use the hoop stress $\sigma_{\theta\theta}$ as the determinant of kink angle, θ_σ^*, and critical load, for initiation of further cracking at a distance away from the main crack tip, the flaws need not be aligned along lines of constant θ. Thus,

---

[3] For the linear-elastic solution, the integral is not defined for ($m > 4$). In nonlinear materials, a similar result depends on m and the strain-hardening coefficient [22].

the principal stress σ_1 is used to determine flaw activation. The mode I elastic stress field possesses maxima in σ_1 at $\theta = \pm 60°$; symmetry dictates that mode I crack advance must occur along $\theta = 0°$, but only statistically.

The expected location of the activated flaws is described from a spatial average of the failure probability. With spatially varying stress, the average of such a coordinate ξ is

$$\bar{\xi} \equiv \frac{\int_{\text{Vol}} \xi \frac{p(\sigma)}{V_0} \, \mathrm{d}V}{\int_{\text{Vol}} \frac{p(\sigma)}{V_0} \, \mathrm{d}V}. \tag{5}$$

When $\xi = x$, $\xi = y$, the mean Cartesian coordinates, $\bar{x}$ and $\bar{y}$, of the fracture initiation are obtained. From these, the angle from the $y = 0$ plane of the average location can be described as $\phi \equiv \tan^{-1}(\bar{y}/\bar{x})$. Eq. (5) also yields the mean distance $\bar{r}$ for $\xi = r$. (This is, in general, distinct from, and usually exceeds, the radial coordinate of the average location, $\sqrt{\bar{x}^2 + \bar{y}^2}$.)

It is possible to describe the relative contribution of stress fields laying at different angles to the total failure probability. The median angle that bisects the angular distribution of failure probability, θ_{50}, is defined through

$$\frac{\int_{-\pi}^{\theta_f} p(\sigma) \, \mathrm{d}\theta}{\int_{-\pi}^{\pi} p(\sigma) \, \mathrm{d}\theta} = f\Phi, \tag{6}$$

with $f = 50\%$. Furthermore, bands of probability can similarly be defined, so that the proportion of the failure probability between two angles is determined, i.e., 90% of failures will be initiated between θ_5 and θ_{95}.

4. Procedures

Calculations for infinite bodies were performed with stresses derived from Eq. (2). Eqs. (5) and (6) were evaluated with the numerical integration package in Mathematica 3.0 (Wolfram Research, Champaign, IL, USA). To evaluate finite geometries, the single edge-crack tension (SE(T)) and middle-crack tension (M(T)) fracture mechanics specimens were modeled by the FEM. A plane-strain linear-elastic finite element code FEAP 4.2 [30] was modified such that during integration of the stiffness matrix, the elastic constants were evaluated at each of the Gauss points (3 per direction, 9 total). This allowed for quadratic variation in modulus to be modeled within a single element.

The crack length, a, was kept constant in the FEM analyses, being equal to half the total width of the sample in the x-direction, W. The mesh for the M(T) sample was the same as for the SE(T) sample (Fig. 3), except it contained an additional row of displacement boundary conditions along $x = -a$, preventing displacements in the x-direction. The mesh consisted of 2300 total elements, 9233 nodes, with extensive refinement using a fan array in the near-tip region, with the smallest element size on the order of $a/10^6$. The fracture mechanics element of Stern and Becker [31] was used in both meshes to model the first ring of elements surrounding the crack tip. A second mesh with 4096 elements was also used to estimate the error via refinement. The difference between the results from the two meshes was negligible, with average initiation angles differing by $<0.1°$.

Stress-intensity factor calibrations were performed for each geometry and gradient. Pointwise evaluation of the stress-intensity factors can be obtain through

$$\begin{Bmatrix} K_{\mathrm{I}} \\ K_{\mathrm{II}} \end{Bmatrix} = \sqrt{2\pi r} \begin{pmatrix} f_{xy}^{\mathrm{I}}(\theta) & f_{xy}^{\mathrm{II}}(\theta) \\ f_{yy}^{\mathrm{I}}(\theta) & f_{yy}^{\mathrm{II}}(\theta) \end{pmatrix}^{-1} \begin{Bmatrix} \sigma_{xy}(r, \theta) \\ \sigma_{yy}(r, \theta) \end{Bmatrix}. \tag{7}$$

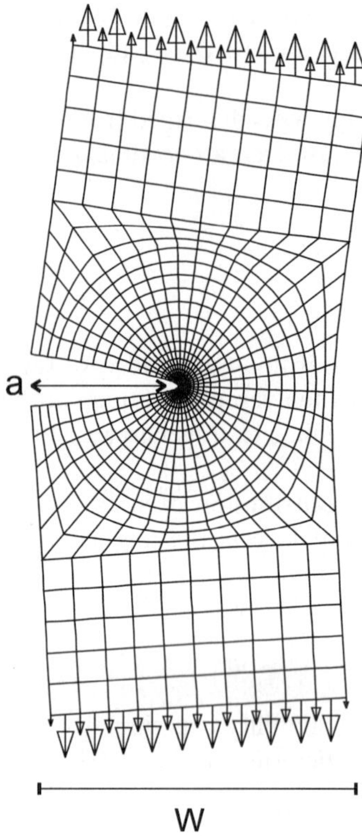

Fig. 3. Finite element mesh used in study of SE(T) fracture mechanics specimen. A high degree of refinement was used, with a focused ring of elements surrounding the crack tip. A similar mesh with different boundary conditions was used for the M(T) sample.

Excluding the crack flanks and the first four elements at the crack tip, these values were averaged over Gauss points within the next five rings of elements (360 elements total, those for $-160° < \theta < 160°$ and $1/10^5 < r/a < 1/10^4$). The precision of these results found to vary weakly with variance of this fitting region. The results of Becker et al. [18] elaborate the need to fit in a range of $rb < 0.01$ in order to avoid effects of higher-order terms (and see Fig. 14).

The statistical modeling used results from each FEM calculation in the probability integral as the sum over the finite elements, where for a variable ξ,

$$\int_{\text{Vol}} \xi \left(\frac{\sigma}{\sigma_0}\right)^m \frac{\mathrm{d}V}{V_0} \approx \frac{B}{\sigma_0^m V_0} \sum_{j}^{\text{elems}} \xi_i \left[\sum_{i}^{\text{Gauss pts}} (\sigma_1^m)_i J w_i \right]_j, \tag{8}$$

where J is the Jacobian of the mapped element (calculated at the Gauss points), w_i's are the weights for Gauss–Legendre quadrature, and B is an out-of-plane thickness. For a physical sample with a large thickness, multiple fracture events are sometimes expected to occur along the crack front, with implications discussed in Section 5.1.

Sigmoidal variation in modulus or strength was employed via a function $g(x)$ or $g(y)$:

$$g\left(\begin{matrix} x \\ y \end{matrix}\right) = (c - 1) \tanh\left(b\left(\begin{matrix} x \\ y \end{matrix}\right)\right) + c. \tag{9}$$

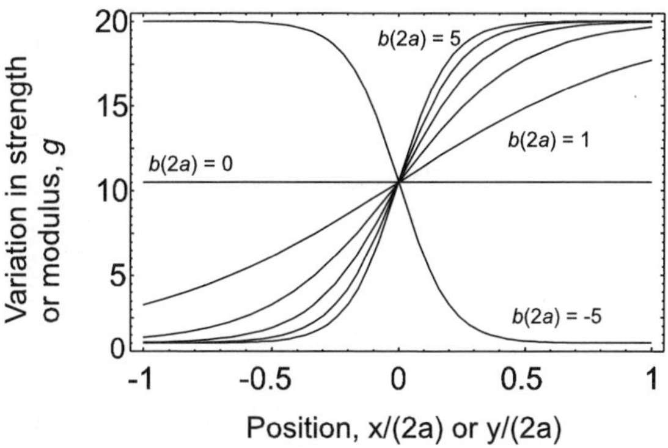

Fig. 4. Sigmoidal gradients modeled. This shape allows for an arbitrarily steep gradient at the crack tip, determined by the parameter b. The total range of variation within the sample is held constant, with $g(\infty)/g(-\infty) = 20$.

At the origin (usually, the crack tip), $g(0) = c$. A value of $c = 10.5$ was chosen for the present analyses, providing a 20-fold change in properties across the sample. As such, b, which was varied, is a measure of the gradient steepness and $(\nabla g/g(0)) = (b(c - 1)/c) \approx b$. In the limit of $b \to \infty$, a sharp interface is formed, with $g(\infty)/g(-\infty) = 2c - 1$. Fig. 4 displays shapes of the gradient for a range of b. The Weibull modulus was investigated as a variable but kept constant in any sample.

5. Results and discussion

5.1. Homogeneous infinite body analyses

Application of the Weibull crack initiation model to material with a mode I crack permits description of the fracture toughness using statistical parameters (σ_u, σ_0 and m) and perhaps other material parameters. Extending the analysis to mixed-mode situations allows predictions about homogeneous material response without specific reference to geometry effects.

5.1.1. Mixed-mode fracture toughness
Combining Eqs. (1), (3) and (4) and inverting gives an expression for the stress intensity to produce a given failure probability, K_Φ:

$$K_\Phi = \frac{\sigma_0(V_0/B)^{1/m}\sqrt{2\pi}(-\ln(1 - \Phi))^{1/m}}{\left[\int_{-\pi}^{\pi} f_1(\theta, \psi)^m \, d\theta\right]^{1/m} F(m, R, \rho)}, \tag{10a}$$

$$F = \begin{cases} \left[\dfrac{R^{2-(m/2)} - \rho^{2-(m/2)}}{2 - \frac{m}{2}}\right]^{1/m} & \text{for } m \neq 4, \\[2em] \left(\ln\left(\dfrac{R}{\rho}\right)\right)^{1/4} & \text{for } m = 4, \end{cases} \tag{10b}$$

where $f_1(\theta, \psi) = \sigma_1\sqrt{2\pi r}/|K|$ and can be inferred from Eq. (1). The integration is conducted over an annulus between the inner cutoff radius, ρ, and the outer radius, $R \gg \rho$, which is still within the K-field (Eq. (1)). Explicit values of K_Φ scale primarily with $\sigma_0(V_0/B)^{1/m}\rho^{(m-4)/2m}$ for large m and $\sigma_0(V_0/B)^{1/m}/R^{(4-m)/2m}$ for small m. That is, for materials with large scatter in strength, toughness is dictated by the large volume of

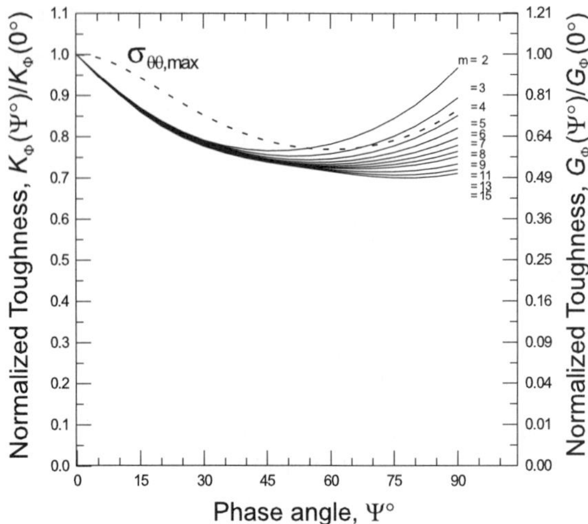

Normalized Toughness, $K_\Phi(\Psi^\circ)/K_\Phi(0^\circ)$

Normalized Toughness, $G_\Phi(\Psi^\circ)/G_\Phi(0^\circ)$

Phase angle, Ψ°

Fig. 5. Prediction for the mixed-mode toughness for an infinite body of a homogeneous material based on statistical fracture model for various values of Weibull modulus (solid lines) compared to the deterministic prediction based on critical hoop stress (dashed line).

low-stress material as $r \rightarrow R$; with little scatter, toughness is determined by the small volume of highly stressed material nearest to the crack tip.

From these results, geometry-independent predictions of the relative fracture toughness for mixed-mode loading are shown in Fig. 5 in terms of $|K_\Phi|$ or G. These toughnesses (being normalized to the mode I toughness and thereby independent of the choice of Φ, for fixed R and ρ) predict (for a nonlocal, stress-controlled failure mechanism) a reduction in fracture toughness under mixed-mode loading. Comparing the isostress contours for $\psi = 0^\circ$ and 45° in Fig. 1 explicates this behavior. The area contained within the 45° contour is plainly larger than that of the 0° contour. This can be interpreted as a larger volume being exposed to a given stress level at a given G, or an elevation of stress at a fixed distance, r, from the crack tip. For a material described by initiation at a distance, this lessens the fracture toughness. This reduction in toughness under mixed-mode conditions at moderate phase angles is stronger than that for a deterministic analysis based on the maximum $\sigma_{\theta\theta}$ [14]. Note these $|K_\Phi|$ and G values describe self similar extension of an unkinked crack; the values plotted for critical $\sigma_{\theta\theta}$ approximate the loading that gives an invariant K_{Ic} or G_{c} at the tip of an optimal kink for continuous crack extension.

5.1.2. Mixed-mode initiation angle and distance

Under mixed-mode conditions, a crack will not, in general, grow in a self-similar manner; rather it will kink off of the $y = 0$ plane. The average initiation angle ϕ calculated via Eq. (5) as $\tan^{-1}(\bar{y}/\bar{x})$ for a crack sustaining mixed-mode stresses in an infinite homogeneous body is shown in Fig. 6. The predicted average crack angles from the statistical formulation based on principal stresses are more negative for $\psi > 0$ and any m than those derived for the continuous growth of the crack tip along the path of maximum $\sigma_{\theta\theta}$, i.e., θ_σ^* [14], or maximum G, i.e., θ_G^* [32], also shown. At higher m, these average initiation angles $-\phi$ increase toward the trend, also plotted, for the angle of maximum principal stress. However, the behavior would only replicate this for extremely large m. Median angles, θ_{50}, exhibit similar trends as does ϕ.

This behavior is easily related to the mixed-mode stress fields plotted in Fig. 1. With pure mode II loading ($\psi = 90^\circ$), the largest principal stress lays along $\theta = 180^\circ$ and locations of average fracture tend toward this extreme. The boundary conditions dictate that this is the σ_{xx} component. Deterministic, continuous-growth analyses predict opening along radial lines and are restricted to the $\sigma_{\theta\theta}$ component, which is zero at the crack flank. This difference requires careful physical interpretation, as discussed in Section 5.1.3.

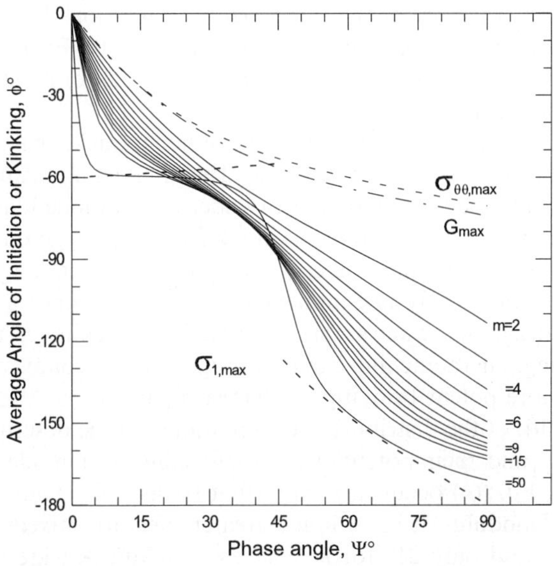

Fig. 6. Average initiation angle for mixed-mode fracture loading for an infinite homogeneous material based on statistical model. Also included are angles of maximum hoop stress and maximum energy release rate criteria (dashed lines).

Table 1
Approximate average distances for fracture initiation

m	2	3	4	7	8	10	16
$\bar{r}$	$R/2$	$R/3$	$R/10$	3ρ	2ρ	1.5ρ	1.2ρ

The mean distances of fracture initiation $\bar{r}$ in an infinite body, also calculated via Eq. (5), are listed in Table 1; these are combined results for $R/\rho = 10^3$ and 10^5. With small values of m (2–4), $\bar{r}$ scales with the outer limit of integration R, $\bar{r} \sim R/2$ to $R/10$. For larger values, $m > 7$, $\bar{r}$ is on the order of the inner cutoff limit of integration ρ, $\bar{r} \sim \rho$ to 3ρ, as is also depicted in Fig. 2. Results are independent of mode mixity.

For small m, the volume is a determinant, and for $m = 2$, the failure probability in a thin annulus at a given r is uniform from the crack tip to infinity in the K-field. Hence, the mean distance is the average of the integration limits $(R + \rho)/2 \approx R/2$. Taking the limit of large m, fracture is dictated simply by the largest stress, that at $\{r, \theta\} = \{\rho, \pm60°\}$ for $\psi = 0$, but even for $m > 11$, $\bar{r}$ measurably exceeds ρ ($\sim1.5\rho$). Thus, the first principal stress σ_1, not the hoop stress, is still the pertinent determinant for failure, as $\sigma_1 \geqslant \sigma_{\theta\theta}$ and flaws are expected to be oriented randomly, not along θ.

5.1.3. Relevance and physical implications

The present model is formally limited to calculating the behavior of the first activated flaw, but this "first failure" may or may not cause catastrophic crack growth. For flaw-intolerant materials such as glasses and fine-grained ceramics, initial flaw growth and catastrophic failure generally are synonymous. However, there is experience with brittle, polycrystalline or composite materials with low-Weibull modulus that near-tip microcracked regions can develop [33,34] or especially that cracks bifurcate extensively [35]. Evidently, flaws exist that are activated by the near-tip stress field, but do not yet possess the driving force necessary to overcome the next higher-length scale fracture resistance (e.g., after cracking a weak or residually stressed grain boundary). The statistical initiation analysis (Fig. 6) shows that such a flaw is unlikely to be located on the trajectory followed by a crack under the same loading conditions but that is continuously moving according to the maximum-$\sigma_{\theta\theta}$ (or G) condition (i.e., $\phi \neq \theta_\sigma^*$ for $\psi > 0$). Even in the mode I case where the statistical and continuous-growth analyses agree on the average angle, most of the microcracks will be initiated at large angles to the crack plane.

When failure ensues from growth of a flaw ahead of the main crack, then further extension will eventually involve connection of this flaw and the main crack. The intervening ligament between the main crack and the microcrack will invariably be fractured or torn under mixed-mode loading requiring much greater dissipation [36]. Moreover, if various flaws along the crack front are triggered prior to criticality, which is more likely with larger B and smaller m, they will each surely be at different angles to the crack plane. Thus, the main crack and the microcracks may not immediately link, but rather a microcrack may next grow forward whilst the uncracked ligaments act as bridges and further enhance the toughness [37]. By such interactions and stress shielding, microcracking can be a toughening mechanism in ceramics [38,39] or more likely serves as a trigger to more potent bridging mechanisms [40,41].

Further analysis of the width of the distribution of initiations can be descriptive of the spatial arrays of microcracks that can develop and their potential for toughening. For mode I loading, only 50% of the fracture initiations ($\theta_{75} - \theta_{25}$, Eq. (6)) occur in a band that spans 120° symmetrically about 0°. Although there is little effect of Weibull modulus, it has a much greater effect for mixed-mode loading, with the 50% band spanning 73° for $m = 3$ and only 21° for $m = 11$ as $\psi \rightarrow 90°$. A wide dispersion in angles may be indicatory of tough material involving rising R-curves, as is $\phi \neq \theta_\sigma^*, \theta_G^*$. However, the problem of treating subsequent failure after first fracture requires an entirely different statistical formulation and detailed simulation of a noncontinuum microstructure with crack–crack interactions [42], which exceeds the present scope.

Thus, very low-m materials, being highly variable locally, can be relatively notch-insensitive globally, with large volumes of low-stress material dominating the probability analysis. For finite geometries, failure would be influenced by stresses outside the K-field, so trends illustrated here are semi-quantitative, even for first fracture, and their application would be geometry dependent. This effect could be somewhat mitigated by the invocation of a cutoff stress σ_u in the Weibull strength distribution (Eq. (4)). Indeed the size of flaws implied by the absence of a cutoff stress becomes unrealistic. This would offset some of the pathological behavior when analyzing cracks in infinite bodies ($K_\Phi \propto R^{-q}$, Eqs. (10a) and (10b)) and may lead to instances where the important behavior occurs within the K-field and more closely reflects trends predicted here, i.e., the pertinent outer cutoff radius should vary as $R \propto 1/\sigma_u^m$.

For large m, one would expect the material to behave in a deterministic manner. Nonetheless, the transition of the fracture mechanism from dictation by σ_1 to $\sigma_{\theta\theta}$ does not follow simply by taking the limit of large m for the present formalism. Thus, this statistical fracture model is unlikely to apply for single phase materials without distributions of internal heterogeneities. Moreover, the nonlocal behavior for intermediate m may be implausible regarding the strength characteristics implied for small volumes under high stress, e.g., near a crack. For example, elementary calculations show that for structural ceramics, the strength implied for small volumes using typical Weibull parameters underestimates the theoretical strength.

For this model to pertain for an intermediate m (>5) material implies that failure does initiate near or beyond some nonatomistic cutoff radius and that $K_\Phi \propto \rho^q$ behavior does occur. Clearly, for semi-brittle materials wherein crack-tip blunting leads to diminished stresses in a region within $r \sim 2K^2/(\sigma_{\text{yield}}E)$ [43], the present type of approximation (perhaps using an HRR stress field) may be robust. For more brittle composites, the lower theoretical strength of impure interfaces [44] and higher internal mismatch stresses compared to a monolithic ceramic could lead to a small interface flaw (off the main tip, but still influenced by the stress concentration of it) reaching criticality before the main crack. Then, it may be possible to conceive ρ as being related to the volume wherein Eqs. (3) and (4) fail, and, moreover, that the quantity $\sigma_0(V_0/B)^{1/m}\rho^{(m-4)/2m}$ can serve as a useful scaling parameter for K_Φ.

5.2. FGM infinite body analyses

In the prior section, analysis was limited to the classical K-field within homogeneous materials. For cracks in graded materials, the stress and/or strength field at a distance away from the crack will vary with the severity of the gradient, and the location and orientation of the crack relative to the gradient. The fracture of an infinite FGM can be modeled similarly, addressing failure probabilities within the context of adjusting a region lying within a K-field by introduction of Eq. (2) into the statistical analysis of Eqs. (3)–(6) and using Eq. (9) to describe gradients in $E(x,y)$ and σ_0. Then, the two-parameter probability function is

$$p(\sigma(r)) = \left\{ \frac{\left[\dfrac{(c_E - 1)\tanh\left(b_E(\xi - \xi_o)\right) + c_E}{(c_E - 1)\tanh(b_E(\xi_o)) + c_E} \right] |K| f_1(\theta, \psi)}{\left[\dfrac{(c_\sigma - 1)\tanh(b_\sigma(\varsigma - \varsigma_o)) + c_\sigma}{(c_\sigma - 1)\tanh(b_\sigma(\varsigma_o)) + c_\sigma} \right] \sigma_0(0)\sqrt{2\pi r}} \right\}^m . \tag{11}$$

This is written to simultaneously allow sigmoidal gradients in both σ_0 and E (characterized by b_σ and c_σ and b_E and c_E, respectively, but with m invariant for simplification); these are aligned along angles θ_σ or θ_E, respectively, and can be offset from the crack tip by ς_o and ξ_o. Situations of gradients in either σ_0 or E aligned parallel or normal to the crack plane with the crack tip centered in the gradient (ς_o, $\xi_o = 0$) are examined first. For these, toughnesses are normalized to that from Eqs. (10a) and (10b) using σ_0 ($b = 0$), which is c_σ with a strength gradient, and as normalized do not depend Φ.

The exact size of the domain of integration in Eq. (3) need not be specified to assess the initiation angle and relative mixed-mode toughness for homogeneous material, as these quantities involving integrations over the classical K-field are independent of the choice of radii. However, determining effects of a strength or modulus gradient on fracture behavior for an FGM requires specification of these dimensions, as it is precisely the gradient-induced variations away from the crack tip that alter the fracture characteristics. Only for sufficiently shallow gradients (low b), or for fracture processes dominated by the very near-tip fields (high m), will the relevant character of the crack fields be the same as in the homogeneous case. Several examples with sigmoidal variations in σ_0 or E are examined with two sets of integration limits in the r-direction, $\rho = R/10^3$ and $R/10^5$, before generalizing results.

5.2.1. Parallel strength gradient, $\sigma_0(x)$

For a graded material with a parallel variation in strength $\sigma_0(x)$ and nominal mode I loading, symmetry dictates a solution with no kinking on average, $\phi = 0$. However, the initiation fracture toughness K_ϕ, i.e., the stress-intensity factor that will result in a stated first failure probability (say 50% or 90%), is affected by the gradient strength.

Results for the normalized toughnesses are displayed in Fig. 7, for m from 2 to 11 and $b(2R) = \{-20, 20\}$. For gradients in strength with $b < 0$, which correspond to cases of cracks growing into weaker material, there is, as expected, a reduction in predicted initiation toughness. The effect is very strong for lower m, as the failure behavior is increasingly volume dependent, and farther ahead of the crack tip larger volumes of weak material are sampled. However, for cracks growing into stronger material ($b > 0$), a similar, but smaller, degradation of initiation toughness trend is also evident, except for very small b. This reduction derives from the sampling of all material in the neighborhood of the crack tip, not just that in front, $x > 0$. The material located outside $-\pi/2 < \theta < \pi/2$ is weaker than that just ahead of the crack tip, and the steeper the gradient, the weaker that material. Given that the maximum principal stress at $\theta = \pm 114°$ is the same as it is at $0°$ (see Fig. 1), this weakness greatly affects the probability integral, although not as much as for a negative gradient.

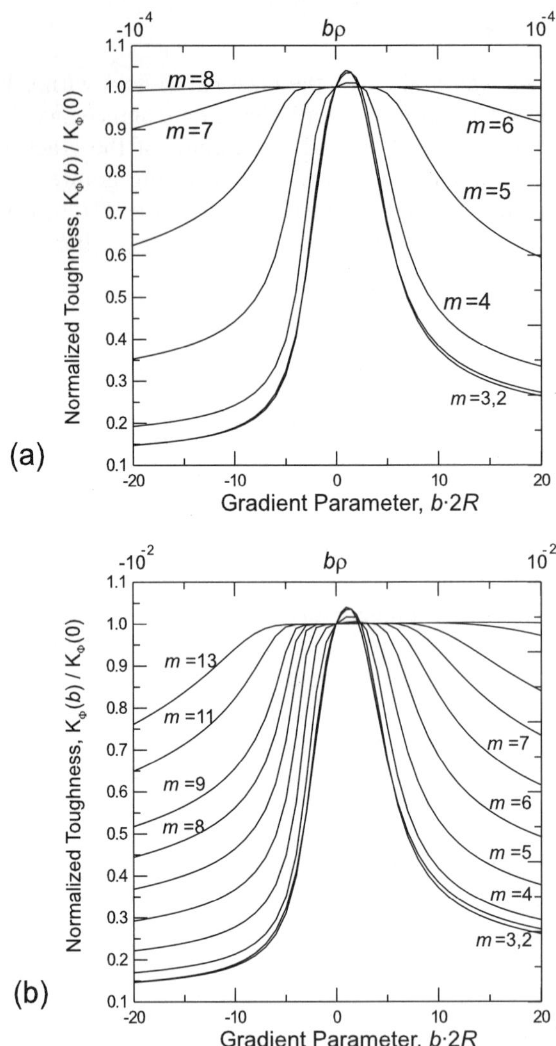

Fig. 7. Mode I fracture toughness for an infinite body FGM with $\sigma_0(x)$ based on statistical model. Results are displayed for two sets of integration limits, ρ and R, with (a) $\rho = R/10^5$ and (b) $\rho = R/10^3$ and m from 2 to 13.

As with homogeneous mixed-mode fracture (Section 5.1), these results must be carefully interpreted. Although material fracture is expected at a lower applied K than in the homogeneous case, a cracking event behind the main crack tip is unlikely to trigger total failure of the body, but rather would lead to a zone of widely distributed damage, or to a discontinuous crack extension that would later produce a bridged crack.

5.2.2. Infinite body with parallel stiffness gradient, $E(x)$

Fig. 8 shows the range of predicted toughness for $E(x)$ gradients. The strength, σ_0, is taken to be constant. Results displayed show a toughening for $b < 0$ and low values of m. That is, for a given crack-tip K, failure is less probable for cracks growing into more compliant material. In contrast, large reductions in toughness are seen for $b > 0$, but again only for relatively low m.

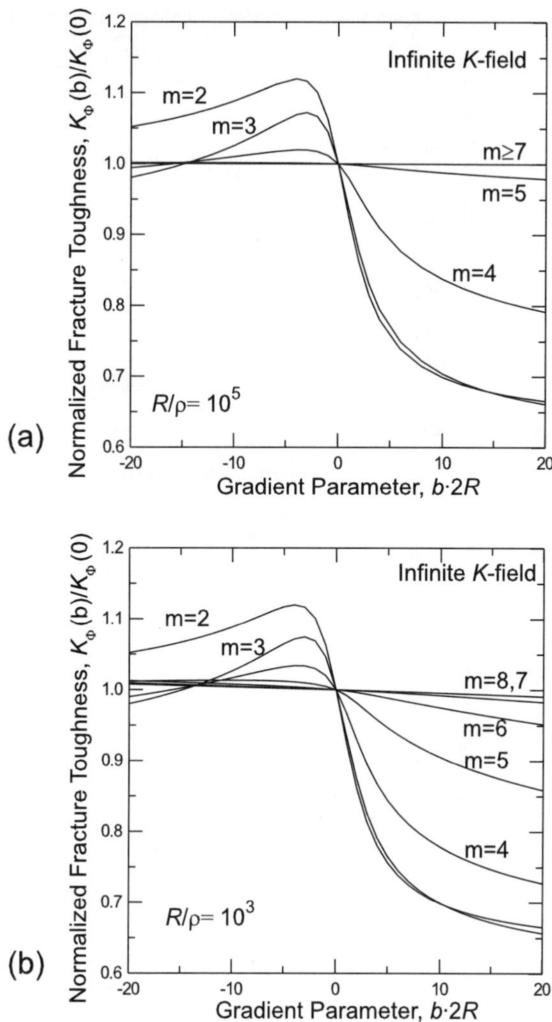

Fig. 8. Mode I fracture toughness for an infinite body FGM with $E(x)$ based on statistical model. Stress field is a one-term modification of the homogeneous field (Eq. (2)). Results are displayed for two sets of integration limits, (a) $\rho = R/10^5$ and (b) $\rho = R/10^3$, and m from 2 to 8.

The effect is clearly a result of the form of the stress field, the FGM modification of the classical field (Eq. (2)). [4] For cases with elevated stresses ahead of the crack, $b > 0$, the failure probability is increased, and thereby the toughness is decreased. The opposite is true of the $b < 0$ case at modest levels of $-b$, where stresses ahead of the crack are lower for a given K.

5.2.3. Infinite body with perpendicular strength gradient, $\sigma_0(y)$

The classical formulae for crack kinking indicate the preferred direction for extension to be along 0° for mode I loading, but for an FGM with a normal gradient in the Weibull strength, $\sigma_0(y)$, the spatial variation

---

[4] Commentary on the accuracy of this approximation can be found in Section 5.3.3.

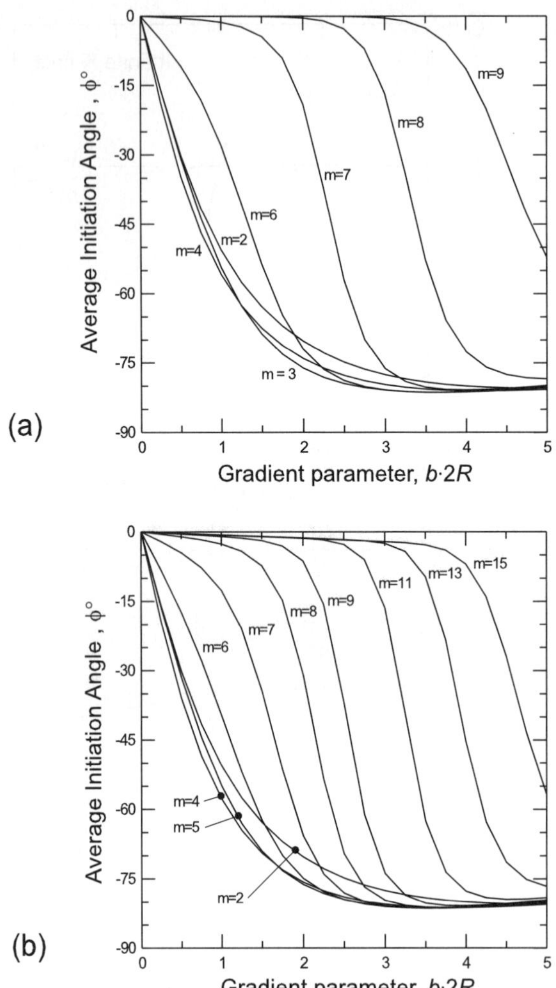

Fig. 9. Average initiation angle ϕ for an infinite body FGM with $\sigma_0(y)$ based on statistical model. Results are displayed for two sets of integration limits, (a) $\rho = R/10^5$ and (b) $\rho = R/10^3$, and m from 2 to 15.

in strength results, on average, in a preferred nonzero initiation angle, ϕ. Fig. 9 shows the range of predicted ϕ, and as results are an odd function of the gradient strength ($\phi(b) = -\phi(-b)$), only for a positive range of $b(2R) = \{0, 5\}$. A strong dependence on $\sigma_0(y)$ is observed, with the average initiation angle approaching 80° for very low Weibull modulus ($m < 5$). Even with higher m, a transition occurs to cracking at 80° although the critical gradient for this transition increases sharply with m.

Note that toughness results for $\sigma_0(y)$ exhibit similar trends as for the case of $\sigma_0(x)$, degradation of toughness. The actual curves are more similar to the $b < 0$ region of Fig. 7. So, not only will the crack tend to grow into the weaker material with a variation in strength present, $\sigma_0(y)$, but fracture will initiate at a lower applied K than in the homogeneous case.

5.2.4. Infinite body with normal stiffness gradient, $E(y)$

Crack initiation was studied for infinite bodies with $E(y)$ in two ways, each with σ_0 held constant. First, the effects of gradient-induced terms in the stress field are examined by using a pure mode I field near the tip

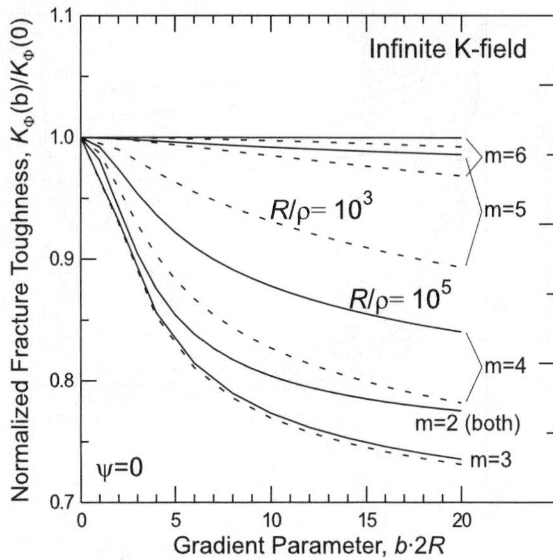

Fig. 10. K_ϕ for infinite body with the modified K-field (Eq. (2)) and a gradient in Young's modulus normal to the crack, $E(y)$. The near-tip region is in pure mode I loading, but the modulus gradient induces an asymmetry in stresses. The elevated stresses in a region $y > 0$ for $b > 0$ cause the failure probability to be increased for a given K.

with the $E(y)$ modification via Eq. (2). Fig. 10 displays the gradient effect on toughness for $b(2R) = \{0, 20\}$. A degradation of over 25% compared to the homogeneous case is evident for low m (2,3), purely due to the elevated stress in the regions of material with higher modulus $E(y)$.

In general, the existence of a stiffness gradient normal to the crack plane causes mixed-mode loading at the crack tip [3]. The FGM stress field depends on the geometry-specific relationship between the modulus gradient and phase angle. The average initiation angles for an infinite sample with a range of Weibull moduli, $m = \{2, 11\}$, are displayed in Fig. 11 for a range of $b(2R) = \{0, 20\}$ (as ϕ is an odd function of b). The stress field is again approximated by Eq. (2); however, the phase angles used are increased with b to match the $\psi(b)$ data of the calibration curve of the SE(T) specimen (subsequently in Fig. 13, Section 5.3.1), which will facilitate later comparison.

For high m, results for this FGM field agree reasonably well with the cracking angles for a homogeneous stress field (Fig. 6) at the same ψ. For instance, for $b = 10$, $\psi = 18.1°$, and for $m = 9–15$, $\phi_{\mathrm{FGM}} = -50°$ to $-60°$, which is similar to that for homogeneous materials at this ψ. At low m, the agreement is poor, with cracking angles being much nearer to zero for graded materials. This difference arises because for the high m, small ρ situation, cracking is dominated by the stresses very near the tip, which, being essentially those of the classical Williams K-field, are affected by the gradient primarily through the phase angle, ψ. As the gradient is made steeper (from $b \sim 10–20$), this phase angle nearly plateaus. However, the assumed multiplicative effect of the modulus field on the stress field, Eq. (2), does not change in form as the gradient becomes steeper. Thus, from Eq. (2), for a fixed phase angle, an increase in modulus will raise all the stresses at $y > 0$ (relative to those in the homogeneous case with the same ψ). The resulting sampling of the whole body will recognize this elevated stress level; for sufficiently low m, it will dominate, as implied by the insensitivity to ρ for small m. Thus, the near-tip stresses for $\psi > 0$ drive the crack in the negative y-direction while the far-field stresses drive the crack in the positive y-direction, because the scaling by $E(y)$ dominates the effect of ψ. This also leads to the average initiation angle being notably smaller (less negative) in some FGMs computed with larger ρ, i.e., for intermediate m.

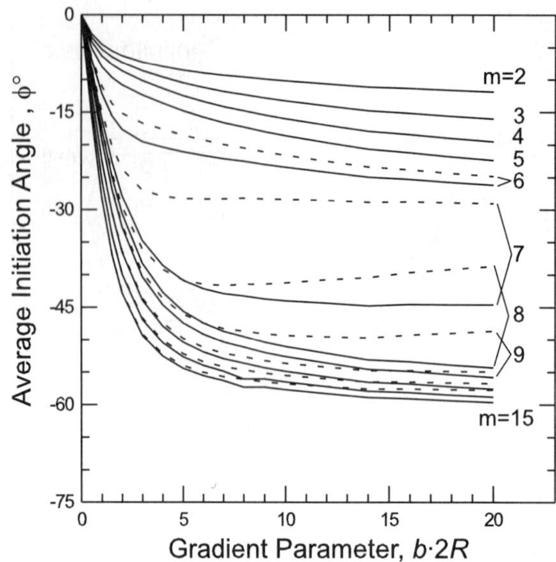

Fig. 11. Average initiation angle ϕ, for an infinite body FGM with $E(y)$ based on statistical fracture model. Stress field is the one-term modification of the homogeneous stress field (Eq. (2)), plus ψ is increased with b (see text). Results are displayed for two sets of integration limits with $\rho = R/10^5$ (solid lines) and $\rho = R/10^3$ (dashed lines) and m from 2 to 15.

5.2.5. Implications

Gradients in material properties affect the toughness in several distinct ways, including the initiation toughness and the crack angle and its dispersion. Both have implications for the final fracture behavior.

The trends depicted previously, as well as the expected behavior for other conditions, can be understood in terms of the most probable radial distance from the crack tip for cracking to initiate. The local cumulative cracking probability is $p(\sigma)\,dV/V_0$, and so the fracture probability in an annulus at a given radial distance from the tip scales with $\int_{-\pi}^{\pi} p(\sigma)r\,d\theta$. Thus, important trends are revealed by the behavior of the weighted probability function $p^*(K,r) \equiv p(\sigma(r))r$, based on $p(\sigma)$ as given in Eq. (11), (for which $\sigma(r)$ is from Eq. (2)) and with the gradient aligned along a direction of high stress for the crack, i.e., high $f_1(\theta,\psi)$, and b selected to give relative weakening (high E/σ_0) ahead of the crack.

In a homogeneous material, cracking frequencies decrease with increasing r for all $m > 2$ in a linear-elastic stress field. However, for a sigmoidal gradient in either σ_0 or E, under certain conditions a local minimum and maximum in p^* exist at radial positions r_n and r_x, respectively, as depicted in Fig. 12. These occur such that $0 < r_n < r_x$, and they scale with $1/b$, as described in Appendix B.

A key issue concerns the strength of the maxima, specifically the relative values of σ/σ_0 and especially of p^* at r_x vs. at the cutoff distance ρ. The cases of $\sigma_0(x)$ with $b(2R) = -20$ and of $E(x)$ with $b(2R) = 20$ for $m = 3$ and 11 provide an illustrative range of examples to examine. Values for $r_x b$ (from Eqs. (B.1) and (B.2)) are listed in Table 2 as are ratios of σ/σ_0 and of p^* (at r_x vs. at ρ) (from Eqs. (B.4a) and (B.4b)) taking $b\rho = 10^{-4}$ and 0.01, the limiting conditions for the plots in Figs. 7–11. For most of these cases, maxima exist for the weighted probability p^*, but they span a range from the local peak in σ/σ_0 (being low but relevant) vs. the value at ρ to $\sigma/\sigma_0(r_x)$ (being too small to influence the cracking probability).

By extending the arguments made regarding the crack locations in homogeneous materials (Table 1) in light of information in Table 2 and Figs. 7 and 8, it is argued that for FGMs the predominant region of failure can be categorized into three situations, at least where the relatively weak region is ahead of the crack, as is described next.

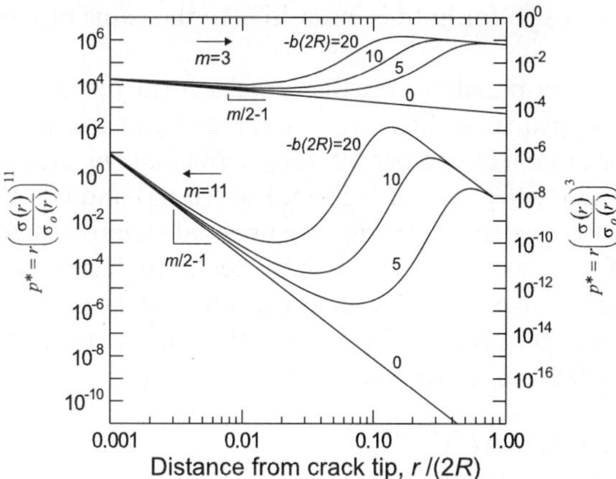

Fig. 12. Variation of the weighted local failure probability $p^*(\sigma)$ for an infinite body with $\sigma_0(x)$ for $m = 3$ and 11. Gradient parameters of $-b(2R) = \{0, 5, 10, 20\}$ are displayed. The tendency for fracture to initiate away from the crack tip is apparent as $p^*(\sigma)$ can reach a local maximum with a value exceeding p^* at ρ.

Table 2
Parameters for local maxima in cracking probability

	$m = 3$		$m = 11$	
For $\sigma_0(x)$, $c = 10$	$\rho b = 0.01$	$\rho b = 0.0001$	$\rho b = 0.01$	$\rho b = 0.0001$
$-r_x b$	3.26	3.26	2.69	2.69
$\sigma/\sigma_0\vert_{r_x}/\sigma/\sigma_0\vert_\rho$	0.55	0.055	0.61	0.061
$p^*(r_x)/p^*(\rho)$	54.4	5.45	1.12	8.65×10^{-10}
For $E(x)$, $c = \infty$	$\rho b = 0.01$	$\rho b = 0.0001$	$\rho b = 0.01$	$\rho b = 0.0001$
$r_x b$	1.7	1.7	0.6^a	0.6^a
$\sigma/\sigma_0\vert_{r_x}/\sigma/\sigma_0\vert_\rho$	0.14	0.014	0.19	0.019
$p^*(r_x)/p^*(\rho)$	0.475	0.0467	5.56×10^{-7}	5.50×10^{-16}

[a] Inflection only for this condition.

If $m < 4$, the cracking sites are spread widely throughout the crack-tip stress field as for a homogeneous material. As can be appreciated from Fig. 12, if $p^*(r_x)/p^*(\rho) \gg 1$, relatively few failures will occur near the crack-tip region itself. Generally, if $m < 4$ and cracking frequencies exhibit a local maxima, cracking will largely be distributed between r_x and R. Following the logic leading to Eqs. (10a) and (10b), the toughness will vary as

$$K_\Phi \approx \frac{\sigma_0\left(\frac{V_0}{B}\right)^{1/m}\sqrt{2\pi}\left(2 - \frac{m}{2}\right)^{1/m}(-\ln(1 - \Phi))^{1/m}}{\left[\int_{-\pi}^{\pi} f_1(\theta, \psi)^m \, d\theta\right]^{1/m}} \frac{\sigma_0(\infty)E(0)}{\sigma_0(0)E(\infty)} \bigg/ R^{(2/m)-(1/2)} \quad (m < 4). \tag{12a}$$

This limiting situation implies that virtually all the cracking occurs at $r > 2/b$, where the tanh functions are saturated. For exact results in the plots of $K_\Phi(b)/K_\Phi(0)$ shown in Fig. 7 for $\sigma_0(x)$ with $b < 0$, and in Fig. 8 for $E(x)$ with $b > 0$, the relative toughnesses for $m = 2$ and 3 are approaching, but have not attained, these limits, $\sim 1/c_\sigma$ and 1/2, respectively, which are insensitive to b and would obtain for $|b(2R)| \gg 20$. Owing to the relative insensitivity to behavior at $r \ll R$, the toughness does not reflect the actual value of $p^*(r_x)/p^*(\rho)$, as is verified by comparing the behavior in Table 2 for $m = 3$, with the exact results in Fig. 7, for $b < 0$ or in

Fig. 8, with $b > 0$, where $K_\Phi(b)/K_\Phi(0)$ for both is insensitive to the value of ρ despite the large differences in $p^*(r_x)/p^*(\rho)$ (Table 2).

For higher m, behavior is determined by whether cracks form just beyond ρ, as in a homogeneous material, with K_Φ nearly unchanged from Eqs. (10a) and (10b) and insensitive to b, or instead are more widely distributed. In behavior more likely to pertain for intermediate m, crack locations would be clustered near r_x if b were large enough that $p^*(r_x) > p^*(\rho)$. From Eqs. (B.4a) and (B.4b), this depends on b and c as well as on the relevant value of ρ needed to describe the fine-scale microstructural variations. For the cases involving $\sigma_0(x)$ illustrated in Fig. 7, where the curves have begun to decline sharply with increasing $-b$, the transition to failure being near r_x has initiated. The transition is largely complete at $-b(2R) = 20$, for $m < 11$ with $b = 0.01$ and for $m < 6$ with $b\rho = 10^{-4}$. This shift cannot occur with a modulus gradient unless $br_0 > 0$. In the limit that most failures occur just at and beyond r_x

$$K_\Phi \propto \sigma_0 \left(\frac{V_0}{B}\right)^{1/m} \frac{\sigma_0(\infty)}{\sigma_0(0)} \frac{E(0)}{E(\infty)} \left(\frac{r_x b}{|b|}\right)^{(1/2)-(2/m)} \qquad (m > 4). \tag{12b}$$

Here, the expected toughness decreases with rising $|b|$ because the stress at the probable cracking location increases with $|b|$ for a given level of K; this follows from Eq. (12b) as $r_x b$ is nearly constant, as shown in Appendix B.

Thus, for FGMs loaded to have a region of high σ/σ_0 ahead of the crack tip, this fracture mechanism involving nonlocal failure is more likely to apply than for homogeneous materials; nonetheless, this applicability still depends, in part, on the value of ρ (or $b\rho$), which reflects other details at the crack tip that preclude direct crack extension being preferable.

The behavior with other gradient alignments or phase angles is more complicated. If a gradient is oriented so that the direction of high E/σ_0 is aligned with the region of high $f_1(\theta, \psi)$ for the stress field, a reduced toughness results with even modest gradient intensities; moreover, the region of damage will be much more focused for both degradation regimes (low and intermediate m) with a gradient present. This is shown in Table 3, which lists the median cracking angles θ_{50} and the dispersion in cracking angles, expressed as $\theta_{75} - \theta_{25}$, for several situations involving gradients in σ_0 with $b(2R) = \{-20, 0, 20\}$ and for $R/\rho = 10^3$. In particular, $\theta_{75} - \theta_{25}$ is smaller if $b < 0$ and $\psi = 0$.

In contrast, if the gradient alignment is opposite, the toughness may rise with modest gradient levels in low-m materials, but diminish for more intense gradients even with higher m values. This is clear in Fig. 7 for variations in $\sigma_0(x)$ and is implied by extrapolating the curves for gradients in $E(x)$ in Fig. 8 to higher levels of $-b$. In essence, material beyond the crack by $r > 1/b(\theta > 0)$ becomes relatively stronger than without a gradient, and so total failure probabilities can be reduced for m low enough that cracking at large r is relevant. However, the degradation in K_Φ with a steeper gradient entails fracture at the crack flanks

Table 3
Effect of gradient on angular dispersion of first fracture locations

$\sigma_0(x)$	$m = 3$		$m = 11$		$\sigma_0(y)$	$m = 3$		$m = 11$	
b	θ_{50}	$\theta_{75} - \theta_{25}$	θ_{50}	$\theta_{75} - \theta_{25}$	b	θ_{50}	$\theta_{75} - \theta_{25}$	θ_{50}	$\theta_{75} - \theta_{25}$
$\psi = 0°$ ($\theta_{\sigma_{\theta\theta|max}} = 0°$)									
-20	$0°$	$70°$	$0°$	$64°$	-20	$73°$	$42°$	$72°$	$24°$
0	$0°$	$124°$	$0°$	$118°$	0	$0°$	$124°$	$0°$	$118°$
20	$0°$	$244°$	$0°$	$124°$	20	$-73°$	$42°$	$-72°$	$24°$
$\psi = 45°$ ($\theta_{\sigma_{\theta\theta|max}} = 52°$)									
-20	$-19°$	$51°$	$-27°$	$28°$	-20	$+46°$	$39°$	$-83°$	$80°$
0	$-74°$	$100°$	$-88°$	$72°$	0	$-74°$	$100°$	$-88°$	$72°$
20	$-146°$	$32°$	$-150°$	$25°$	20	$-90°$	$57°$	$-90°$	$38°$

(which is normally unlikely), as shown in Table 3 for $\sigma_0(x)$ at $b > 0$ for which $\theta_{75} - \theta_{25}$ has become very large (244°). From inspection of Fig. 1, it can be seen that if a cracking condition that depended upon the hoop stress, $\sigma_{\theta\theta}$, were pertinent, the tendency of weakening effects to outweigh local strengthening would be lessened owing to the narrower angular range of singular crack-tip stresses.

Based on the tendency for weakening to predominate, it is not surprising that for nonparallel gradient alignments, a decline in initiation toughness generally follows, as for $E(y)$ (Fig. 10), for which the toughness declines with b almost as rapidly as when the gradient is aligned along $E(x)$ with $b > 0$ (Fig. 8).

The probable angle of crack initiation is obviously shifted by the presence of a gradient. It should be noted that the angular location of cracking even when it occurs near ρ can be markedly changed by relatively small levels of b, i.e., at levels having only minimal affects on K_Φ. This emerges from comparing the levels of b that markedly change the crack angle with a gradient in $\sigma_0(y)$ (Fig. 9) with the much larger levels of $-b$ needed to alter the toughness with $\sigma_0(x)$ (Fig. 7). This obtains because the angular dependence of the crack-tip stresses, $f_1(\theta, \psi)$, especially for σ_1, is rather weak over a wide interval, Fig. 1a, and so the angle of the highest stress can be shifted even by a shallow material property gradient.

An equally important issue concerns whether the range in probable crack initiation angles, or the difference in expected initiation angle, ϕ, and stable extension angle, θ^*, is made larger or narrower by the presence of a gradient. Expected trends are again illustrated in Table 3. For situations in which the range of initiation angles is raised with increasing gradient, a microcrack or bifurcation zone would be wide, or more diffuse, especially for low m. This should raise the difference between the loads for first cracking and final failure, i.e., steepen the R-curve. Where the zone of probable cracking is narrower, i.e., focused, the fracture surface will be less rough and the R-curve relatively flatter, e.g., for σ_0 $(b < 0)$.

For mixed-mode loading, the expected microcrack zone tends to be less wide (Table 3). However, the average location depends strongly on m and is unlikely to be on the path of the eventual main crack $(\phi \neq \theta^*)$. It is assumed that subsequent cracking would statistically tend to be along the path of maximum $G/G_c(x, y, \theta)$, and the resistance could be heterogeneous as well. When these angles differ, strongly rising R-curve effects could also be very significant.

Thus, the presence of a strength gradient would almost invariably lower the toughness for first cracking; however, unless the weak material were largely ahead of the crack, the corollary effect would be to markedly widen the damage zone perhaps enough to induce a net increase in final toughness if a substantial trend develops toward first cracking occurring near the crack flanks. For modest modulus gradients, the effects on first cracking and final cracking would tend to be the same rather than offsetting unless the gradient were nearly normal to the crack. Thus, situations with some toughening both for first cracking and from a rising R-curve can exist.

Finally, it should be emphasized that if the position of the crack tip is shifted such that $br_0 \sim 1$, the qualitative differences between gradients in σ_0 and E will be switched (as follows from Eqs. (B.1)–(B.3), (B.4a), (B.4b)). That is more toughening would be accessible with σ_0 gradients and more degradation in K_Φ for initiation could derive with modulus gradients, but accompanied by opportunities for R-curve behavior.

5.3. FEA of finite-sized specimens

Although dependencies of the toughness and initiation angle on the modulus gradient emerge for the infinite body FGM (Section 5.2) due to the gradient-induced modification of the K-field, at finite r, other deviations from the classical K-field can also be expected. Even in homogeneous materials for cracks in finite bodies, boundary condition-induced higher-order terms in the Williams expansion will dictate the stress field at increasing distance from the crack tip. Thus, it should be appreciated that with a nonlocal fracture criterion, fracture toughness could depend upon the actual specimen size and geometry, especially for an FGM. By comparing the results for an infinite body (Sections 5.2) and those derived from FEM results for a specific specimen geometry, it can be determined how effects of geometry and gradient are likely to interact or dominate.

After describing the calibration of the single-edge crack tension SE(T) sample, several examples are given in which failure probabilities are computed based on stresses in the entire graded sample excluding a zone within either $\rho = a/10^5$ or $a/10^3$. These toughnesses are normalized to that of a homogeneous body ($b = 0$) for the finite sample, which differs in detail from that in Eqs. (10a) and (10b), and are, thereby, insensitive to the failure probability, Φ, chosen as a basis for comparison. Samples having gradients in either σ_0 or E aligned with the crack tip centered in the gradient ($\varsigma_\mathrm{o}, \xi_\mathrm{o} = 0$) are assessed.

5.3.1. Specimen calibration

A modulus gradient parallel to the direction of the crack, $E(x)$, retains the symmetry of the homogeneous problem so that $K_{\mathrm{II}} = 0$. For homogeneous materials with load boundary conditions, the stress-intensity factor is not influenced by Young's modulus; however, for an FGM, the magnitude of K_{I} is increased for $b < 0$ and decreased for $b > 0$ as shown in Fig. 13 for the SE(T) geometry with $a = W/2$ (Fig. 3). This behavior is consistent with other observations relating the far-field load and K_{I} for an FGM with $E(x)$ [3].

Deterministic failure comparisons based on applied loads can be directly inferred from the results of Fig. 13. This indicates that cracks growing into positive b gradients are expected to carry higher loads than homogeneous or negative b materials when fracture is determined by an invariant K_{Ic} or near tip $\sigma_{\theta\theta}$.

The effect of a modulus gradient normal to the crack plane, $E(y)$, is to take a nominally mode I geometry and load the crack tip in both tension and shear. The mixed-mode calibrations for the stress-intensity factors, K_{I} and K_{II}, are also shown in Fig. 13. K_{I} is an even function of b, $K_{\mathrm{I}}(b) = K_{\mathrm{I}}(-b)$ and K_{II} is odd. Both stress-intensity factors are amplified with increasing $|b|$, but with a resulting monotonic growth in the tip phase angle, $\psi = \tan^{-1}(K_{\mathrm{II}}/K_{\mathrm{I}})$. This trend seems to plateau for $b(2a) > 10$, as ψ shifts from 0° to 18° while $b(2a)$ goes from 0 to 10 but by only an additional 2° while $b(2a)$ goes from 10 to 20.

The subsequent results depend on the stress field of the entire sample, not just aspects pertinent to the asymptotic crack-tip field, Eq. (1). The deviations of the stresses for a homogeneous material from those of

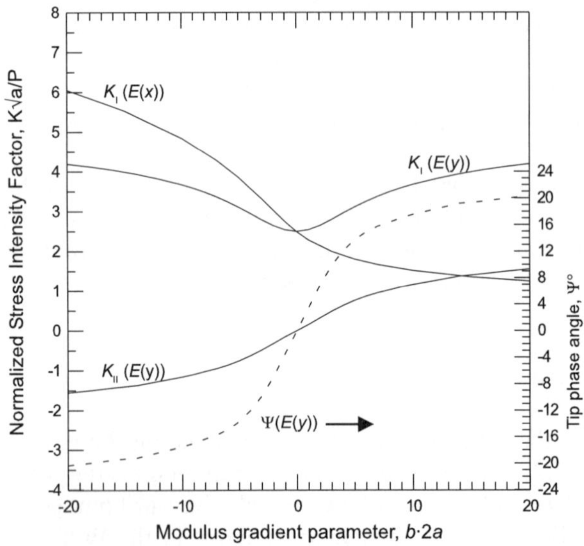

Fig. 13. Stress-intensity factors, K_{I} and K_{II} for the SE(T) fracture mechanics sample for modulus gradients $E(x)$ and $E(y)$ (solid lines). The gradient in the y-direction results in a shearing of the crack tip, such that that the tensile geometry results in a mixed-mode loading with ψ increasing in magnitude for increasing gradient (dashed line).

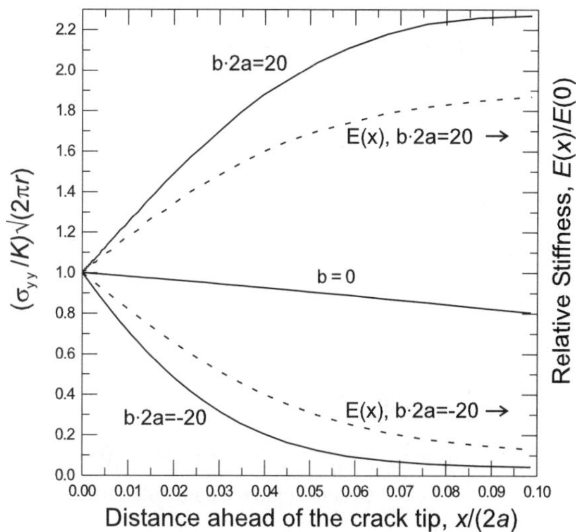

Fig. 14. Deviation of yy-stresses ahead of the crack tip from classical K-field values for an SE(T) specimen with $E(x)$ gradient for cases with $b(2a) = \{-20, 0, 20\}$. Normalized moduli $E(x)$ are plotted for comparison. The approximation in Eq. (2) would appear to better describe the near-tip stresses, but underestimates the magnitude of the gradient-induced adjustment.

the K-field are displayed in Fig. 14 (curve labeled $b = 0$). These stresses deviate weakly from those of the K-field compared to those with an $E(x)$ gradient present, also shown for $b(2a) = \{20, -20\}$.

5.3.2. SE(T) with parallel strength gradient, $\sigma_0(x)$

For an SE(T) sample with strength gradients $\sigma_0(x)$ but homogeneous modulus, E, the variation in the relative fracture toughness, K_Φ, is shown in Fig. 15 as a function of strength gradient with $b(2a) = \{-20, 20\}$. Fig. 15a displays a reduction in toughness for all gradients when $m < 7$. With the larger cutoff region (Fig. 15b), the degradation in toughness occurs for $m \leqslant 11$. The $\rho = a/10^5$ case (Fig. 15a) biases the results toward the homogeneous situation, which lessens the gradient effects on toughness. It is worth noting the general agreement between results in Figs. 7 and 15, the latter of which only utilize the modulus-corrected Williams expansion near the crack tip (but have an outer cutoff R). However, discernable differences exist even for intermediate m values for which most cracking occurs between ρ and r_x.

5.3.3. SE(T) with parallel modulus gradient, $E(x)$

The calculated statistical fracture toughness, K_Φ, for a range of modulus gradients $E(x)$ and uniform strength is plotted in Fig. 16. Clearly, for a crack growing into a stiffer material, the toughness of the structure is decreased for material with low m. Conversely, toughening occurs for cracks growing into a more compliant material.

These results, obtained via FEM analysis of the SE(T) geometry, utilize the complete stress field, and effects are due to both the modulus gradient and the finite geometry. Prior calculations with a linear $E(x)$, rather than a sigmoidal variation, exhibited similar trends, which differed in detail [29]. For either geometry, it can be seen that the loss of toughness with $b > 0$, based on the local K, is offset by the modulus shielding for comparisons based on applied load.

The trends of a large influence of a gradient occurring with low m correspond to the probable location of fracture initiation shifting away from the crack tip, to where the stress solution deviates most from the classical singular form, as depicted in Fig. 14 for the yy-component. These results indicate that although K is a valid scaling parameter for the stress field very near the crack tip for an FGM, the dependence of the

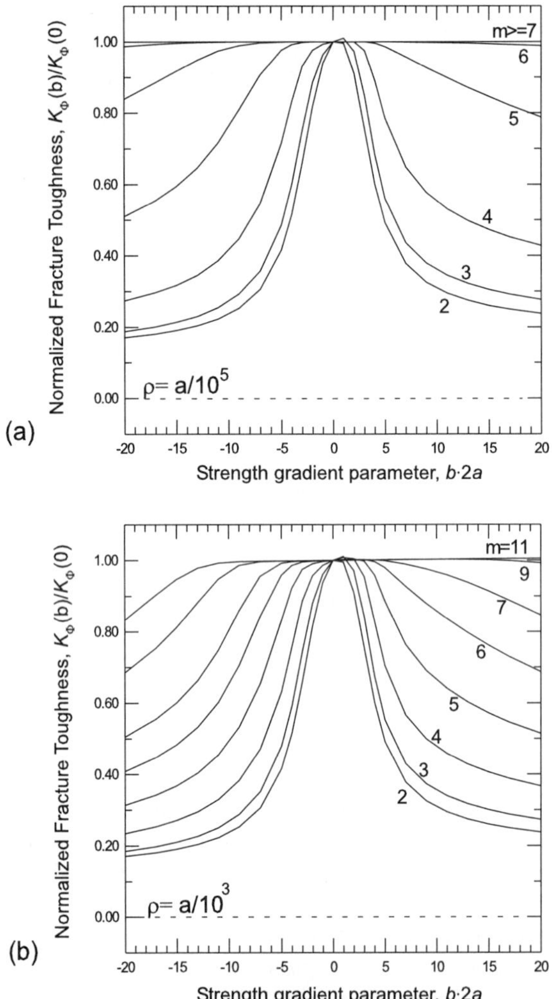

Fig. 15. Effect of variable strength $\sigma_0(x)$ on normalized fracture toughness of an SE(T) specimen. Two near-tip cutoff regions are used, $\rho = a/10^5$ (a) and $a/10^3$ (b).

stresses away from the crack on the gradient in E renders K_Φ to be an inaccurate predictor of failure if it initiates away from the tip. These effects arise because stresses ahead of the tip, and hence the cracking probability, are higher with a positive gradient than would exist for a homogeneous material at the same K.

The trends roughly agree in Figs. 8 and 16; however, the infinite body analysis only utilizes one correction to the singular term in the asymptotic expansion near the crack tip and underestimates the magnitude of gradient effects. The reason can again be seen in Fig. 14. Here the σ_{yy} stress ahead of the crack tip (along $y = 0$) normalized by the K-field value is plotted for $b = \{-20, 0, 20\}$; also plotted are the normalized variations in the elastic modulus for the same gradients. The homogeneous stress solution deviates slowly from the classical K-field, with exact agreement as $x \to 0$ (by nature of the fitting routine for K) and 10% deviation at $x/a = 0.1$. In contrast, the exact solutions with gradients deviate from the K-field by 10% at $x/a = 0.007$, meaning $x|b| = 0.13$. Eq. (2) describes the stress field in the FGM as that of the homogeneous field multiplied by the pointwise variation in modulus. Fig. 14 demonstrates that although this method-

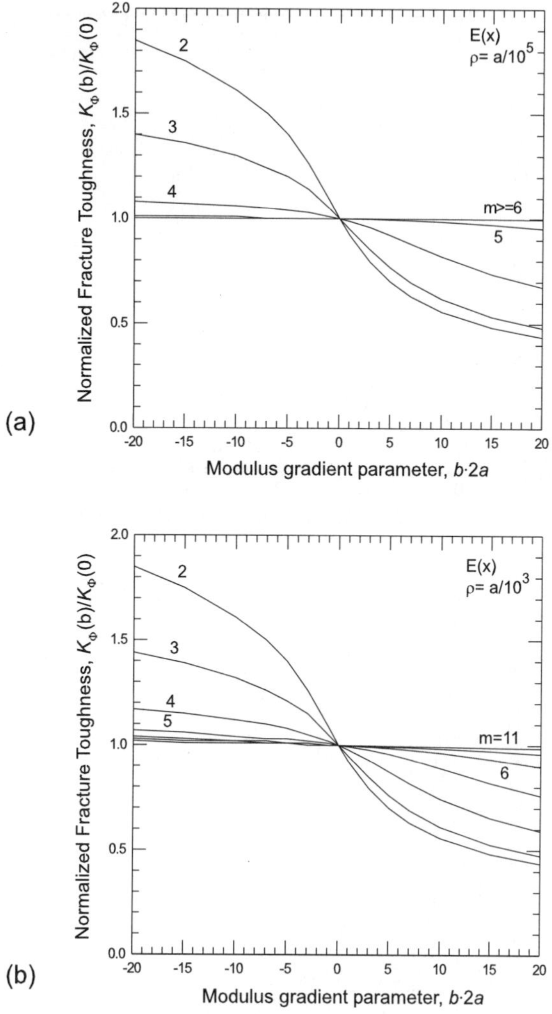

Fig. 16. Effect of variable modulus $E(x)$ on normalized fracture toughness of an SE(T) specimen. Two near-tip cutoff regions are used, $\rho = a/10^5$ (a) and $a/10^3$ (b).

ology yields a better estimate of the stress field than does the Williams K-field, it underestimates the true change for FGMs, even at $x|b| \ll 1$. Evidently, the infinite body analysis of Section 5.2.2 uses a stress field which varies too weakly with b and, especially for $b < 0$, the overall effect on fracture behavior is underestimated. For example, for $m = 3$ and $b = -10$ in the infinite body with the larger cutoff region, $K_\Phi(b)/K_\Phi(0) = 1.03$, and for the SE($T$) geometry, $K_\Phi(b)/K_\Phi(0) = 1.35$.

5.3.4. SE(T) with perpendicular strength gradient, $\sigma_0(y)$

For the SE(T) specimen with normal gradients in strength $\sigma_0(y)$ varying over $b = \{0, 5\}$, the average initiation angle is displayed in Fig. 17. The results agree well with those in Fig. 9, with a large negative initiation angle predicted for a positive gradient, $\phi b < 0$. The lower the value of m, the larger the average initiation angle will be, until $\phi = 90$ at sufficient b. As greater distances from the crack tip participate in the fracture event at low m, the region with lower strength material will predominate. For high m, (>9 for

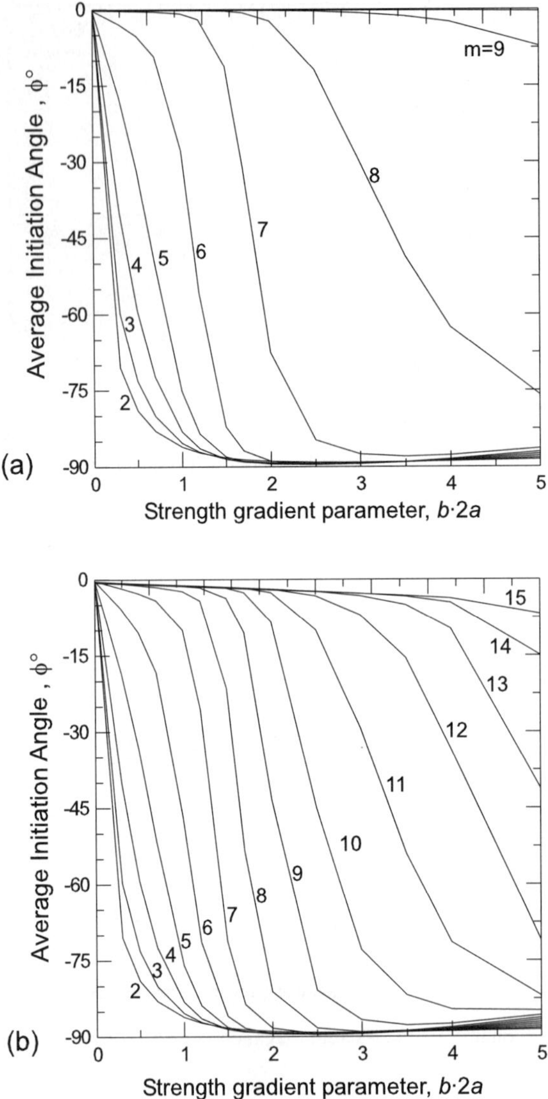

Fig. 17. Average initiation angle for the case of pure mode I loading and $\sigma_0(y)$ in an SE(T) specimen. Strength gradient skews average location of fracture such that nonzero initiation angle is preferred. Two near-tip cutoff regions are used, $\rho = a/10^5$ (a) and $a/10^3$ (b).

$\rho = a/10^5$, >15 for $\rho = a/10^3$), the very near-tip region dominates the calculation for the b levels explored, where the strength variation is limited and the symmetry of the stress field dictates that $\phi = 0$. Overall, the results are similar to those in Fig. 9; however, the sensitivity to increasing m is notably greater for the finite sample, indicating a role for the classical T-stresses (as E is uniform here), which are compressive and not particularly strong for this geometry [45].

5.3.5. SE(T) with perpendicular modulus gradient, E(y)

The average crack initiation angles for the SE(T) specimen with a modulus gradient $E(y)$ for a range of m from 2 to 15 are shown in Fig. 18(a). As ϕ is an odd function of gradient strength, only a range of

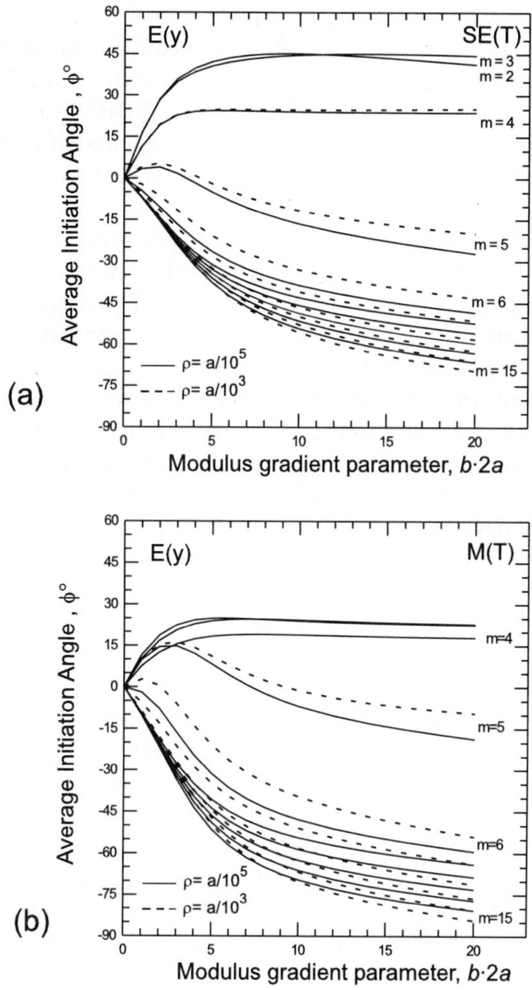

Fig. 18. Average initiation angle with a modulus gradient $E(y)$ in (a) single-edge notched tension, SE(T), and (b) middle-cracked tension, M(T), geometries. The gradient induces a positive crack-tip phase angle as seen in Fig. 13, but the gradient-induced higher-order stresses cause a positive initiation angle for very low m in both geometries.

$b = \{0, 20\}$ is plotted. The dramatic effect of m is evident. First, for high $m(>10)$, positive gradients yield negative initiation angles, i.e., $\phi b < 0$. The high-stress near-tip region of the sample, which experiences a shift in phase angle, dominates these results. As seen for the infinite body analysis for both homogeneous material and FGMs, positive phase angles lead to crack initiation angling downward, $\phi < 0$ for $\psi > 0$.

In contrast, for very low values of $m(\leqslant 4)$, the average direction of initiation is actually positive for positive gradients, $\phi b > 0$. This is, in part, in response to higher stresses associated with the increasing elastic modulus in the positive y-direction, which are only important when cracking is widely distributed. In comparison with the results in Fig. 11, this effect of a gradient with low m is more drastic. This is again due to the fact that Eq. (2) underestimates the effect of $E(y)$ on stresses at finite distances from the crack tip, including treating all components in proportion.

5.3.6. M(T) specimen with perpendicular modulus gradient, E(y)

To further elucidate the effect of geometry in the presence of a modulus gradient, a middle-cracked tension (M(T)) fracture mechanics specimen was also analyzed with a variation $E(y)$. When homogeneous, this geometry is known to develop a markedly different stress distribution outside of the K-dominant region. The nonsingular compressive T-stress in the M(T) sample is nearly five times that in the SE(T) sample [45]. Also, there is a change in sign in the deviation of the σ_{yy} stress from the K-field due to higher-order geometry terms [46].

The two salient trends, namely tip-dominated, negative angles for $m > 7$, and positive average initiation angles for $m = 2$–4 are evident for both sample geometries. However, the initiation angles for large b are shifted to about 20° lower values for all m for the M(T) sample. The other differences are mainly expressed for the intermediate m where neither tip nor very far-field stresses dominate. It is interesting to note that the phase angle–gradient relationship for these two samples was found to be very similar, thus indicating that the differences in Fig. 18 arise on the basis of b through higher-order stress terms and not simply through $\psi(b)$ at the tip. At low m, the crack initiation trends in Fig. 18(a) and (b) for the two samples are more similar, compared to the infinite case, Fig. 11, but differences are notable.

Note that recent FEM calculations for a small crack show that the shift in phase angle induced by a normal modulus gradient only develops at a dimension of $rb \sim 1$ [18]. If the effect is similar for these large cracks, then for $m < 4$ material, where most cracking initiates at $rb > 1$, the relevant phase angle would be small and the effect of higher stresses in the stiffer material, at $y > 0$, would appear to dictate the crack locations giving $\phi > 0$.

5.3.7. Implications for fracture experiments and application to FGMs

Effects with far-field mode I loading of FGMs with variable $\sigma_0(x)$, $\sigma_0(y)$, $E(x)$ and $E(y)$ and the crack tip centered in the gradient have been explored. For more general cases with far-field mixed-mode loading and an FGM with arbitrary $\sigma_0(x,y)$ and $E(x,y)$, the space of possible combinations is prohibitively large to survey completely. The results thus far presented for $m > 6$ indicated that behavior is somewhat similar between the finite and infinite bodies. For both situations, the gradient is more influential with larger ρ, and the general similarities indicate that the trends regarding crack location discussed in Section 5.2.5 pertain for finite geometries. Therefore, Eq. (2) for analysis of infinite bodies can be used as in Eq. (11) to anticipate some results of cases not explored here; however, the geometry-induced shifts in phase angle with elastic modulus gradient must be known. Using Eq. (11), it can be observed that as a plausible approximation for real materials, $\sigma_0 \propto E$, and so some of the gradient effects could tend to compensate; nonetheless, large effects of induced ψ, Figs. 5 and 6, still pertain, which may ultimately prove to be more influential at higher m (Figs. 11 and 18).

For lower and intermediate m, the fracture behavior may not be adequately determined by the stresses even within the gradient-corrected K-field of a finite geometry. Even when failure occurs near the crack tip, gradient-induced perturbations influence crack angles (see discussion concerning, Figs. 9 and 11, Section 5.2.5). For these cases, the combined effects of near crack tip, gradient induced and geometry determined stresses must be taken into account. This requires numerical determination of the stress field, e.g., through FEM analyses, which is a more expensive procedure than the semi-analytic modeling of infinite bodies. It is reiterated that when such effects yield first fracture locations that are well away from the eventual crack path for complete failure, various mechanisms may be triggered that can raise the ultimate toughness.

Note that for ductile materials, the nonlinear fracture mechanics is most simply characterized using a single parameter, J; however, this still leads to an appreciable geometry-to-geometry variability in measured crack-growth resistance for many materials. In those materials, the T-stress (or, nearly equivalently, the constraint factor, Q) has been evoked to account for the effects of the next term in the nonlinear stress field expansion about the crack tip [19,47]. This two-parameter characterization allows the description of the material not only at the crack tip, but also at finite distances, and thereby is more successful in determining critical conditions for crack initiation and growth.

The characterization of the stress field in an FGM by the single parameter K has similar limitations. The approximate formulation (Eq. (2)) alludes to the form of the next higher-order term in the expansion. Although in the asymptotic near-tip limit one could also assume the existence of a parallel T-stress, this suffers from the same incompatibility discussed in Section 2. The next-order stress term will therefore not be strictly constant, but using near-tip stresses in an FGM to calculate an effective T-stress may prove to allow for comparison with known effects in homogeneous materials. In a recent study by the authors [18], a kink-length effect was found for the preferred (deterministic) kink angle in an FGM. For kink lengths shorter than the gradient dimension, $1/b$, this effect can almost entirely be accounted for in terms of this effective T-stress, which may depend more on modulus gradient than finite geometry.

It seems clear that the addition of a gradient-induced T-stress can be used to more accurately describe the stress field in an FGM and also provide a more powerful two-parameter scheme to predict statistical crack behavior in different geometries and gradients, at least for the intermediate and high-m material. For longer kink lengths in the above mentioned study, other factors came strongly into play, in part owing to the reduction in induced phase angle for $rb > 1$ already mentioned. Thus, for materials with $m < 4$, in which cracking occurs over dimensions far exceeding $1/b$, further corrections will also be needed. (Actually, ignoring the modulus induced phase shift in conjunction with using Eq. (2) may be a useful approximation at lower m.)

Finally, there are situations in which the stress fields decay rapidly as occurs under an indenter. Then, effects of nonlocal cracking on fracture criteria and causing distributed damage could be great, but using a description based on an asymptotic stress field will be of limited utility.

6. Summary and conclusions

The application of two-parameter Weibull statistics to near-crack fracture problems has been elaborated and extended to the case of mixed-mode loading, with predictions made for both the toughness and average initiation angle of a crack in a brittle material. Conditions have been explored to justify the use of a nonlocal fracture model for brittle materials. This fracture model allows for the statistical correlation between near-tip stresses and first fracture for heterogeneous materials, and possibly for predicting the emergent shape of microcrack damage zones. This is especially interesting for problems in composite-type graded materials (FGMs), where stress fields vary from the homogeneous form away from the crack tip and the strengths can be highly variable throughout. A combination of infinite-body and finite geometry analyses reveal that

- For statistically homogeneous materials, the mixed-mode loading of an elastic crack subject to a nonlocal stress-controlled failure criteria yields a reduction of the fracture toughness by up to 30%, depending on the phase angle ψ and Weibull modulus m. This is roughly 150% of the effect predicted by mechanisms dictated by $\sigma_{\theta\theta}$, usually invoked for continuous kinking.
- The location of microcracks initiated by the main crack stress field do not coincide with the trajectory determined by maximum energy criteria, usually being at larger angles. Thus, diffuse damage zones and crack bridging may develop, thereby, expectedly leading to R-curve toughening for materials that fail by crack initiation away from the main crack.
- Gradients in Weibull scaling stress σ_0 lead to a dramatic decrease in initiation toughness for most alignments of the gradient. However, for cracks growing into stronger, $b > 0$, materials or with gradients normal to the crack $\sigma_0(y)$, weakening for first fracture will involve crack growth initiating at angles that may induce R-curve toughening that will offset this drop in toughness for final failure.

- Calibration of effects of gradients in modulus $E(x)$ for a single-edge notched fracture mechanics specimen show an increased stress-intensity factor for cracks growing into more compliant materials, and crack-tip shielding when growing into a stiffer material.
- When comparing FGMs with gradients in Young's modulus, E, and the same crack-tip stress intensity, the crack growing into the steeper negative gradient, $b < 0$, will be tougher for moderate gradients; whereas most other alignments and very steep gradients lead to a loss of initial toughness.
- Gradients in modulus $E(y)$ for a SE(T) or M(T) fracture mechanics sample lead to initiation behavior qualitatively similar to that in homogeneous materials having the same phase angle for $m > 5$, with the average angle $\phi < 0$ for $b > 0$; for lower Weibull modulus though, further effects of an elastic modulus gradient on the stress field cause the average fracture initiation site to be skewed toward the stiffer (higher stress) region, $\phi > 0$ for $b > 0$.
- Infinite-body analyses predict many of the gradient-induced features seen in a finite geometry. When comparing two fracture-mechanics specimen geometries, additional gradient effects on crack-initiation angle were found to be important. There is scope for further analysis that explores the several higher-order terms indicated to be important in the stress fields for FGMs.
- The evolution of damage near a crack tip will vary strongly with the Weibull modulus, m, of the material. For low m materials, a distribution of damage is expected before catastrophic crack advance. The effects of considering "action-at-a-distance" are many-fold. Most importantly, the first flaws to be activated in a mixed-mode loading are most likely not in line with what is considered to be the energetically most favorable path. Therefore, it is expected that materials with wide distributions of strength will develop diffuse damage zones around the crack tip. At elevated loads, one expects that these microcracks will link up, or the main crack will jump beyond this damage zone. Under either scenario, the tearing or bridging that results will act as an R-curve mechanism. These effects could be greater in FGMs than for nominally homogeneous materials.

Acknowledgements

This work was supported by Director, Office of Science, Office of Basic Energy Sciences, Materials Sciences Division of the US Department of Energy under Contract no. DE-AC03-76SF00098. The authors would like to thank Prof. Panos Papadopoulos for supplying the finite element code FEAP, and Dr. James M. McNaney and Prof. Daryl Chrzan for helpful discussions regarding Section 3.

Appendix A. Monte Carlo simulations

To illustrate the effects of the elastic near-tip stress field on a material with strength described by Weibull statistics, a Monte Carlo simulation was performed. A domain, 200 units square, was considered with a crack tip at the center $\{0,0\}$. For the reasons discussed in Section 3.2, a region (with radius 1 unit) was excluded surrounding the crack tip. For each loop in the simulation algorithm:

- Select random location within the domain (outside of the region of exclusion).
- For element with size $\Delta V = r\,\delta r\,\delta\theta$, centered at $\{x,y\}$, compute $\Phi^{\Delta V}$ from Eq. (3) and Williams stress field (Eq. (1)), $\delta r = 0.1$ unit and $\delta\theta = 10°$.
- Select random $\{\Phi_{\text{test}}^{\Delta x}\}$ (between 0 and 1)

  $\Phi_{\text{test}}^{\Delta x} < \Phi^{\Delta x} \rightarrow$ "hit"; the value of $\{x,y\}$ is stored as $\{x,y\}_i^{\text{hit}}$; $N^{\text{hits}} = N^{\text{hits}} + 1$

  $\Phi_{\text{test}}^{\Delta x} > \Phi^{\Delta x} \rightarrow$ "miss"; the value of $\{x,y\}$ is discarded

Fig. 1 displays the results of 5×10^6 runs. From these results, averages can be calculated from the location of the "hits".

$$x^{\text{ave}} = \frac{1}{N^{\text{hits}}} \sum_{i=1}^{N^{\text{hits}}} x_i^{\text{hit}} \qquad y^{\text{ave}} = \frac{1}{N^{\text{hits}}} \sum_{i=1}^{N^{\text{hits}}} y_i^{\text{hit}}.$$

For this average location, the angle is $\tan^{-1}(y^{\text{ave}}/x^{\text{ave}})$. For the mode I case, this is 4.3°. For a phase angle of $\psi = 45°$ case, the angle of the average location is 80.7°. The different mode loading also resulted in a greater number of "hits" (fracture initiations), with 111 hits for the mode-case and 764 for the mixed-mode case.

Appendix B. Derivation of the weighted probability function maxima and minima

As discussed in Section 5.2.5, the incremental fracture probability in an annulus at a given radial distance from the tip depends on $\int_{-\pi}^{\pi} p(\sigma) r \, d\theta$. As such, important trends are revealed by the behavior of the weighted probability function $p^*(K, r) \equiv p(\sigma r) r$, where $p(\sigma)$ is given in Eq. (11).

When the gradient is aligned along a direction of high $f_1(\theta, \psi)$ with b selected to give relative weakening ahead of the crack, local minima, at r_n, and maxima, at r_x, in failure probability can occur. The locations can be found based on $\mathrm{d}p^*/\mathrm{d}r = 0$, and it can be ascertained by inspection that these occur such that $0 < r_n < r_x$.

With only a gradient in σ_0 present, along the ray of greatest weakening ($\theta = -\theta_\sigma$ and $\varsigma \to -r$), the local minimum and maximum in $p^*(K, r)$ are at positions that satisfy

$$br = \left(\frac{m-2}{8m} \right) \left(\frac{c_\sigma}{c_\sigma - 1} \right) \left[\frac{e^{2b(r-r_0)}}{c_\sigma} + 2 + \frac{(2c_\sigma - 1)}{c_\sigma} e^{-2b(r-r_0)} \right]. \tag{B.1}$$

By inspection, it can be shown that for $c_\sigma > 2.3$, such a maximum exists for nearly all $m > 2$; if $r_0 = 0$, the maximum typically occurs where $r_x b \approx 2\text{–}3$, being larger at lower m. Instead, with a gradient in E, along a ray aligned toward the stiffer direction ($\theta = \theta_E$ and $\xi \to r$), which corresponds to that of rising stress, then r_n and r_x, satisfy

$$br = \left(\frac{m-2}{8m} \right) \left(\frac{c_E}{c_E - 1} \right) \left[\frac{e^{2b(r-r_0)}}{c_E} + 2 + \frac{(2c_E - 1)}{c_E} e^{-2b(r-r_0)} \right]. \tag{B.2}$$

The existence condition for a maximum, derived assuming $b(r - r_0) \equiv b\Delta r \gg 0$ (so that $\exp -b\Delta r \sim 0$), can be expressed as

$$br > \left(\frac{m-2}{4m} \right) \left(\frac{c_E}{c_E - 1} \right) + \frac{1}{2} - \frac{1}{2} \ln \left[\left(\frac{4m}{m-2} \right) \left(\frac{c_E - 1}{2c_E - 1} \right) \right]. \tag{B.3}$$

For large c_E and $r_0 = 0$, a maximum exists only if $m < 4.51$; usually $r_x b \approx 1\text{–}2$, being larger at smaller m.

For either type gradient, if $m = 2$, then $r_n \to 0$ and $r_x \to \infty$. The rise in $\sigma(r)$ owing to the change in $E(\infty)/E(0)$ is only $(2c_E - 1)/c_E$ if $r_0 = 0$; so an actual maximum is less likely to occur than with a gradient in σ_0 where the corresponding ratio is $1/c_\sigma$. However, with a crack tip displaced back from the gradient center, $br_0 > 0$, the modulus change ahead of the crack tip is greater and the behavior would be more like that with a gradient in σ_0 and $r_0 = 0$, as can be seen better in the following.

One important issue concerns the strength of the maxima, in particular the relative values of σ/σ_0 and especially of p^* at r_x vs. at the cutoff distance ρ. The latter is described by

$$\frac{p^*(r_x)}{p^*(\rho)} = \left(\frac{\rho b}{r_x b} \right)^{(m/2)-1} \left(\frac{c_\sigma}{-(c_\sigma - 1)\tanh(r_x b) + c_\sigma} \right) \approx \left(\frac{\rho b}{r_x b} \right)^{(m/2)-1} c_\sigma^m, \tag{B.4a}$$

$$\frac{p^*(r_x)}{p^*(\rho)} = \left(\frac{\rho b}{r_x b}\right)^{(m/2)-1} \left(\frac{-(c_E - 1)\tanh(r_x b) + c_E}{c_E}\right) \approx 2^m \left(\frac{\rho b}{r_x b}\right)^{(m/2)-1}, \tag{B.4b}$$

respectively, for a gradient in σ_0 and for a gradient in E with c_E being large.

From Fig. 1a, it can be seen that the major contributions to the angular integrations occur over an interval often exceeding $-2\pi/3 < \theta < 2\pi/3$ for the principal stress, but this does depend on the gradient strength. Thus, log–log plots of $\int_{-\pi}^{\pi} p(\sigma) r \, d\theta$ vs. r would be similar to that of p^* vs. r in Fig. 12, but adjusted owing to the r- and b-dependence of the angular integration. Specifically, with the gradient aligned parallel to the crack so as to give relative weakening, these adjustments would reduce the effects of stress variations at very high angles from the crack plane and so flatten the local maximum somewhat. They would also slightly increase the distance needed to reach the second asymptotic portion of the curve and would reduce the level of that portion of the curve for low m by a factor near unity from that expected based on the ratio of $\sigma/\sigma_0|_\infty / \sigma/\sigma_0|_0$ along the gradient. Owing to this rather weak sensitivity of the angular term, the results from Eqs. (10a) and (10b) can be applied to give estimates of the limiting toughness for large b for two situations; these are given in Eqs. (12a) and (12b) based upon values from Eqs. (B.1) and (B.2).

It is unclear, however, how these relations pertain in the limit as $b \to \infty$ where other pathologies described for interface cracks would develop, such as singularities in K (e.g., Refs. [16,48]). However, the form of the phase shifts for cracking parallel to an interface are quite different than that for the graded material and a T-stress develops at dimensions small compared to the gradient change that has no counterpart in the solutions for sharp interface cracks which pertain only for $rb \gg 1$ [18].

References

[1] Parsons WB. Engineers and engineering in the renaissance. Baltimore, MD: Williams and Wilkers; 1939.

[2] Weibull W. A statistical distribution function of wide applicability. J Appl Mech 1951;18:293.

[3] Erdogan F. Fracture mechanics of functionally graded materials. Compos Engng 1995;5:753–70.

[4] Ishibashi H, Tobimatsu H, Matsumoto T, Hayashi K, Tomsia AP, Saiz E. Characterization of Mo–SiO$_2$ functionally graded materials. Metall Mater Trans A 2000;31A:299–308.

[5] Lee CS, DeJonghe LC, Thomas G. Mechanical properties of polytypoidally joined Si$_3$N$_4$/Al$_2$O$_3$. Acta Mater 2001;49:3767–73.

[6] Sand Ch, Adler J, Lenk R. A new concept for manufacturing sintered materials with a three-dimensional composition gradient using a silicon carbide–titanium carbide composite. Mater Sci Forum 1999;308–311:65–70.

[7] Krell T, Schulz U, Peters M, Kayseer WA. Graded EB-PVD alumina–zirconia thermal barrier coatings—an experimental approach. Mater Sci Forum 1999;308-311:396–401.

[8] Erdogan F. The crack problem for bonded nonhomogeneous materials under antiplane shear loading. J Appl Mech 1985;52:823–8.

[9] Jin Z, Noda N. Crack-tip singular fields in nonhomogeneous materials. J Appl Mech 1994;61:738–40.

[10] Williams ML. On the stress distribution at the base of a stationary crack. J Appl Mech 1957;79:109–14.

[11] Konda N, Erdogan F. The mixed mode crack problem in a nonhomogeneous elastic medium. Engng Fract Mech 1994;47:533–45.

[12] Jin Z-H, Batra RC. Some basic fracture mechanics concepts in functionally graded materials. J Mech Phys Solids 1996;44:1221–35.

[13] Becker TL, Cannon RM, Ritchie RO. An approximate method for residual stress calculation in functionally graded materials. Mech Mater 2000;32:85–97.

[14] Erdogan F, Sih GC. On the crack extension in plates under plane loading and transverse shear. J Basic Engng 1963;85:519–27.

[15] Cotterell B, Rice JR. Slightly curved or kinked cracks. Int J Fract 1980;16:155–69.

[16] Hutchinson JW, Suo Z. Mixed mode cracking in layered materials. Adv Appl Mech 1992;29:64–191.

[17] Gu P, Asaro RJ. Crack deflection in functionally graded materials. Int J Solids Struct 1997;34:3085–98.

[18] Becker TL, Cannon RM, Ritchie RO. Finite crack kinking and T-stresses in functionally graded materials. Int J Solids Struct 2001;38:5545–63.

[19] Larsson SG, Carlsson AJ. Influence of non-singular stress terms and specimen geometry on small-scale yielding at crack tips in elastic–plastic materials. J Mech Phys Solids 1973;21:263–77.

[20] Ishibashi H, Saiz E, Tomsia AP. Powder processing of Mo/SiO$_2$ FGMs fabricated by slip casting. Am Ceram Soc Bull 1997;76:231.

[21] Ritchie RO, Rice JR, Knott JF. On the relationship between critical stress and fracture toughness in mild steel. J Mech Phys Solids 1973;21:395–410.

[22] Lin T, Evans AG, Ritchie RO. A statistical model of brittle fracture by transgranular cleavage. J Mech Phys Solids 1986;34:477–97.

[23] Lei Y, O'Down NP, Busso EP, Webster GA. Weibull stress solutions for 2D cracks in elastic and elastic–plastic materials. Int J Fract 1998;89:245–68.

[24] Lin T, Evans AG, Ritchie RO. Statistical analysis of cleavage fracture ahead of sharp cracks and rounded notches. Acta Metall 1986;34:2205–16.

[25] Hutchinson JW. Singular behavior at the end of a tensile crack in a hardening material. J Mech Phys Solids 1968;16:13–31.

[26] Rice JR, Rosengren GF. Plane strain deformation near a crack tip in a power law hardening material. J Mech Phys Solids 1968;16:1–12.

[27] Beremin FM. A local criterion for cleavage fracture of a nuclear pressure vessel steel. Metall Trans A 1983;14A:2277–87.

[28] Evans AG. Statistical aspects of cleavage fracture in steel. Metall Trans A 1983;14A:1349–455.

[29] Becker TL, Cannon RM, Ritchie RO. A statistical RKR fracture model for the brittle fracture of functionally graded materials. Mater Sci Forum 1999;308–311:957–62.

[30] Zienkiewicz OC, Taylor RL. The finite element method. New York, NY: McGraw-Hill; 1987.

[31] Stern M, Becker EB. A conforming crack tip element with quadratic variation in the singular fields. Int J Numer Meth Engng 1978;12:279–88.

[32] He M-H, Hutchinson JW. Kinking of a crack out of an interface. J Appl Mech 1989;56:270–8.

[33] Rühle M, Claussen N, Heuer AH. Transformation and microcrack toughening as complementary processes in ZrO_2-toughened Al_2O_3. J Am Ceram Soc 1986;69:195–9.

[34] Rühle M, Evans AG, McMeeking RM, Charalambides PG, Hutchinson JW. Microcrack toughening in zirconia toughened alumina. Acta Metall 1987;35:2701–10.

[35] Wu CCm, Freiman SW, Rice RW, Mecholsky JJ. Microstructural aspects of crack propagation in ceramics. J Mater Sci 1978;13:2659–70.

[36] McClintock FA. A three-dimensional model for polycrystalline cleavage and problems in cleavage after extended plastic flow or cracking. In: Chan KS, editor. Cleavage fracture. George R Irwin Symp. Warrendale, PA: TMS; 1997. p. 81–94.

[37] Mai Y-W, Lawn BR. Crack stability and toughness characteristics in brittle materials. Ann Rev Mater Sci 1986;16:415–39.

[38] Hutchinson JW. Crack tip shielding by micro-cracking in brittle solids. Acta Metall 1987;35:1605–19.

[39] Evans AG. High toughness ceramics. Mater Sci Engng 1989;A105–106:65–75.

[40] Becher PF. Microstructural design of toughened ceramics. J Am Ceram Soc 1991;74:255–69.

[41] Rödel J. Interaction between crack deflection and crack bridging. J Eur Ceram Soc 1992;10:143–50.

[42] McClintock FA, Zaveral F. An analysis of the mechanics and statistics of brittle crack initiation. Int J Fract 1978;25:107–18.

[43] Rice JR, Johnson MA. The role of large crack tip geometry changes in plane strain fracture. In: Kanninen M, Adler W, Rosenfield AR, Jaffee R, editors. Inelastic behavior of solids. NY: McGraw-Hill; 1970. p. 641–72.

[44] Hong T, Smith JR, Srolovitz DJ. Theory of metal-ceramic adhesion. Acta Metall Mater 1995;43:2721–30.

[45] Sherry AH, France CC, Goldthorpe MR. Compendium of *T*-stress solutions for two- and three-dimensional cracked geometries. Fatigue Fract Engng Mater Struct 1995;18:141–55.

[46] Knott JF. Fundamentals of fracture mechanics. London: Butterworth; 1973.

[47] Du ZZ, Hancock JW. The effect of non-singular stresses on crack-tip constraint. J Mech Phys Solids 1991;39:555–67.

[48] Bleeck O, Munz D, Schaller W, Yang YY. Effect of a graded interlayer on the stress intensity factor of cracks in a joint under thermal loading. Engng Fract Mech 1998;60:615–23.

PERGAMON

Engineering Fracture Mechanics 69 (2002) 1557–1586

Engineering Fracture Mechanics

www.elsevier.com/locate/engfracmech

Mixed-mode fracture of orthotropic functionally graded materials using finite elements and the modified crack closure method

Jeong-Ho Kim, Glaucio H. Paulino *

Department of Civil and Environmental Engineering, University of Illinois at Urbana-Champaign, Newmark Laboratory, 205 North Mathews Avenue, Urbana, IL 61801, USA

Received 26 September 2001; received in revised form 20 December 2001; accepted 21 December 2001

Abstract

A finite element methodology is developed for fracture analysis of orthotropic functionally graded materials (FGMs) where cracks are arbitrarily oriented with respect to the principal axes of material orthotropy. The graded and orthotropic material properties are smooth functions of spatial coordinates, which are integrated into the element stiffness matrix using the isoparametric concept and special graded finite elements. Stress intensity factors (SIFs) for mode I and mixed-mode two-dimensional problems are evaluated and compared by means of the modified crack closure (MCC) and the displacement correlation technique (DCT) especially tailored for orthotropic FGMs. An accurate technique to evaluate SIFs by means of the MCC is presented using a simple two-step (predictor–corrector) process in which the SIFs are first predicted (e.g. by the DCT) and then corrected by Newton iterations. The effects of boundary conditions, crack tip mesh discretization and material properties on fracture behavior are investigated in detail. Many numerical examples are given to validate the proposed methodology. The accuracy of results is discussed by comparison with available (semi-) analytical or numerical solutions.
© 2002 Elsevier Science Ltd. All rights reserved.

Keywords: Functionally graded material; Stress intensity factor; Modified crack closure; Displacement correlation technique; Finite element method

1. Introduction

Functionally graded materials (FGMs) are special composites in which the volume fraction of constituent materials vary gradually, giving a non-uniform microstructure with continuously graded macroproperties [1]. Functional gradation opens new possibilities for optimizing both material and component structures to achieve high performance and material efficiency. However, at the same time, it also poses challenging mechanics problems including the understanding of damage and fracture behavior of such materials,

---

* Corresponding author. Tel.: +1-217-333-3817; fax: +1-217-265-8041.
E-mail address: paulino@uiuc.edu (G.H. Paulino).

especially when preferential directions of orthotropy develop due to the material processing technique utilized, as discussed below. Thus the goal of this investigation is to develop a general-purpose finite element framework for fracture of orthotropic FGMs.

With the introduction of the FGM concept, extensive research on various aspects of fracture of *isotropic* FGMs under mechanical [2–4] or thermal [5–11] loads has been conducted. Mode I [12,13] and mixed-mode [14,15] crack problems have been investigated by means of the finite element method (FEM) and the path-independent J_k^*-integral, which includes an extra term for the explicit derivative of strain energy density to account for the material variation. Recently, a simplified method has been proposed in the limit of very small integration domains so that the extra term may be neglected [16]. Multiple cracking [17] and delamination cracking and buckling [18] in isotropic functionally graded ceramic/metal coatings have also been investigated under mechanical and thermal loads using the FEM.

On the other hand, the nature of processing techniques of ceramic–metal FGMs may lead to loss of isotropy. For example, graded materials processed by a *plasma spray* technique generally have a lamellar structure [19]. Flattened splats and relatively weak splat boundaries create an oriented material with higher stiffness and weak cleavage planes parallel to the boundary. Furthermore, graded materials processed by the *electron beam physical vapor deposition* technique could have a columnar structure [20], which result in a higher stiffness in the thickness direction and weak fracture planes perpendicular to the boundary. Thus, such materials would not be isotropic, but orthotropic with material directions that can be considered perpendicular to each other as an initial approximation. Gu and Asaro [21] performed theoretical studies in a four point bending specimen consisting of orthotropic FGMs with varying Poisson's ratio. Ozturk and Erdogan [22,23] used integral equations to investigate Mode I and mixed-mode crack problems in an infinite non-homogeneous orthotropic medium with a crack aligned with one of the material directions considering constant Poisson's ratio. One of the goals of this study is to compare the numerical results (FEM) for stress intensity factors (SIFs) in FGMs with the semi-analytical solutions by Ozturk and Erdogan [22,23].

This paper presents numerical techniques to evaluate SIFs in orthotropic FGMs by means of the FEM. To this end, the modified crack closure (MCC) and the displacement correlation technique (DCT) are specifically tailored for orthotropic FGMs. A general-purpose implementation is presented which is able to handle multiple, interacting, arbitrarily shaped and oriented cracks. The remainder of this paper is organized as follows. Section 2 reviews crack tip fields in orthotropic FGMs. Section 3 presents the DCT, and Section 4 addresses the MCC. Section 5 discusses some aspects of the FEM implementation. In order to assess and validate the above development, Section 6 presents several numerical results which are compared to available numerical and/or semi-analytical solutions. Finally, Section 7 concludes the present investigation.

2. Crack tip fields in orthotropic functionally graded materials

The most general anisotropic form of linear elastic stress–strain relations is given by

$$\varepsilon_{ij} = S_{ijkl}\sigma_{kl} \quad (i,j,k,l = 1,2,3), \tag{1}$$

where σ_{ij} is the linear stress tensor, ε_{ij} is the linear strain tensor, and S_{ijkl} is the fourth-order compliance tensor. The compliance tensor has 81 independent components, but, because of the symmetry of σ_{ij} and ε_{ij}, the number of independent components reduces to 36. Furthermore, the existence of a strain energy function provides a reduction of the number of independent components to 21 ($S_{ijkl} = S_{klij}$). In order to represent S_{ijkl} in compact form, a contracted notation a_{ij} may be introduced as follows:

$$\varepsilon_i = a_{ij}\sigma_j, \quad a_{ij} = a_{ji} \quad (i,j = 1,2,\ldots,6), \tag{2}$$

where the compliance coefficients, a_{ij}, are related to the elastic properties of the material and

$$\varepsilon_1 = \varepsilon_{11}, \quad \varepsilon_2 = \varepsilon_{22}, \quad \varepsilon_3 = \varepsilon_{33}, \quad \varepsilon_4 = 2\varepsilon_{23}, \quad \varepsilon_5 = 2\varepsilon_{13}, \quad \varepsilon_6 = 2\varepsilon_{12}$$

$$\sigma_1 = \sigma_{11}, \quad \sigma_2 = \sigma_{22}, \quad \sigma_3 = \sigma_{33}, \quad \sigma_4 = \sigma_{23}, \quad \sigma_5 = \sigma_{13}, \quad \sigma_6 = \sigma_{12}. \tag{3}$$

For the special case of plane stress or plane strain problems in transversely isotropic materials (where at each point through the thickness there is a plane of material symmetry parallel to the plane of the problem), Eq. (2) can be reduced to depend upon six independent elastic parameters for plane stress:

$$a_{ij} \quad (i,j = 1,2,6) \tag{4}$$

and a set of corresponding constants for plane strain:

$$b_{ij} = a_{ij} - \frac{a_{i3}a_{j3}}{a_{33}} \quad (i,j = 1,2,6). \tag{5}$$

Fig. 1 shows a crack tip referred to the Cartesian coordinate systems in orthotropic FGMs. Two dimensional anisotropic elasticity problems can be formulated in terms of the analytic functions, $\phi_k(z_k)$, of the complex variable, $z_k = x_k + iy_k$ $(k = 1,2)$, where

$$x_k = x + \alpha_k y, \quad y_k = \beta_k y \quad (k = 1,2). \tag{6}$$

The parameters α_k and β_k are the real and imaginary parts of $\mu_k = \alpha_k + i\beta_k$, which can be determined from [24]

$$a_{11}\mu^4 - 2a_{16}\mu^3 + (2a_{12} + a_{66})\mu^2 - 2a_{26}\mu + a_{22} = 0. \tag{7}$$

The roots μ_k are always complex or purely imaginary in conjugate pairs as $\mu_1, \bar{\mu}_1, \mu_2, \bar{\mu}_2$, of which μ_1 *and* μ_2 *must be calculated at the location of a crack tip for FGMs.*

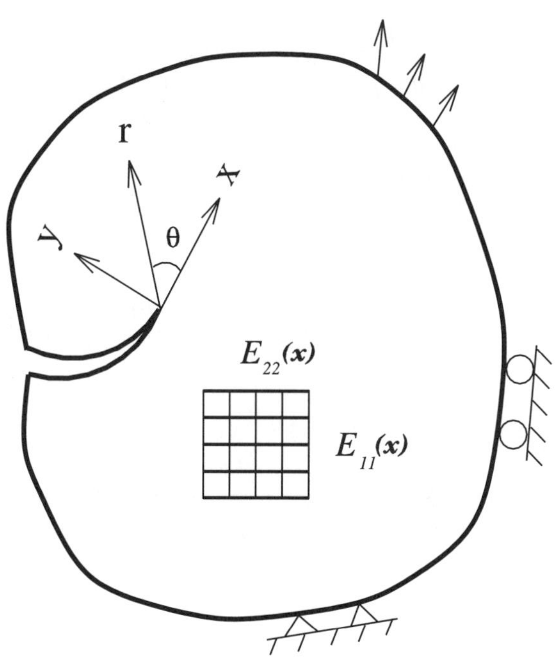

Fig. 1. Coordinate systems at the crack tip in orthotropic FGMs.

For pure mode I, the stresses in the vicinity of the crack tip are

$$\sigma_{11} = \frac{K_{\mathrm{I}}}{\sqrt{2\pi r}} \mathrm{Re}\left[\frac{\mu_1^{\mathrm{tip}}\mu_2^{\mathrm{tip}}}{\mu_1^{\mathrm{tip}} - \mu_2^{\mathrm{tip}}}\left\{\frac{\mu_2^{\mathrm{tip}}}{\sqrt{\cos\theta + \mu_2^{\mathrm{tip}}\sin\theta}} - \frac{\mu_1^{\mathrm{tip}}}{\sqrt{\cos\theta + \mu_1^{\mathrm{tip}}\sin\theta}}\right\}\right]$$

$$\sigma_{22} = \frac{K_{\mathrm{I}}}{\sqrt{2\pi r}} \mathrm{Re}\left[\frac{1}{\mu_1^{\mathrm{tip}} - \mu_2^{\mathrm{tip}}}\left\{\frac{\mu_1^{\mathrm{tip}}}{\sqrt{\cos\theta + \mu_2^{\mathrm{tip}}\sin\theta}} - \frac{\mu_2^{\mathrm{tip}}}{\sqrt{\cos\theta + \mu_1^{\mathrm{tip}}\sin\theta}}\right\}\right] \tag{8}$$

$$\sigma_{12} = \frac{K_{\mathrm{I}}}{\sqrt{2\pi r}} \mathrm{Re}\left[\frac{\mu_1^{\mathrm{tip}}\mu_2^{\mathrm{tip}}}{\mu_1^{\mathrm{tip}} - \mu_2^{\mathrm{tip}}}\left\{\frac{1}{\sqrt{\cos\theta + \mu_1^{\mathrm{tip}}\sin\theta}} - \frac{1}{\sqrt{\cos\theta + \mu_2^{\mathrm{tip}}\sin\theta}}\right\}\right]$$

and the displacements are

$$u_1 = K_{\mathrm{I}}\sqrt{\frac{2r}{\pi}}\mathrm{Re}\left[\frac{1}{\mu_1^{\mathrm{tip}} - \mu_2^{\mathrm{tip}}}\left\{\mu_1^{\mathrm{tip}}p_2\sqrt{\cos\theta + \mu_2^{\mathrm{tip}}\sin\theta} - \mu_2^{\mathrm{tip}}p_1\sqrt{\cos\theta + \mu_1^{\mathrm{tip}}\sin\theta}\right\}\right]$$

$$u_2 = K_{\mathrm{I}}\sqrt{\frac{2r}{\pi}}\mathrm{Re}\left[\frac{1}{\mu_1^{\mathrm{tip}} - \mu_2^{\mathrm{tip}}}\left\{\mu_1^{\mathrm{tip}}q_2\sqrt{\cos\theta + \mu_2^{\mathrm{tip}}\sin\theta} - \mu_2^{\mathrm{tip}}q_1\sqrt{\cos\theta + \mu_1^{\mathrm{tip}}\sin\theta}\right\}\right]. \tag{9}$$

Similarly, for pure mode II, the stresses in the vicinity of the crack tip are

$$\sigma_{11} = \frac{K_{\mathrm{II}}}{\sqrt{2\pi r}} \mathrm{Re}\left[\frac{1}{\mu_1^{\mathrm{tip}} - \mu_2^{\mathrm{tip}}}\left\{\frac{(\mu_2^{\mathrm{tip}})^2}{\sqrt{\cos\theta + \mu_2^{\mathrm{tip}}\sin\theta}} - \frac{(\mu_1^{\mathrm{tip}})^2}{\sqrt{\cos\theta + \mu_1^{\mathrm{tip}}\sin\theta}}\right\}\right]$$

$$\sigma_{22} = \frac{K_{\mathrm{II}}}{\sqrt{2\pi r}} \mathrm{Re}\left[\frac{1}{\mu_1^{\mathrm{tip}} - \mu_2^{\mathrm{tip}}}\left\{\frac{1}{\sqrt{\cos\theta + \mu_2^{\mathrm{tip}}\sin\theta}} - \frac{1}{\sqrt{\cos\theta + \mu_1^{\mathrm{tip}}\sin\theta}}\right\}\right] \tag{10}$$

$$\sigma_{12} = \frac{K_{\mathrm{II}}}{\sqrt{2\pi r}} \mathrm{Re}\left[\frac{1}{\mu_1^{\mathrm{tip}} - \mu_2^{\mathrm{tip}}}\left\{\frac{\mu_1^{\mathrm{tip}}}{\sqrt{\cos\theta + \mu_1^{\mathrm{tip}}\sin\theta}} - \frac{\mu_2^{\mathrm{tip}}}{\sqrt{\cos\theta + \mu_2^{\mathrm{tip}}\sin\theta}}\right\}\right]$$

and the displacements are

$$u_1 = K_{\mathrm{II}}\sqrt{\frac{2r}{\pi}}\mathrm{Re}\left[\frac{1}{\mu_1^{\mathrm{tip}} - \mu_2^{\mathrm{tip}}}\left\{p_2\sqrt{\cos\theta + \mu_2^{\mathrm{tip}}\sin\theta} - p_1\sqrt{\cos\theta + \mu_1^{\mathrm{tip}}\sin\theta}\right\}\right]$$

$$u_2 = K_{\mathrm{II}}\sqrt{\frac{2r}{\pi}}\mathrm{Re}\left[\frac{1}{\mu_1^{\mathrm{tip}} - \mu_2^{\mathrm{tip}}}\left\{q_2\sqrt{\cos\theta + \mu_2^{\mathrm{tip}}\sin\theta} - q_1\sqrt{\cos\theta + \mu_1^{\mathrm{tip}}\sin\theta}\right\}\right]. \tag{11}$$

In the above equations, μ_1^{tip} and μ_2^{tip} denote the crack tip parameters calculated as the roots of Eq. (7), which are taken such that $\beta_k > 0$ ($k = 1, 2$) and p_k and q_k are given by

$$p_k = a_{11}(\mu_k^{\mathrm{tip}})^2 + a_{12} - a_{16}\mu_k^{\mathrm{tip}}$$

$$q_k = a_{12}\mu_k^{\mathrm{tip}} + \frac{a_{22}}{\mu_k^{\mathrm{tip}}} - a_{26}. \tag{12}$$

Moreover, K_{I} and K_{II} denote the mode I and mode II SIFs, respectively.

3. Displacement correlation technique for orthotropic functionally graded materials

The DCT is one of the simplest methods to evaluate SIFs. It consists of correlating numerical results for displacement at specific locations on the crack with available analytical solutions. For quarter point singular elements, the crack opening displacement (COD) and crack sliding displacement (CSD) at $x = -r$ are given by [25]

$$\text{COD}(-r) = (4u_{2,i-1} - u_{2,i-2})\sqrt{\frac{r}{\Delta a}}, \tag{13}$$

$$\text{CSD}(-r) = (4u_{1,i-1} - u_{1,i-2})\sqrt{\frac{r}{\Delta a}}, \tag{14}$$

respectively, where $u_{1,i-1}$, $u_{1,i-2}$, $u_{2,i-1}$, and $u_{2,i-2}$ are the relative displacements with respect to the crack tip in the x_i $(i = 1, 2)$ direction at locations $(i - 1)$ and $(i - 2)$, r is the distance from the crack tip along the local x_1 direction, and Δa is the characteristic length of the crack tip elements (see Fig. 2).

For $\theta = 180°$, by combination of the two modes, Eqs. (9) and (11) reduce to

$$u_1 = K_I\sqrt{\frac{2r}{\pi}}\text{Re}\left[\frac{i}{\mu_1^{\text{tip}} - \mu_2^{\text{tip}}}(\mu_1^{\text{tip}}q_2 - \mu_2^{\text{tip}}q_1)\right] + K_{II}\sqrt{\frac{2r}{\pi}}\text{Re}\left[\frac{i}{\mu_1^{\text{tip}} - \mu_2^{\text{tip}}}(q_2 - q_1)\right], \tag{15}$$

$$u_2 = K_I\sqrt{\frac{2r}{\pi}}\text{Re}\left[\frac{i}{\mu_1^{\text{tip}} - \mu_2^{\text{tip}}}(\mu_1^{\text{tip}}p_2 - \mu_2^{\text{tip}}p_1)\right] + K_{II}\sqrt{\frac{2r}{\pi}}\text{Re}\left[\frac{i}{\mu_1^{\text{tip}} - \mu_2^{\text{tip}}}(p_2 - p_1)\right]. \tag{16}$$

Equating Eq. (13) with Eq. (16), and Eq. (14) with Eq. (15), one obtains

$$K_I = \frac{1}{4}\sqrt{\frac{2\pi}{\Delta a}}\frac{D(4u_{1,i-1} - u_{1,i-2}) - B(4u_{2,i-1} - u_{2,i-2})}{AD - BC}, \tag{17}$$

$$K_{II} = \frac{1}{4}\sqrt{\frac{2\pi}{\Delta a}}\frac{A(4u_{2,i-1} - u_{2,i-2}) - C(4u_{1,i-1} - u_{1,i-2})}{AD - BC}, \tag{18}$$

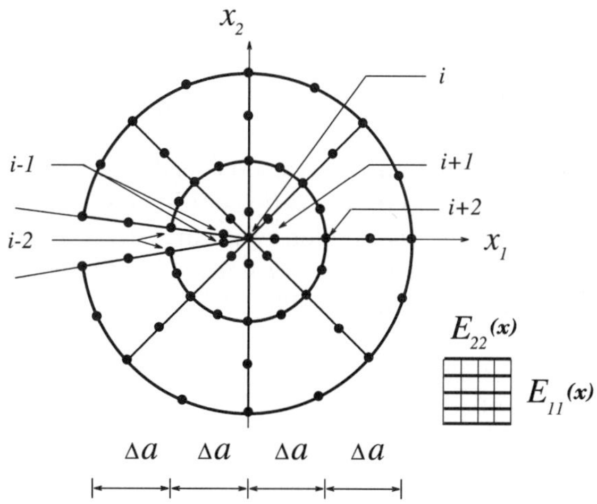

Fig. 2. Crack tip rosette of singular quarter-point (first ring) and regular (second ring) finite elements.

where

$$A = \mathrm{Re}\left[\frac{i}{\mu_1^{\mathrm{tip}} - \mu_2^{\mathrm{tip}}}(\mu_1 p_2 - \mu_2 p_1)\right]$$

$$B = \mathrm{Re}\left[\frac{i}{\mu_1^{\mathrm{tip}} - \mu_2^{\mathrm{tip}}}(p_2 - p_1)\right]$$

$$C = \mathrm{Re}\left[\frac{i}{\mu_1^{\mathrm{tip}} - \mu_2^{\mathrm{tip}}}(\mu_1 q_2 - \mu_2 q_1)\right]$$

$$D = \mathrm{Re}\left[\frac{i}{\mu_1^{\mathrm{tip}} - \mu_2^{\mathrm{tip}}}(q_2 - q_1)\right],$$

$$(19)$$

in which all the material parameters and related coefficients must be considered at the crack tip location for FGMs.

4. Modified crack closure integral for orthotropic functionally graded materials

The MCC integral method is based on Irwin's virtual crack closure approach [26], which uses the stresses ahead of the crack tip and the displacements behind the crack tip. Rybicki and Kanninen [27] used this method to obtain a formula for strain energy release rate with four-node quadrilateral non-singular elements, and Raju [28] extended the method to a set of quarter-point singular elements and provided a modified formulation for the case of crack faces loaded with uniform pressure.

Because no assumption of isotropy or homogeneity around the crack is made, the method is ideally suited for orthotropic FGMs. The energy release rate is estimated only in terms of the work done by the stresses (or equivalent nodal forces) over the displacements behind the crack tip produced by a virtual crack extension.

The expression for $\mathscr{G}_{\mathrm{I}}$ (strain energy release rate for mode I) and $\mathscr{G}_{\mathrm{II}}$ (strain energy release rate for mode II) may be obtained according to Irwin as

$$\mathscr{G}_{\mathrm{I}} = \lim_{\delta a \to 0} \frac{2}{\delta a} \int_{x_1=0}^{x_1=\delta a} \tfrac{1}{2}\sigma_{22}(r = x_1, \theta = 0, a)\, u_2(r = \delta a - x_1, \theta = \pi, a + \delta a)\, \mathrm{d}x_1, \qquad (20)$$

$$\mathscr{G}_{\mathrm{II}} = \lim_{\delta a \to 0} \frac{2}{\delta a} \int_{x_1=0}^{x_1=\delta a} \tfrac{1}{2}\sigma_{12}(r = x_1, \theta = 0, a)\, u_1(r = \delta a - x_1, \theta = \pi, a + \delta a)\, \mathrm{d}x_1, \qquad (21)$$

where $\sigma_{12} \equiv \sigma_{xy}$ and $\sigma_{22} \equiv \sigma_{yy}$ are shear and normal stresses ahead of the crack tip, and $u_1 \equiv u_x$ and $u_2 \equiv u_y$ are the relative displacements with respect to the crack tip coordinates, respectively. Fig. 3 illustrates a self-similar virtual crack extension δa and the distribution of normal stress ahead of the crack tip.

For the *particular* case in which the material is orthotropic with the crack on a plane of material symmetry, the following relationships are obtained [29]:

$$\mathscr{G}_{\mathrm{I}} = K_{\mathrm{I}}^2 \sqrt{\frac{a_{11}a_{22}}{2}}\left[\sqrt{\frac{a_{22}}{a_{11}}} + \frac{2a_{12} + a_{66}}{2a_{11}}\right]^{1/2}$$

$$\mathscr{G}_{\mathrm{II}} = K_{\mathrm{II}}^2 \frac{a_{11}}{\sqrt{2}}\left[\sqrt{\frac{a_{22}}{a_{11}}} + \frac{2a_{12} + a_{66}}{2a_{11}}\right]^{1/2},$$

$$(22)$$

where K_{I} and K_{II} are uncoupled.

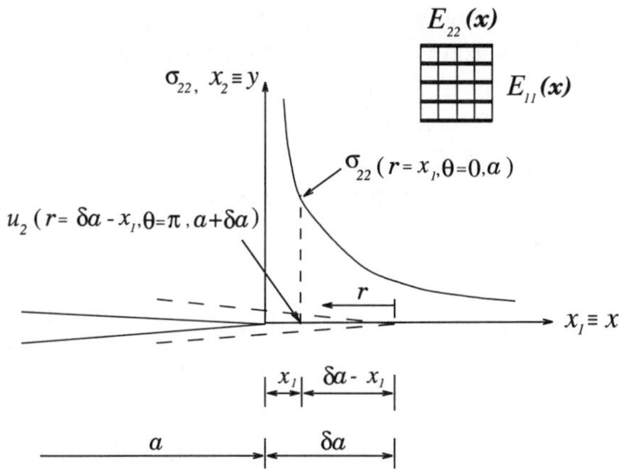

Fig. 3. Self-similiar crack extension and normal stress distribution.

For the *general* case of mixed-mode fracture in orthotropic materials where the crack is arbitrarily oriented with respect to the principal material directions, the SIFs are related to the values of the potential energy release rates through the following expressions [29]:

$$\mathscr{G}_{\mathrm{I}} = -\frac{K_{\mathrm{I}}}{2} a_{22} \mathrm{Im} \left[\frac{K_{\mathrm{I}}(\mu_1^{\mathrm{tip}} + \mu_2^{\mathrm{tip}}) + K_{\mathrm{II}}}{\mu_1^{\mathrm{tip}} \mu_2^{\mathrm{tip}}} \right]$$

$$\mathscr{G}_{\mathrm{II}} = \frac{K_{\mathrm{II}}}{2} a_{11} \mathrm{Im} \left[K_{\mathrm{II}}(\mu_1^{\mathrm{tip}} + \mu_2^{\mathrm{tip}}) + K_{\mathrm{I}} \mu_1^{\mathrm{tip}} \mu_2^{\mathrm{tip}} \right].$$

(23)

For this case, the SIFs are coupled and they may be solved by means of the Newton iteration algorithm summarized below. Define a system of non-linear equations as follows:

$$\mathbf{F}(\mathbf{K}) = (F_1(\mathbf{K}), F_2(\mathbf{K}))$$

$$F_1(\mathbf{K}) = \mathscr{G}_{\mathrm{I}} + \frac{K_{\mathrm{I}}}{2} a_{22} \mathrm{Im} \left[\frac{K_{\mathrm{I}}(\mu_1^{\mathrm{tip}} + \mu_2^{\mathrm{tip}}) + K_{\mathrm{II}}}{\mu_1^{\mathrm{tip}} \mu_2^{\mathrm{tip}}} \right]$$

$$F_2(\mathbf{K}) = \mathscr{G}_{\mathrm{II}} - \frac{K_{\mathrm{II}}}{2} a_{11} \mathrm{Im} \left[K_{\mathrm{II}}(\mu_1^{\mathrm{tip}} + \mu_2^{\mathrm{tip}}) + K_{\mathrm{I}} \mu_1^{\mathrm{tip}} \mu_2^{\mathrm{tip}} \right],$$

(24)

where $\mathbf{F}$ is a vector-valued function of F_1 and F_2, and $\mathbf{K}$ is a vector of the unknowns $(K_{\mathrm{I}}, K_{\mathrm{II}})$.

Determination of $(K_{\mathrm{I}}, K_{\mathrm{II}})$ using Newton iteration method:

1. Select $\mathbf{K}^{(0)}(K_{\mathrm{I}}, K_{\mathrm{II}})$ and initialize counter $i = 0$.
2. Compute gradient: $\nabla \mathbf{F}(\mathbf{K})$
3. Perform iteration:

$$\mathbf{K}^{(i+1)} = \mathbf{K}^{(i)} - \frac{\mathbf{F}(\mathbf{K}^{(i)})}{\nabla \mathbf{F}(\mathbf{K}^{(i)})}$$

4. Check convergence: If $|\mathbf{F}(\mathbf{K}^{(i)})| > \mathrm{TOL}$, then $i \leftarrow i + 1$, and GOTO Step 1.

For the initial values of $\mathbf{K}^{(0)}(K_{\mathrm{I}}, K_{\mathrm{II}})$ in Step 1, we may use the SIFs obtained by the DCT because it provides physically reasonable SIF values and accelerates the iterative procedure. *Essentially, this procedure is a two-step (predictor–corrector) process in which the SIFs are predicted by the DCT and corrected by Newton iterations for the MCC.* However, any judicious choice for the initial values of $\boldsymbol{K}^{(0)}(K_{\mathrm{I}}, K_{\mathrm{II}})$ may also be acceptable, as discussed subsequently in Section 6.

Ramamurthy et al. [30], and Raju [28] have shown that the values of $\mathscr{G}_{\mathrm{I}}$ and $\mathscr{G}_{\mathrm{II}}$ can be written in terms of the equivalent nodal forces $F_2 \equiv F_y$ and $F_1 \equiv F_x$, and the relative nodal displacements u_2 and u_1 when employing quarter-point singular elements around the crack tip (see Fig. 2). For mixed-mode problems, the general expressions for $\mathscr{G}_{\mathrm{I}}$ and $\mathscr{G}_{\mathrm{II}}$ are given by [28]

$$
\begin{aligned}
\mathscr{G}_{\mathrm{I}} = \frac{1}{2\Delta a} & [F_{2,i}(t_{11}u_{2,i-2} + t_{12}u_{2,i-1}) + F_{2,i+1}(t_{21}u_{2,i-2} + t_{22}u_{2,i-1}) + F_{2,i+2}^{\mathrm{T}}(t_{31}\bar{u}_{2,i-2} + t_{32}\bar{u}_{2,i-1}) \\
& + F_{2,i+2}^{\mathrm{B}}(t_{31}\hat{u}_{2,i-2} + t_{32}\hat{u}_{2,i-1})] \\
\mathscr{G}_{\mathrm{II}} = \frac{1}{2\Delta a} & [F_{1,i}(t_{11}u_{1,i-2} + t_{12}u_{1,i-1}) + F_{1,i+1}(t_{21}u_{1,i-2} + t_{22}u_{1,i-1}) + F_{1,i+2}^{\mathrm{T}}(t_{31}\bar{u}_{1,i-2} + t_{32}\bar{u}_{1,i-1}) \\
& + F_{1,i+2}^{\mathrm{B}}(t_{31}\hat{u}_{1,i-2} + t_{32}\hat{u}_{1,i-1})],
\end{aligned}
\tag{25}
$$

where the first subscript in F or u refers to the Cartesian coordinate ($x_1 \equiv x$ or $x_2 \equiv y$), the second subscript refers to the nodal point, the parameters t_{kl} ($k = 1, 2, 3;\ l = 1, 2$) are given by

$$
\begin{aligned}
t_{11} &= 14 - \frac{33\pi}{8}, \quad t_{12} = -52 - \frac{33\pi}{2}, \\
t_{21} &= -\frac{7}{2} + \frac{21\pi}{16}, \quad t_{22} = 17 - \frac{21\pi}{4}, \\
t_{31} &= 8 - \frac{21\pi}{8}, \quad t_{32} = -32 + \frac{21\pi}{2}
\end{aligned}
\tag{26}
$$

and Δa is the characteristic length of singular elements around the crack tip which are six-node quarter-point triangular elements as illustrated in Fig. 2. The superscripts T and B indicate the top and bottom regions of the crack with respect to x_1-axis, and thus F^{T} and F^{B} indicate the forces at the top and bottom surfaces, respectively. The fields $\bar{u}$ and $\hat{u}$ represent the relative displacement of the top and bottom surfaces with respect to the crack tip. For example, at location $(i - 1)$ in Fig. 2, $\bar{v}_{y,i-1} = v_{y,i-1}^{\mathrm{T}} - v_{y,i}$ and $\hat{v}_{y,i-1} = v_{y,i-1}^{\mathrm{B}} - v_{y,i}$ where v represent absolute displacements.

Eq. (25) is valid for calculating strain energy release rates for self-similar crack growth when the crack faces are traction free. When the crack faces are pressure loaded, Eq. (25) must be modified. Fig. 4(a) shows a crack tip under a uniform crack face pressure loading, denoted by p, and Fig. 4(b) shows the idea of superposition of computed nodal forces and equivalent nodal forces corresponding to the opposite of the applied pressure when six-node triangular quarter-point (T6qp) singular elements are used. The closure of the crack faces from $a + \Delta a$ to a can be decomposed into two parts. In the first part, the applied pressure is "erased" and, in the second part, the stress free crack faces between $a + \Delta a$ and a are "closed". In the FEM, the first part is equivalent to addition of nodal forces corresponding to the opposite of the applied pressure at all nodes between $a + \Delta a$ and a. For the second part, the general procedure described earlier for the MCC method is used. Thus, the resulting forces at nodes i, $i + 1$, and $i + 2$ are

$$
F_{y,i} = F_{y,i}^{\mathrm{c}}, \quad F_{y,i+1} = F_{y,i+1}^{\mathrm{c}} - \frac{2p}{3}\Delta a, \quad F_{y,i+2} = F_{y,i+2}^{\mathrm{c}} - \frac{p}{3}\Delta a.
\tag{27}
$$

The corrected nodal forces are used to calculate the strain energy release rates.

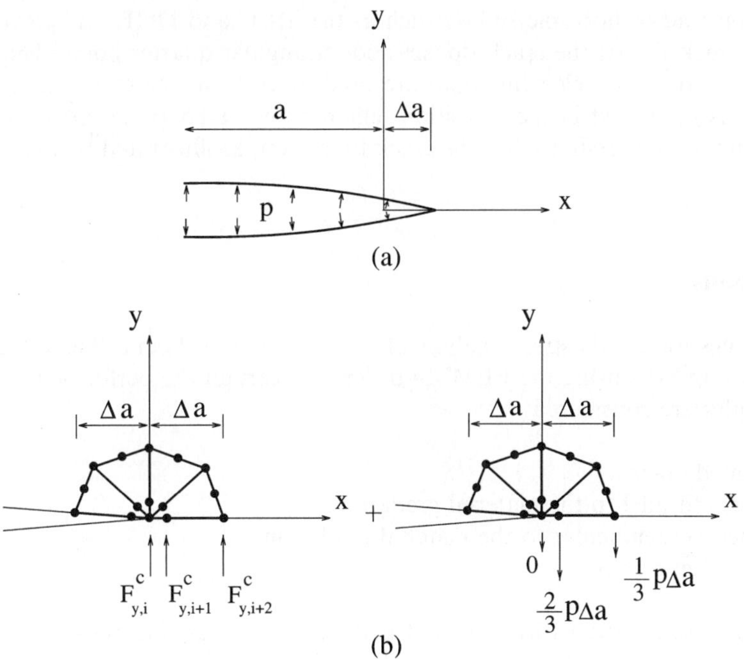

Fig. 4. Modification of nodal forces for uniform crack face pressure loading: (a) crack face pressure loading; (b) superposition of total computed forces and forces needed to account for the pressure loading.

5. Finite element implementation

Graded elements (rather than homogeneous elements) need to be introduced to discretize FGM properties [15,31]. The material properties at Gauss quadrature points are interpolated from the nodal material properties via isoparametric interpolation functions, as illustrated by Fig. 5. The behavior of such graded elements in fracture mechanics of isotropic FGMs has been studied by Paulino and Kim [32]. In order to develop the orthotropic graded elements and perform fracture analyses, the public domain FEM code FRANC2D (FRacture ANalysis Code 2D) [33,34] has been used as the basis for implementing fracture capabilities in FGMs. The source code of FRANC2D is fully accessible and thus it is well-suited for research and for new developments. The code with extended fracture capabilities for FGMs is called I_FRANC2D (Illinois FRacture ANalysis Code 2D). The extensions include special techniques to evaluate

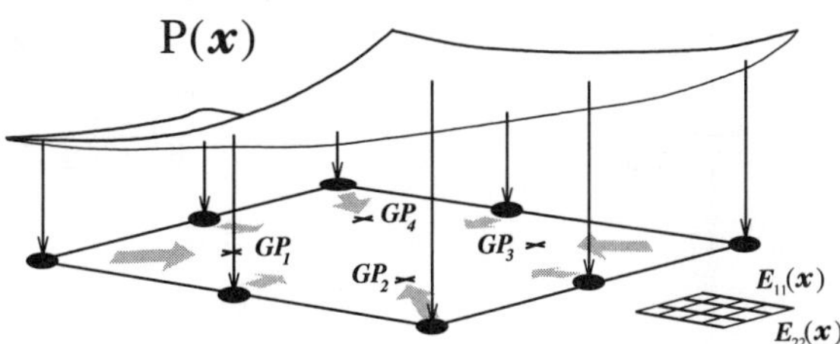

Fig. 5. Generalized isoparametric formulation [15,31] using graded finite elements. The above figure illustrates a graded Q8 element and $P(x)$ denotes a generic material property. The Gauss point properties are obtained as $P_{GP} = \sum N_i P_i$ where N are element shape functions.

SIFs in both isotropic and orthotropic FGMs such as the MCC and DCT, and creation of many sectors and rings around a crack tip. At the crack tip, six-node triangular quarter-point (T6qp) singular elements are used. Eight-node serendipity elements (Q8) are used away from the crack(s), and regular triangular quadratic elements (T6) are used in the transition region between T6qp and Q8 elements. Thus the code allows a careful design of the mesh around the crack tip region, as illustrated by the examples given in the next section.

6. Computational results

This paper examines the elastic stress analysis of orthotropic FGMs and the performance of the MCC and DCT on computing SIFs using the FEM. In order to ascertain the performance of the two methods, the following examples are considered:

(1) Plate with a slanted crack.
(2) Plate with a crack parallel to the material gradation.
(3) Plate with a crack perpendicular to the material gradation.
(4) Two interacting offset cracks.

This set of problems assesses the FEM code and performance of the methods for evaluation of SIFs in orthotropic FGMs. The examples have either numerical (e.g. finite element), semi-analytical (e.g. integral equation) or experimental results available. Thus, the solutions obtained with the I_FRANC2D code are compared with those available results. Moreover, the present solutions can also be verified against other methods for evaluating SIFs in FGMs, e.g. the J_k^*—integral formulation of Kim and Paulino [35].

For the sake of comparison with the semi-analytical solutions by Ozturk and Erdogan [22,23], the following average parameters (originally proposed by Krenk [36]) are used in the second and third examples, as illustrated by Fig. 6(a) and (b), respectively. The independent engineering constants E_{ii}, G_{ij} and v_{ij} $((v_{ij}/E_{ii}) = (v_{ji}/E_{jj}))$ $(i,j = 1,2,3)$ can be replaced by the averaged Young's modulus E, the effective Poisson's ratio v, the stiffness ratio δ^4 and the shear parameter κ_0 [36] defined by

$$E = \sqrt{E_{11}E_{22}}, \quad v = \sqrt{v_{12}v_{21}}, \quad \delta^4 = \frac{E_{11}}{E_{22}} = \frac{v_{12}}{v_{21}}, \quad \kappa_0 = \frac{E}{2G_{12}} - v \tag{28}$$

for generalized plane stress and

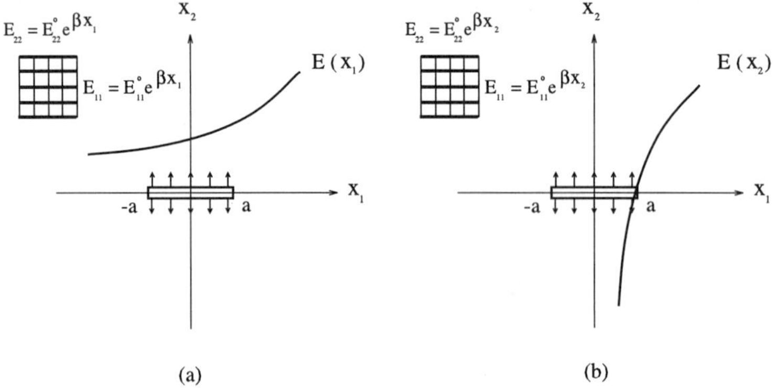

Fig. 6. Geometry and notation for crack problems in an orthotropic non-homogeneous medium: (a) Mode I [22]; (b) mixed-mode [23].

$$E = \sqrt{\frac{E_{11}E_{22}}{(1 - v_{13}v_{31})(1 - v_{23}v_{32})}}, \quad v = \sqrt{\frac{(v_{12} + v_{13}v_{32})(v_{21} + v_{23}v_{31})}{(1 - v_{13}v_{31})(1 - v_{23}v_{32})}},$$

$$\delta^4 = \frac{E_{11}(1 - v_{23}v_{32})}{E_{22}(1 - v_{13}v_{31})}, \quad \kappa_0 = \frac{E}{2G_{12}} - v \tag{29}$$

for plane strain.

It is worth mentioning that the SIF normalization factor used by Ozturk and Erdogan [22,23] is $k_0 = \sigma\sqrt{a}$ (or $k_0 = \tau\sqrt{a}$) while in this paper $K_0 = \sigma\sqrt{\pi a}$ (or $K_0 = \tau\sqrt{\pi a}$). However, the ratios K_I/K_0 or K_I/k_0 (and K_{II}/K_0 or K_{II}/k_0) are comparable quantities in both pieces of work (i.e. Refs. [22,23] and here) because the $\sqrt{\pi}$ factor cancels out.

6.1. Plate with a slanted crack

Fig. 7(a) shows a slanted crack of length $2a$ located in a finite two-dimensional plate under constant applied tension and Fig. 7(b) shows the complete finite element mesh configuration. Fig. 7(c) shows a detail of the mesh with 4 rings (R4) and 16 sectors (S16) around the crack tips. The mesh has 879 Q8, 430 T6, and 32 T6qp crack tip elements with a total of 1341 elements and 3694 nodes. All the elements are graded and orthotropic [31]. The applied load corresponds to $\sigma_{22}(-10 \leqslant x_1 \leqslant 10, \pm 20) = \pm\sigma = \pm 1.0$ along the top and bottom edges. The displacement boundary condition is prescribed such that $u_1 = u_2 = 0$ for the node in the middle of the left edge and $u_2 = 0$ for the node in the middle of the right edge.

With reference to the example of Fig. 7, the following data were used for the FEM analysis:

$$2a = 2\sqrt{2}, \quad L/W = 2.0,$$
$$E_{11}(x_1) = E_{11}^0 e^{\alpha x_1}, \quad E_{22}(x_1) = E_{22}^0 e^{\beta x_1}, \quad G_{12}(x_1) = G_{12}^0 e^{\gamma x_1},$$
$$E_{11}^0 = 3.5 \times 10^6, \quad E_{22}^0 = 12 \times 10^6, \quad G_{12}^0 = 3 \times 10^6, \quad v_{12} = 0.204$$

generalized plane stress, $\quad 2 \times 2$ Gauss quadrature.

This example is investigated with respect to three different material variations for E_{11}, E_{22} and G_{12}, which are assumed to be exponential functions; while the Poisson's ratio v_{12} is assumed to be constant. The cases examined are the following:

(1) Homogeneous orthotropic material: $(\alpha, \beta, \gamma) = (0, 0, 0)$.
(2) Orthotropic FGM with proportional material variation: $(\alpha, \beta, \gamma) = (0.2, 0.2, 0.2)$.
(3) Orthotropic FGM with non-proportional material variation: $(\alpha, \beta, \gamma) = (0.5, 0.4, 0.3)$.

In the first case, the solutions obtained are compared with those available in the literature. However, for the other two cases, there are no other solutions available for comparison.

For the homogeneous orthotropic material case with $(\alpha, \beta, \gamma) = (0, 0, 0)$, Table 1 shows the SIFs using the MCC and DCT in comparison with the reference solutions given by Sih et al. [29] (complex variable approach), Atluri et al. [37] (FEM–hybrid displacement model), and Wang et al. [38] (FEM–M integral). The MCC and DCT estimate the SIFs to within 1.3% and 2.2% errors, respectively, with respect to the reference solution by Sih et al. [29].

The orthotropic FGM case with proportional material variation is investigated considering $(\alpha, \beta, \gamma) = (0.2, 0.2, 0.2)$. The importance of the initial values for SIFs (K_I, K_{II}) are elaborated in this example. Fig. 8 shows four possible solutions (four intersection points) for SIFs at the right crack tip, i.e. $(K_I(+a), K_{II}(+a))$:

$$(\mathbf{1.762}, \mathbf{1.439}), \quad (1.3959, -0.9565), \quad (-1.762, -1.439), \quad (-1.3959, 0.9565).$$

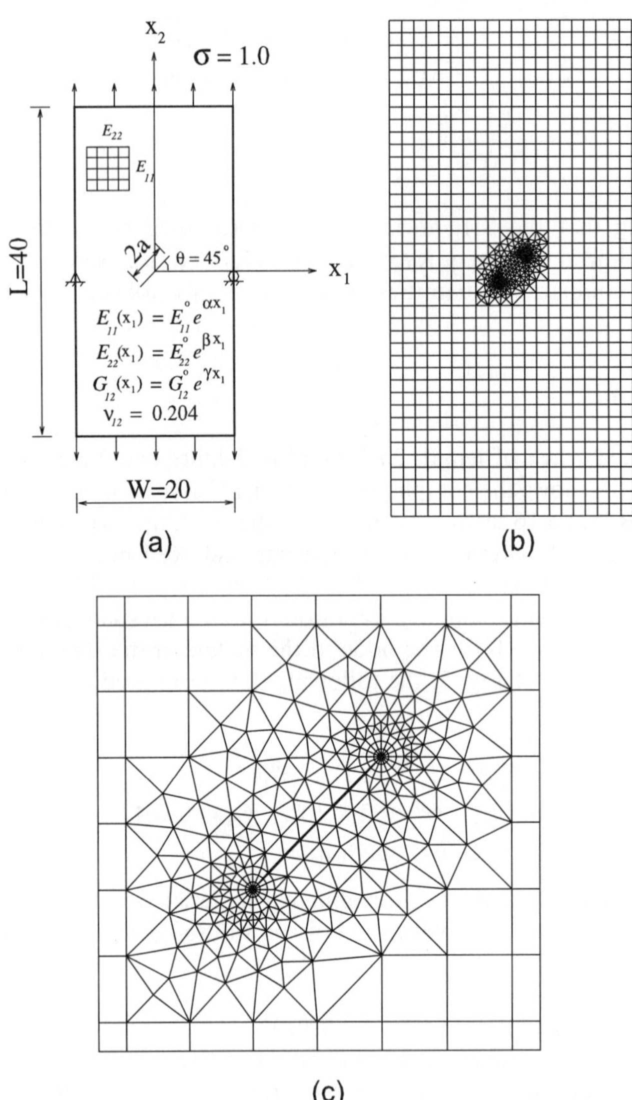

Fig. 7. Plate with a slanted crack: (a) geometry and BCs under remote tension; (b) complete finite element mesh; (c) mesh detail using 4 rings (R4) and 16 sectors (S16) around crack tips.

Initial values for SIFs must be carefully chosen for this problem because there are two sets of admissible solutions for this problem (the first two results above). If we use initial values from the SIFs obtained by the DCT, i.e. $(K_{\mathrm{I}}, K_{\mathrm{II}}) = (1.769, 1.419)$, the final SIFs $(K_{\mathrm{I}}, K_{\mathrm{II}}) = (1.762, 1.439)$ are obtained after two iterations, and are indicated in bold above. However, initial guesses other than the one provided by the DCT may also lead to the correct solution. For instance, we have tried the following initial guesses $(K_{\mathrm{I}}^{(0)}(+a), K_{\mathrm{II}}^{(0)}(+a))$:

$$(1,1), \quad (0.7, 0.2), \quad (0.5, 0.5) \quad \text{and} \quad (0.2, 0.7)$$

and after four, seven, six, and eight iterations, respectively, all of them conducted to the correct solution, i.e. $(K_{\mathrm{I}}(+a), K_{\mathrm{II}}(+a)) = (1.762, 1.439)$. The SIF results for this case are also summarized in Table 1.

Table 1
SIFs in a non-homogeneous orthotropic plate with a slanted crack under uniform remote tension loading (see Fig. 7)

Material	Method	$K_I^+(a)$	$K_{II}^+(a)$	$K_I^-(a)$	$K_{II}^-(a)$
Homogeneous $(\alpha, \beta, \gamma) = (0,0,0)$	Sih et al. [29]	1.0539	1.0539	1.0539	1.0539
	Atluri et al. [37]	1.0195	1.0795	1.0195	1.0795
	Wang et al. [38]	1.023	1.049	1.023	1.049
	MCC	1.067	1.044	1.067	1.044
	DCT	1.077	1.035	1.077	1.035
FGM (proportional) $(\alpha, \beta, \gamma) = (0.2, 0.2, 0.2)$	MCC	1.762	1.439	1.403	1.288
	DCT	1.769	1.419	1.419	1.284
FGM (non-proportional) $(\alpha, \beta, \gamma) = (0.5, 0.4, 0.3)$	MCC	2.384	1.581	1.437	1.225
	DCT	2.387	1.553	1.456	1.229

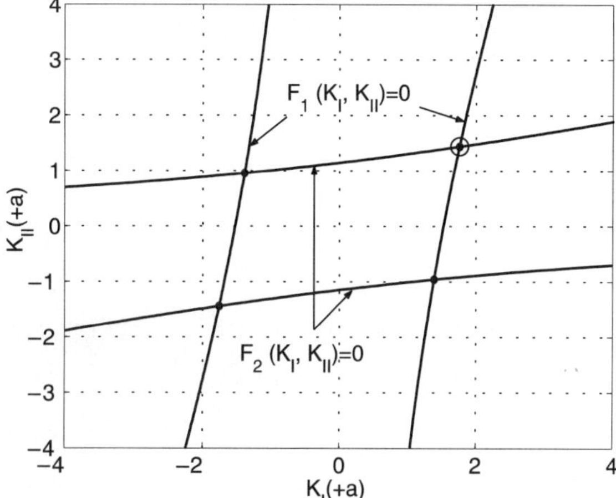

Fig. 8. SIF solutions of non-linear system of equations by Newton's iteration method considering $(\alpha, \beta, \gamma) = (0.2, 0.2, 0.2)$ (see Fig. 7). The circle around the bullet indicates the converged solution.

The situation depicted by Fig. 8 can change significantly depending on the selection of the material parameters. For instance, if $(\alpha, \beta, \gamma) = (1.0, 1.0, 1.0)$, the plot in Fig. 9 is obtained, which shows that the roots can be quite close to each other. If we use the SIFs provided by the DCT as initial estimates, i.e. $(K_I(+a), K_{II}(+a)) = (0.7274, 0.1553)$, then the final SIFs $(K_I(+a), K_{II}(+a)) = (0.7387, 0.1611)$ are obtained after two iterations.

Finally, the orthotropic FGM case with non-proportional material variation is investigated considering $(\alpha, \beta, \gamma) = (0.5, 0.4, 0.3)$. The SIF results are also summarized in Table 1. Notice that both MCC and DCT provide symmetric SIF responses at the left and right tips for the homogeneous case, while such symmetry is lost for the FGM case (as expected).

6.2. Plate with a crack parallel to material gradation

Ozturk and Erdogan [22] have investigated the Mode I crack problem for an infinite non-homogeneous orthotropic medium as illustrated by Fig. 6(a). Here the infinite medium is approximated by a square plate

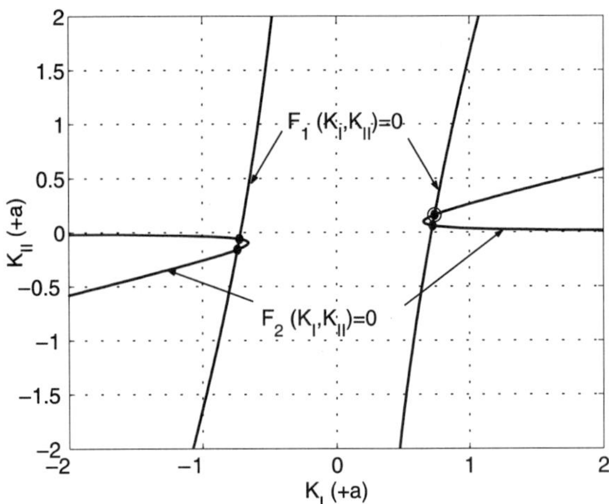

Fig. 9. SIF solutions of non-linear system of equations by Newton's iteration method considering $(\alpha, \beta, \gamma) = (1.0, 1.0, 1.0)$ (see Fig. 7). The circle around the bullet indicates the converged solution.

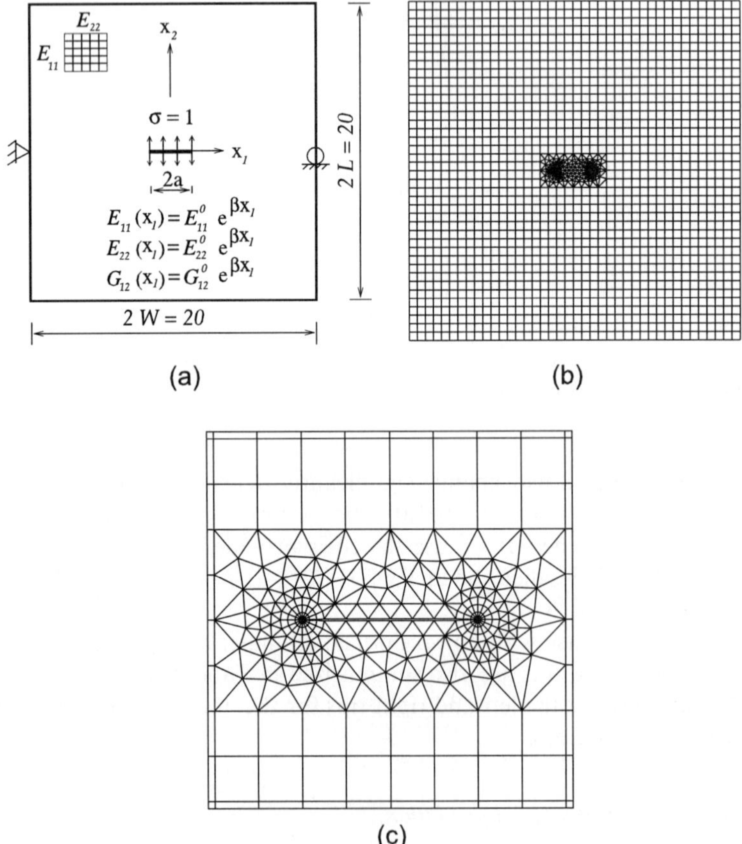

Fig. 10. Plate with a center crack parallel to the material gradation: (a) geometry and BCs under crack face loading; (b) complete finite element mesh; (c) mesh detail using 4 rings (R4) and 16 sectors (S16) around crack tips.

where the ratio of the plate edge length over the crack length is equal to 10. Fig. 10(a) shows a crack of length $2a$ located in a finite two-dimensional plate under uniform crack pressure loading, and Fig. 10(b) shows the corresponding mesh configuration. Notice that the crack is parallel to the material gradation as the average Young's modulus $E \equiv E(x_1) = E^0 e^{\beta x_1}$ where $E^0 = \sqrt{E^0_{11} E^0_{22}}$. Fig. 10(c) shows a detail with 4 rings (R4) and 16 sectors (S16) around the crack tips. The displacement boundary condition is prescribed such that $u_1 = u_2 = 0$ for the node in the middle of the left edge and $u_2 = 0$ for the node in the middle of the right edge. The applied load corresponds to $\sigma_{22}(-1 \leqslant x_1 \leqslant 1, \pm 0) = \pm \sigma = \pm 1.0$ along the crack faces, as illustrated by Fig. 10(a).

As illustrated by Fig. 10(a), the variations of E_{11}, E_{22}, and G_{12} are assumed to be exponential functions of x_1 and proportional to one another, while the Poisson's ratio v_{12} is constant. The mesh has 1666 Q8, 303 T6, and 32 T6qp singular elements with a total of 2001 elements and 5851 nodes. The following data were used for the FEM analysis:

$$a/W = 0.1, \quad L/W = 1.0,$$
$$E_{11}(x_1) = E^0_{11} e^{\beta x_1}, \quad E_{22}(x_1) = E^0_{22} e^{\beta x_1}, \quad G_{12}(x_1) = G^0_{12} e^{\beta x_1},$$
$$\beta a = (0.0\text{--}1.0), \quad \kappa_0 = (-0.25, 5.0), \quad v = 0.3$$

generalized plane stress, 2×2 Gauss quadrature.

The average modulus indicated above can be rewritten as $E = E^0 e^{(\beta a)(x_1/a)}$, and thus the non-homogeneity parameter β enters the analysis through the dimensionless constant βa. Table 2 reports the normalized SIFs using the MCC and DCT in an orthotropic plate under crack face pressure loading for various βa and for two different values of the shear parameter κ_0, which are compared with the results reported by Ozturk and Erdogan [22]. Both the MCC and the DCT estimate the SIFs within 4% of the results by Ozturk and Erdogan [22]. The influence of κ_0 on K_I (and u_2) appears to be less significant than that of βa. Notice that, in Table 2, the SIFs on the stiffer side of the plate are always greater than those on the less stiff side. Although this result may seem counter intuitive, it can be explained by comparing the displacement profiles of the non-homogeneous medium with those of the homogeneous medium [22]. Fig. 11 plots the crack opening displacements in a non-homogeneous medium with $E(x_1) = E^0 e^{x_1/a}$ and also in homogeneous materials with $E(-a) = E^0 e^{-1.0}$, $E = E^0$, and $E(a) = E^0 e^{1.0}$. When comparing the solution of a non-homogeneous medium with that of a homogeneous medium with the same material properties at the right crack tip $(x_1 = a)$, one can observe that because the crack opening displacement in the non-homogeneous medium is greater than that in the corresponding homogeneous medium $(E(a) = E^0 e^{1.0})$, the SIF (K_I) in the non-homogeneous medium is greater than that in the homogeneous medium. Similarly, for the same material properties at the left crack tip $(x_1 = -a)$, the SIF (K_I) in the non-homogeneous medium is lower than that in the corresponding homogeneous medium $(E(-a) = E^0 e^{-1.0})$. In Fig. 11, the crack opening displacement considering $E = E^0$ serves as a reference curve between those curves for $E = E^0 e^{1.0}$ and $E = E^0 e^{-1.0}$.

6.3. Plate with a crack perpendicular to material gradation

Ozturk and Erdogan [23] have also investigated the mixed-mode crack problem for an infinite non-homogeneous orthotropic medium (see Fig. 6(b)). Similarly to the previous example, the infinite medium is approximated by a square plate where the ratio of the plate edge length over the crack length is equal to 10. The material parameters $E_{11}(\boldsymbol{x})$, $E_{22}(\boldsymbol{x})$ and $G_{12}(\boldsymbol{x})$ are exponentially graded as functions of x_2 and proportional, while the Poisson's ratio v_{12} is constant. The following data were used for the FEM analysis:

Table 2
The normalized SIFs in a non-homogeneous orthotropic plate under uniform crack face loading for mode I problem $(K_0 = \sigma\sqrt{\pi a})$

Method	βa	$\kappa_0 = -0.25$		$\kappa_0 = 5.0$	
		$K_I^+(a)/K_0$	$K_I^-(a)/K_0$	$K_I^+(a)/K_0$	$K_I^-(a)/K_0$
Ozturk and Erdogan [23]	0.0	1.0	1.0	1.0	1.0
	0.01	1.0025	0.9975	1.0025	0.9975
	0.1	1.0246	0.9747	1.0231	0.9733
	0.25	1.0604	0.9364	1.0531	0.9306
	0.50	1.1177	0.8740	1.0946	0.8594
	0.75	1.1720	0.8154	1.1281	0.7932
	1.00	1.2235	0.7616	1.1556	0.7339
	1.50	1.3184	0.6701	1.1979	0.6367
	2.00	1.4043	0.5979	1.2290	0.5636
MCC S16, $\Delta a = a/24$	0.0	1.005	1.005	1.012	1.012
	0.01	1.007	1.003	1.015	1.009
	0.1	1.034	0.9816	1.041	0.9873
	0.25	1.081	0.9467	1.082	0.9484
	0.50	1.1390	0.8761	1.1255	0.8688
	0.75	1.1819	0.8073	1.1486	0.792
	1.00	1.2293	0.7503	1.1706	0.7278
	1.50	1.3275	0.6572	1.2135	0.6268
	2.00	1.4206	0.5839	1.2496	0.5524
DCT S16, $\Delta a = a/24$	0.0	0.9895	0.9895	1.020	1.020
	0.01	0.9918	0.9873	1.022	1.017
	0.1	1.015	0.9685	1.046	0.9978
	0.25	1.057	0.9372	1.084	0.9613
	0.50	1.107	0.8720	1.120	0.8869
	0.75	1.1419	0.8105	1.1362	0.8141
	1.00	1.1808	0.7577	1.1501	0.7533
	1.50	1.3444	0.6979	1.2017	0.6606
	2.00	1.4307	0.6262	1.2259	0.5884

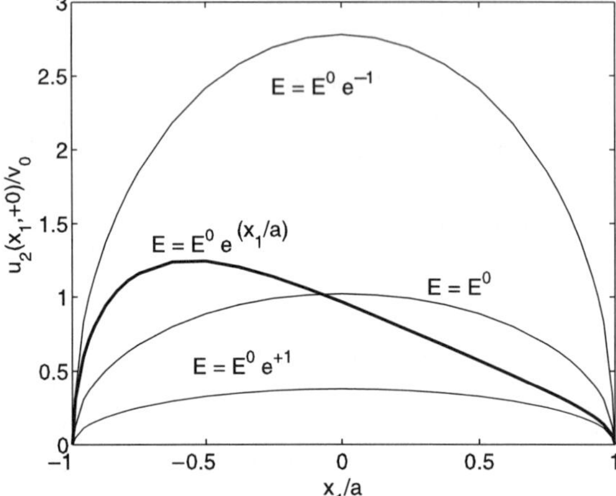

Fig. 11. COD u_2 in a non-homogeneous orthotropic medium under uniform crack face pressure loading where $v_0 = (s_1 + s_2)a\sigma\delta/E^0$ with $s_1 = \sqrt{\kappa_0 + \kappa_1}$, $s_2 = \sqrt{\kappa_0 - \kappa_1}$, $\kappa_1 = \sqrt{\kappa_0^2 - 1}$, $\kappa_0 = 0.5$, $v = 0.3$, and $\beta a = 1.0$. The COD for the crack in the FGM is indicated by a thicker line.

$$a/W = 0.1, \quad L/W = 1.0,$$

$$E_{11}(x_2) = E_{11}^0 e^{\beta x_2}, \quad E_{22}(x_2) = E_{22}^0 e^{\beta x_2}, \quad G_{12}(x_2) = G_{12}^0 e^{\beta x_2},$$

dimensionless non-homogeneity parameter: $\beta a = (0.0–2.0)$,

$$\delta^4 = E_{11}/E_{22} = (0.25, 0.5, 1.0, 3.0, 10.0),$$

$$\kappa_0 = (-0.25, 0.0, 0.5, 1.0, 2.0, 5.0), \quad v = 0.15, 0.30, 0.45,$$

generalized plane stress, $\quad 2 \times 2$ Gauss quadrature.

Notice that the average Young's modulus indicated above can be rewritten as $E = E^0 e^{(\beta a)(x_2/a)}$, and thus the non-homogeneity parameter β enters the analysis through the dimensionless constant βa. Therefore $1/\beta$ is the length scale of non-homogeneity. This problem (i.e. crack perpendicular to material gradation) will be investigated for two different boundary conditions (BCs) (see Fig. 12(a) and (b)) and two loading cases. For the first set of BCs (Fig. 12(a)), $\beta a = (0.0–2.0)$ and for the second set of BCs (Fig. 12(b)), $\beta a = (0.0–1.0)$.

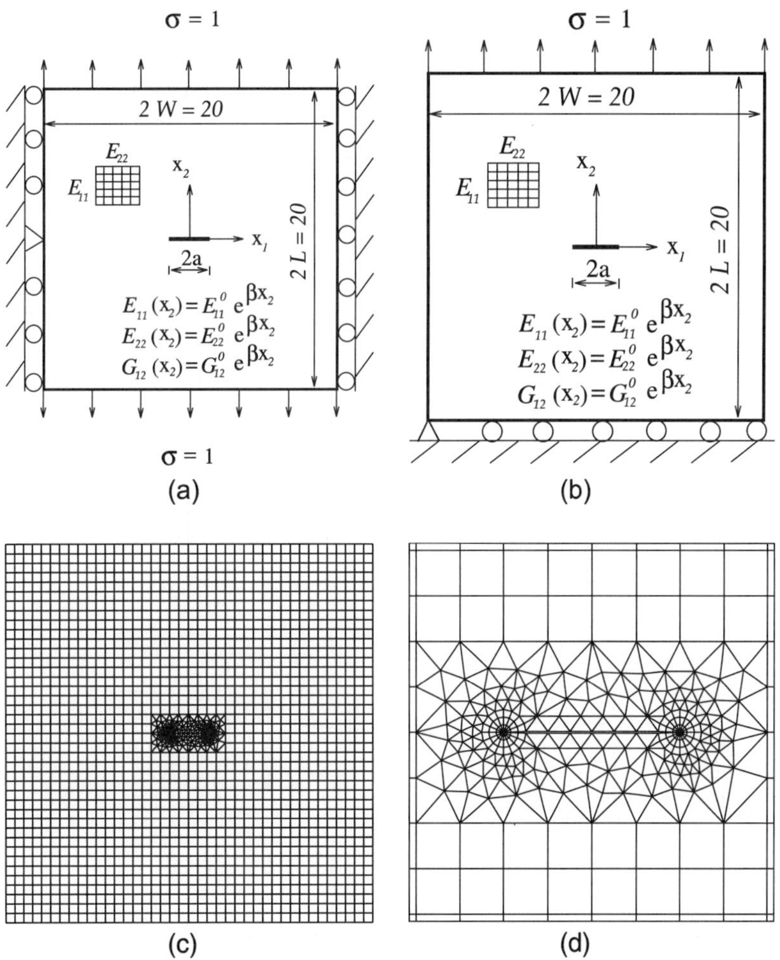

Fig. 12. Plate with a center crack perpendicular to the material gradation: (a) first set of BCs; (b) second set of BCs; (c) complete finite element mesh; (d) mesh detail using 4 rings (R4) and 16 sectors (S16) around crack tips.

6.3.1. Remote tension loading

Fig. 12(a) and (b) show a crack of length $2a$ located in a finite two-dimensional plate under remote uniform tension loading for two different boundary conditions. The complete mesh used in both cases is shown in Fig. 12(c), which is the same as that used for the example for a plate with a crack parallel to the material gradation (see Section 6.2 and Fig. 10(b)). Fig. 12(d) shows a detail with 4 rings (R4) and 16 sectors (S16) around the crack tips. The boundary conditions are prescribed such that, for Fig. 12(a), $u_1 = 0$ along the left and right edges and, in addition, $u_2 = 0$ for the node in the middle of the left edge, while for Fig. 12(b), $u_2 = 0$ along the bottom edge and in addition, $u_1 = 0$ at the left corner node of the bottom edge. The mesh has 1666 Q8, 303 T6, and 32 T6qp singular elements with a total of 2001 elements and 5851 nodes. The applied load corresponds to $\sigma_{22}(x_1, \pm 10) = \pm\sigma = \pm 1.0$ for the BCs in Fig. 12(a) and $\sigma_{22}(x_1, 10) = \sigma = 1.0$ for the BCs in Fig. 12(b).

Some representative results for the strain energy release rate, calculated from Eq. (25), for a non-homogeneous orthotropic plate under uniform tension considering two different BCs (cf. Fig. 12(a) and (b)) are plotted in Figs. 13–19. Notice in these figures that the normalized strain energy release rate $\mathscr{G}_0 = \pi\sigma^2 a/E^0$ corresponds to a homogeneous isotropic medium ($\beta a = 0$, $\kappa_0 = 1$, $\delta^4 = 1$).

Fig. 13 compares the FEM results for two different BCs and a fixed stiffness ratio $\delta^4 = 10$ with varying material non-homogeneity βa and shear parameter κ_0. Notice that the BCs and the Poisson's ratio (ν_{12}) (see Fig. 12(a) and (b)) have much influence on the strain energy release rates and, consequently, SIFs. The BCs of Fig. 12(a) prohibits the Poisson's effect of contraction in the x_1 direction. For these BCs (see Fig. 12(a)), the FEM results agree with the strain energy release rate ($\mathscr{G}$) obtained by Ozturk and Erdogan [23], which is a monotonically increasing function of κ_0 and βa. However, for the other BCs (see Fig. 12(b)), the results (dashed lines in Fig. 13) differ significantly from the previous ones (solid lines in Fig. 13). Notice that, for the case illustrated by Fig. 12(b), although $\mathscr{G}$ is still an increasing function of κ_0, it is a decreasing function of βa.

Figs. 14 and 15 show plots of the FEM results with varying βa and δ^4 for a fixed $\kappa_0 = 1$ considering the two BCs of Fig. 12(a) and (b), respectively. For the BCs of Fig. 12(a), Fig. 14 shows that $\mathscr{G}$ is an increasing function of δ^4 for $0 \leqslant \beta a \leqslant \approx 1.5$ and loses such behavior for $\beta a \geqslant \approx 1.5$, which agrees well with the results by Ozturk and Erdogan [23]. Moreover, $\mathscr{G}$ is a monotonically increasing function of βa. However, for the

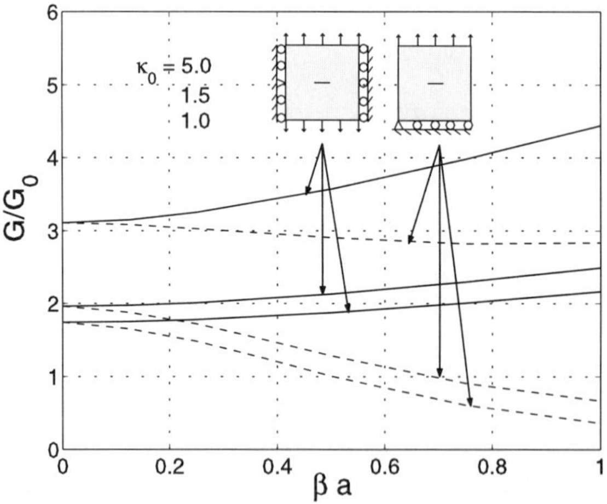

Fig. 13. Normalized strain energy release rate versus the non-homogeneity parameter βa and the shear parameter κ_0 considering uniformly applied tension ($\sigma_{22}(x_1, \pm L) = \pm\sigma$ for the left side BC and $\sigma_{22}(x_1, L) = \sigma$ for the right side BC) and $\delta^4 = E_{11}/E_{22} = 10.0$, $\nu = 0.3$, $\mathscr{G}_0 = \pi\sigma^2 a/E^0$. The solid lines correspond to the first set of BCs (see Fig. 12(a)) and the dashed lines correspond to the second set of BCs (see Fig. 12(b)).

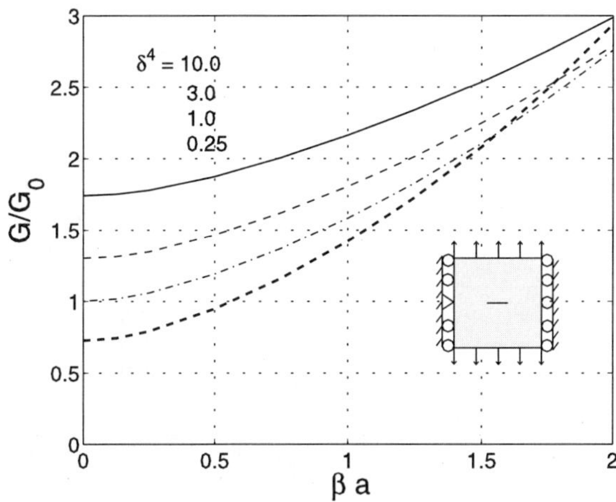

Fig. 14. Normalized strain energy release rate versus the non-homogeneity parameter βa and the stiffness parameter δ^4 considering uniformly applied tension with BCs of Fig. 12(a) and $\sigma_{22}(x_1, \pm L) = \pm\sigma$, $\kappa_0 = 1$, $v = 0.3$, $\mathcal{G}_0 = \pi\sigma^2 a/E^0$.

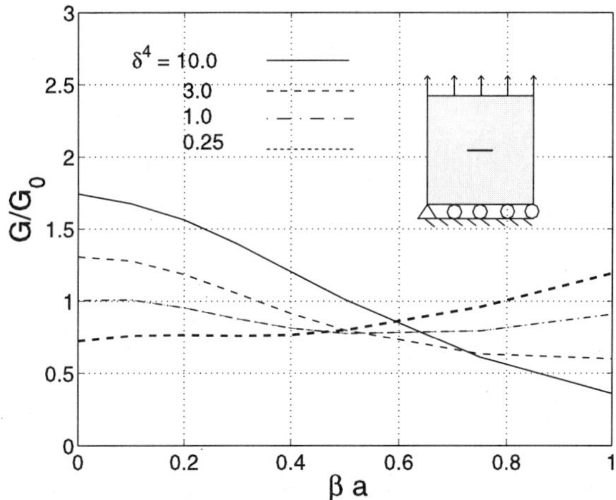

Fig. 15. Normalized strain energy release rate versus the non-homogeneity parameter βa and the stiffness parameter δ^4 considering uniformly applied tension with BCs of Fig. 12(b) and $\sigma_{22}(x_1, L) = \sigma$, $\kappa_0 = 1$, $v = 0.3$, $\mathcal{G}_0 = \pi\sigma^2 a/E^0$.

BCs of Fig. 12(b), the overall behavior is quite different as shown by Fig. 15, which indicates a change in the overall trends for $\beta a \approx 0.5$. The energy release rate $\mathcal{G}$ increases with δ^4 to the left of this point and it decreases to the right of this point. Note that $\mathcal{G}$ is a monotonically increasing function of βa for $\delta^4 = 0.25$ and a monotonically decreasing function of βa for $\delta^4 = 10$.

Figs. 16 and 17 show the FEM results with varying βa and δ^4 for the two BCs of Fig. 12(a) and (b), respectively. For the BCs of Fig. 12(a), Fig. 16 shows that $\mathcal{G}$ is an increasing function of βa, which matches well with the results by Ozturk and Erdogan [23], while for the BCs of Fig. 12(b), $\mathcal{G}$ loses such trends for βa as illustrated by Fig. 17.

Figs. 18 and 19 show the variation of $\mathcal{G}$ with βa and κ_0 for fixed $\delta^4 = 9$ considering the two BCs of Fig. 12(a) and (b), respectively. For the BCs of Fig. 12(a), Fig. 18 shows that $\mathcal{G}$ is an increasing function of βa and κ_0, which agrees with the results by Ozturk and Erdogan [23]. However, for the BCs of Fig. 12(b),

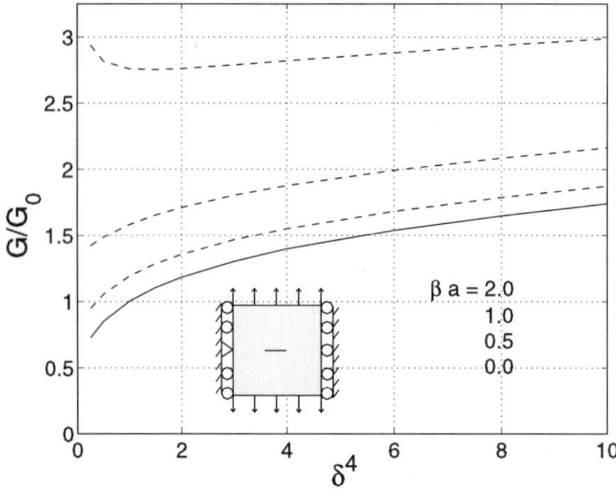

Fig. 16. Normalized strain energy release rate versus the stiffness parameter $\delta^4 = E_{11}/E_{22}$ and the non-homogeneity parameter βa considering uniformly applied tension with BCs of Fig. 12(a) and $\sigma_{22}(x_1, \pm L) = \pm \sigma$, $\kappa_0 = 1$, $v = 0.3$, $\mathscr{G}_0 = \pi\sigma^2 a/E^0$.

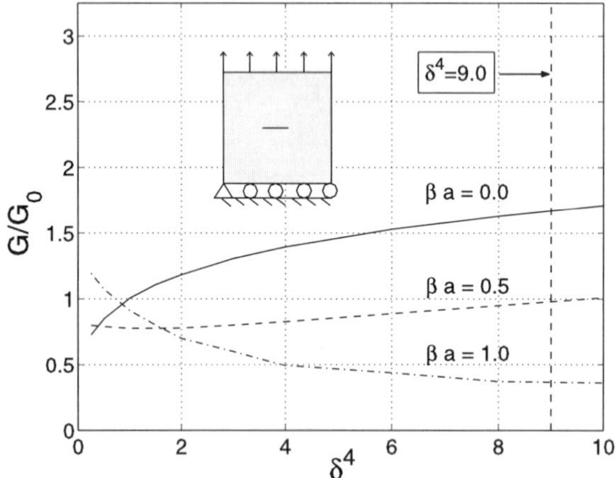

Fig. 17. Normalized strain energy release rate versus the stiffness parameter $\delta^4 = E_{11}/E_{22}$ and the non-homogeneity parameter βa considering uniformly applied tension with BCs of Fig. 12(b) and $\sigma_{22}(x_1, L) = \sigma$, $\kappa_0 = 1$, $v = 0.3$, $\mathscr{G}_0 = \pi\sigma^2 a/E^0$.

Fig. 19 shows that $\mathscr{G}$ is a decreasing function of βa and increasing function of κ_0. Moreover, consider the vertical line $\kappa_0 = 1$ in Fig. 19, and note that for increasing $\mathscr{G}$, the values of βa are $(1.0, 0.5, 0.0)$. This is consistent with the results of Fig. 17 (see vertical line in the graph).

6.3.2. Crack face loading

Fig. 20(a) shows a crack of length $2a$ located in a finite two-dimensional plate under either uniform crack face pressure (normal) loading or crack face shear loading. For uniform crack face pressure loading, the applied load corresponds to

$$\sigma_{22}(-1 \leqslant x_1 \leqslant 1, \pm 0) = \pm \sigma = \pm 1.0,$$

and for uniform crack face shear loading, the applied load corresponds to

$$\sigma_{12}(-1 \leqslant x_1 \leqslant 1, \pm 0) = \pm \tau = \pm 1.0.$$

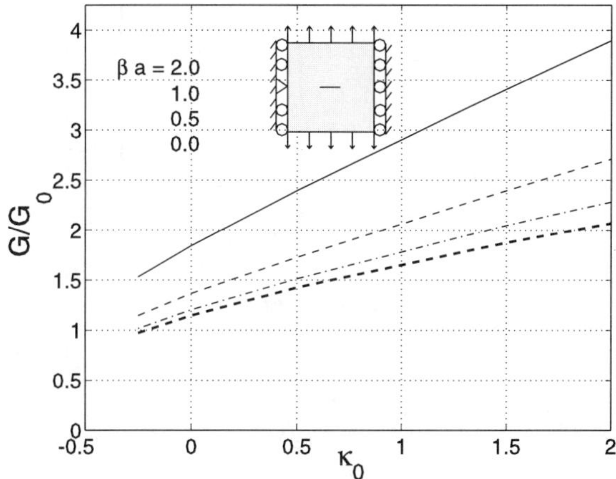

Fig. 18. Normalized strain energy release rate versus the shear parameter κ_0 and the non-homogeneity parameter βa considering uniformly applied tension with BCs of Fig. 12(a) and $\sigma_{22}(x_1, \pm L) = \pm\sigma$, $\delta^4 = 9$, $v = 0.3$, $\mathcal{G}_0 = \pi\sigma^2 a/E^0$.

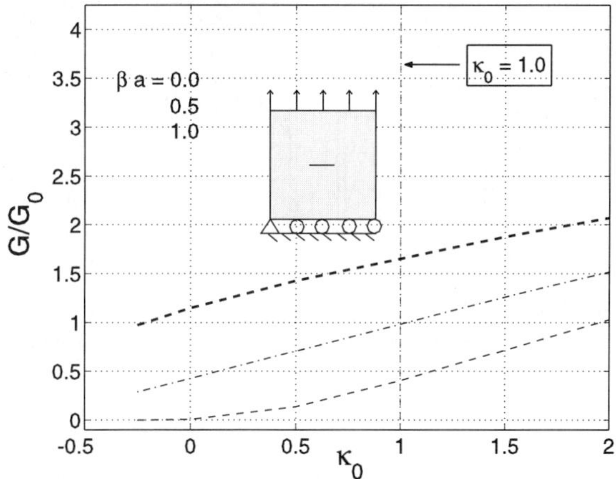

Fig. 19. Normalized strain energy release rate versus the shear parameter κ_0 and the non-homogeneity parameter βa considering uniformly applied tension with BCs of Fig. 12(b) and $\sigma_{22}(x_1, L) = \sigma$, $\delta^4 = 9$, $v = 0.3$, $\mathcal{G}_0 = \pi\sigma^2 a/E^0$.

The displacement boundary condition is prescribed such that $u_1 = u_2 = 0$ for the node in the middle of the left edge, and $u_2 = 0$ for the node in the middle of right edge. To address the influence of the crack tip discretization, two mesh configurations are used: one has 5 rings and 8 sectors (S8) around the crack tips (Fig. 20(b)) and the other has 4 rings (R4) and 16 sectors (S16) around the crack tips (Fig. 20(c)). The latter configuration is the same as those shown in Figs. 10(c) and 12(d). The results for this example are summarized in Figs. 21–25 and Tables 3–6.

Fig. 21 shows the increase in $K_I(a)$ with increasing βa and κ_0 under crack face pressure loading, which agrees with the results by Ozturk and Erdogan [23]. The influence of mesh discretization on fracture behavior is shown in Figs. 22 and 23, which illustrate the variation of SIFs with the non-homogeneity parameter βa and the shear parameter κ_0 under uniform crack face loading. Fig. 22 refers to the discretization with 8 sectors (S8) and Fig. 23 refers to the discretization with 16 sectors (S16) around the crack tips. Comparing Figs. 22 and 23, one notices that the results for $\kappa_0 = 1.0$, 1.5 and 5.0 are similar in both figures,

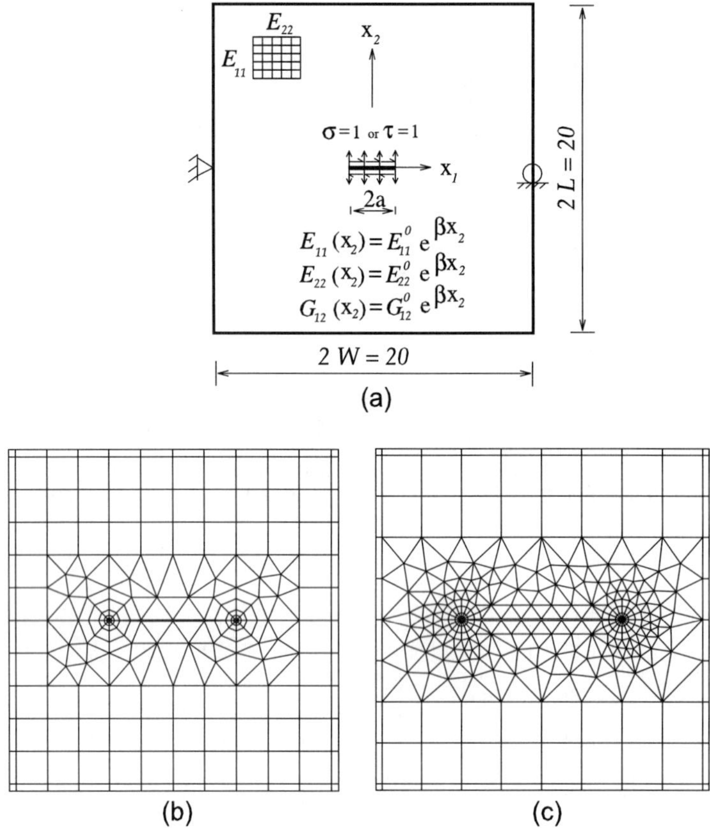

Fig. 20. Plate with a center crack perpendicular to the material gradation: (a) geometry and BCs; (b) mesh detail using 5 rings (R5) and 8 sectors (S8) around crack tips; (c) mesh detail using 4 rings (R4) and 16 sectors (S16) around crack tips.

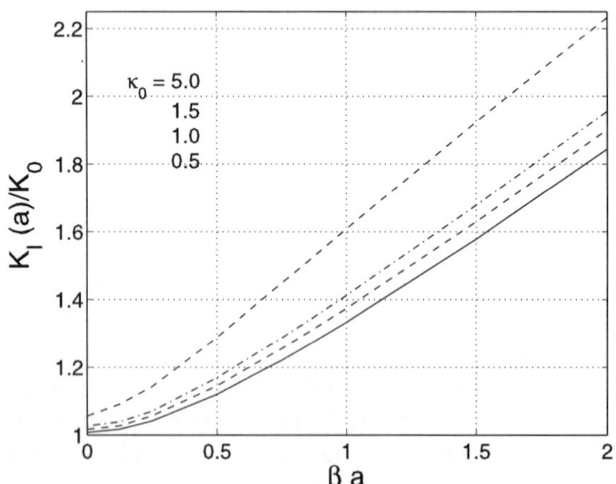

Fig. 21. Normalized strain energy release rate versus the non-homogeneity parameter βa and the shear parameter κ_0 under uniform crack face pressure loading, $\sigma_{22}(-a \leqslant x_1 \leqslant a, \pm 0) = \pm\sigma$, $\delta^4 = 0.25$, $v = 0.3$, $K_0 = \sigma\sqrt{\pi a}$.

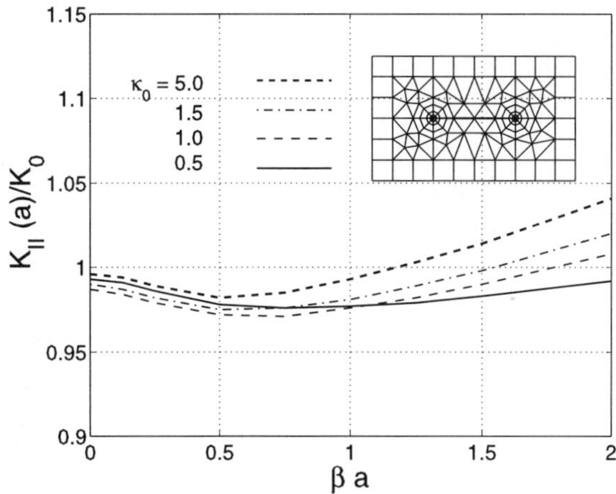

Fig. 22. Variation of the normalized SIFs versus the non-homogeneity parameter βa and the shear parameter κ_0 under uniform crack face shear loading, $\sigma_{12}(-a \leqslant x_1 \leqslant a, \pm 0) = \pm\tau$, $\delta^4 = 0.25$, $v = 0.3$, $K_0 = \tau\sqrt{\pi a}$ using 8 sectors (S8) around the crack tip.

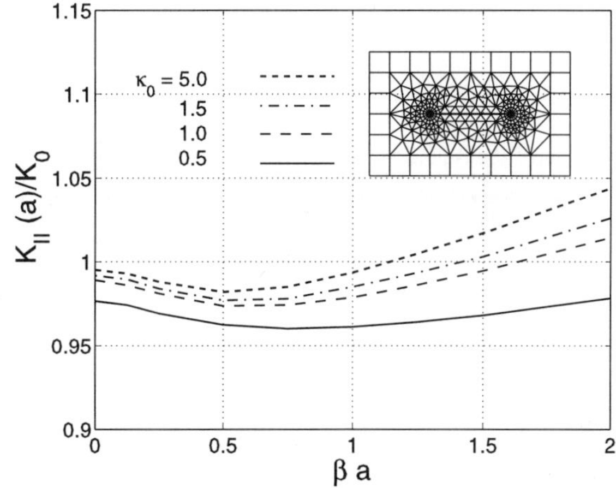

Fig. 23. Variation of the normalized SIFs versus the non-homogeneity parameter βa and the shear parameter κ_0 under uniform crack face shear loading, $\sigma_{12}(-a \leqslant x_1 \leqslant a, \pm 0) = \pm\tau$, $\delta^4 = 0.25$, $v = 0.3$, $K_0 = \tau\sqrt{\pi a}$ using 16 sectors (S16) around the crack tip.

but not the results for $\kappa_0 = 0.5$. In fact, the results in Fig. 23 (finer discretization: S16) are more reliable than those in Fig. 22 (coarser discretization: S8) and also agree with the results by Ozturk and Erdogan [23]. It is observed in Figs. 21–23 that K_{I} and K_{II} approach $K_0 = \sigma\sqrt{\pi a}$ and $K_0 = \tau\sqrt{\pi a}$, respectively, as βa tends to zero, while exact solutions are obtained by Ozturk and Erdogan [23]. This is due to numerical errors in the FEM which are absent in the integral equation method. However this type of method (i.e. based on integral equations) is much less general than the FEM. Fig. 24 shows a comparison of CODs by the FEM with those by Ozturk and Erdogan [23] for $v = 0.3$, $\kappa_0 = 0.5$, and $\delta^4 = 10$. Both results agree quite well. Fig. 25 shows the CSDs obtained by the FEM. It is interesting to observe that CSDs by the FEM are almost two times those by Ozturk and Erdogan [23]. The excellent agreement of the CODs and the factor of 2 for the CSDs seem to suggest a typo in the CSD results of reference [23]. The crack profiles in Fig. 25 are significantly influenced by the mesh discretization. A finer discretization along the crack surfaces (away from the tips) leads to a smoother profile (cf. Fig. 20(c)).

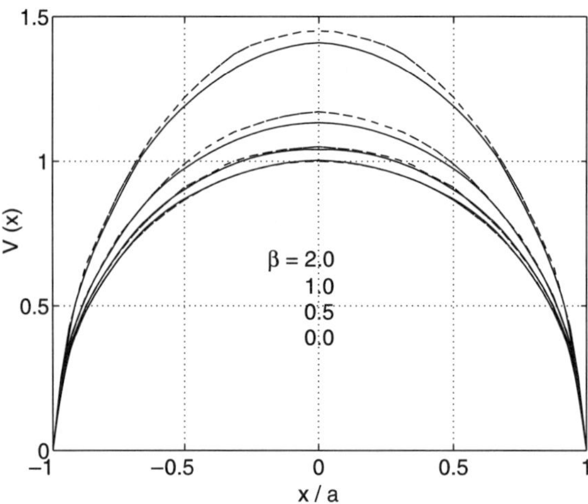

Fig. 24. Influence of the non-homogeneity parameter βa on the COD considering uniform crack face pressure loading, $\sigma_{22}(-a \leqslant x_1 \leqslant a, \pm 0) = \pm\sigma$, $\delta^4 = 10.0$, $v = 0.3$, $\kappa_0 = 0.5$, $V = (u_2(x_1, +0) - u_2(x_1, -0))/v_0$, $v_0 = a\sigma/E_2$, $(E_2 = E^0\delta/(2\sqrt{2}(\delta^2 + (E_{11}/2G_{12} - v_{12}))^{1/2}))$. The dotted lines represent the Ozturk and Erdogan's [23] analytical results and the solid lines show the FEM results.

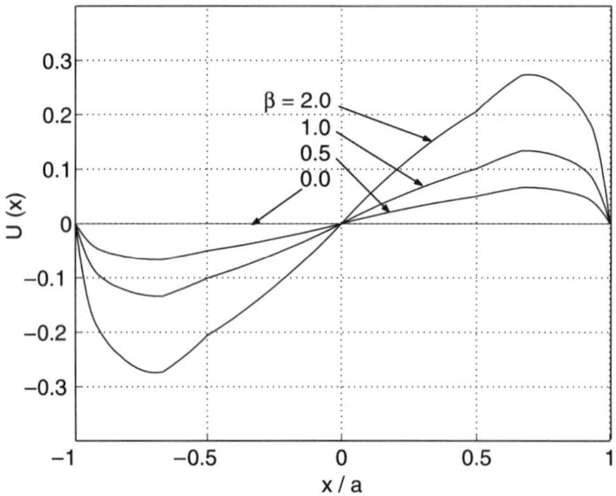

Fig. 25. Influence of the non-homogeneity parameter βa on the CSD considering uniform crack face pressure loading, $\sigma_{22}(-a \leqslant x_1 \leqslant a, \pm 0) = \pm\sigma$, $\delta^4 = 10.0$, $v = 0.3$, $\kappa_0 = 0.5$, $U = (u_1(x_1, +0) - u_1(x_1, -0))/u_0$, $u_0 = a\sigma/E_1$, $(E_1 = E^0\delta^2/(2\sqrt{2}(\delta^2 + (E_{11}/2G_{12} - v_{12}))^{1/2}))$.

Several SIF results are presented in Tables 3–6 and are also compared with the results obtained by Ozturk and Erdogan [23]. Table 3 shows the effect of loading conditions and the non-homogeneity parameter βa on the normalized SIFs for uniform crack face pressure loading and for uniform crack face shear loading using both the MCC and DCT. Tables 4 and 5 show the effect of the Poisson's ratio on the normalized SIFs in a non-homogeneous orthotropic plate with the uniform crack face pressure loading and shear loading, respectively. To show the influence of mesh discretization on the SIFs, Table 6 presents mixed-mode SIFs in a non-homogeneous orthotropic plate with crack face pressure loading considering $\beta = 0.5$, $\kappa_0 = 5.0$, and $\delta^4 = 10$. Notice that the mesh with 16 sectors (S16) leads to results that are closer to

Table 3
The effect of the loading conditions on the normalized SIFs in a non-homogeneous orthotropic plate for the mixed-mode problem considering $\kappa_0 = 0.5$ ($K_0 = \sigma\sqrt{\pi a}$ for uniform crack face pressure loading and $K_0 = \tau\sqrt{\pi a}$ for uniform crack face shear loading)

Method	βa	σ		τ	
		$K_{\mathrm{I}}(a)/K_0$	$K_{\mathrm{II}}(a)/K_0$	$K_{\mathrm{I}}(a)/K_0$	$K_{\mathrm{II}}(a)/K_0$
Ozturk and Erdogan [23]	0.0	1.0	0.0	0.0	1.0
	0.1	1.0115	0.0250	−0.0494	0.9989
	0.25	1.0489	0.0627	−0.1191	0.9968
	0.50	1.1351	0.1263	−0.2217	0.9965
	1.0	1.3494	0.2587	−0.3682	1.0071
	2.0	1.8580	0.5529	−0.5725	1.0499
MCC S16, $\Delta a = a/24$	0.0	1.021	0.0	0.0	0.992
	0.1	1.0273	0.0246	−0.0484	0.9907
	0.25	1.0550	0.0601	−0.1150	0.9850
	0.50	1.1351	0.1204	−0.2126	0.9788
	1.0	1.3495	0.2458	−0.3685	0.9884
	2.0	1.8663	0.5226	−0.5940	1.0350
DCT S16, $\Delta a = a/24$	0.0	1.0302	0.0	0.0	0.9811
	0.1	1.0352	0.0242	−0.0492	0.9799
	0.25	1.0618	0.0609	−0.1196	0.9749
	0.50	1.1390	0.1227	−0.2219	0.9687
	1.0	1.3444	0.2515	−0.3857	0.9777
	2.0	1.8240	0.5377	−0.6245	1.0189

Table 4
The effect of the Poisson's ratio on the normalized SIFs in a non-homogeneous orthotropic plate with crack face (normal) pressure loading for the mixed-mode problem ($\kappa_0 = 5.0$, $K_0 = \sigma\sqrt{\pi a}$)

Method	βa	δ^4	0.25			10.0		
		ν	0.15	0.30	0.45	0.15	0.30	0.45
Ozturk and Erdogan [23]	0.5	$K_{\mathrm{I}}(a)/K_0$	1.2516	1.2596	1.2674	1.0748	1.0776	1.0804
		$K_{\mathrm{II}}(a)/K_0$	0.1259	0.1259	0.1259	0.1252	0.1252	0.1251
	1.0	$K_{\mathrm{I}}(a)/K_0$	1.5589	1.5739	1.5884	1.1892	1.1955	1.2017
		$K_{\mathrm{II}}(a)/K_0$	0.2555	0.2557	0.2558	0.2511	0.2512	0.2512
MCC S16, $\Delta a = a/24$	0.5	$K_{\mathrm{I}}(a)/K_0$	1.2536	1.2615	1.2694	1.073	1.076	1.079
		$K_{\mathrm{II}}(a)/K_0$	0.1196	0.1196	0.1196	0.1297	0.1297	0.1298
	1.0	$K_{\mathrm{I}}(a)/K_0$	1.5611	1.5763	1.5910	1.1881	1.1949	1.2011
		$K_{\mathrm{II}}(a)/K_0$	0.2402	0.2403	0.2404	0.2415	0.2415	0.2414
DCT S16, $\Delta a = a/24$	0.5	$K_{\mathrm{I}}(a)/K_0$	1.267	1.275	1.282	1.082	1.084	1.086
		$K_{\mathrm{II}}(a)/K_0$	0.1166	0.1167	0.1169	0.1143	0.1144	0.1144
	1.0	$K_{\mathrm{I}}(a)/K_0$	1.563	1.578	1.591	1.1960	1.1994	1.2051
		$K_{\mathrm{II}}(a)/K_0$	0.2363	0.2366	0.2370	0.2273	0.2274	0.2274

those by Ozturk and Erdogan [23] than the mesh with 8 sectors (S8), which confirms the need for careful crack tip mesh discretization in order to obtain satisfactory SIF results by the FEM. In summary, for this example, the MCC and DCT give accurate SIF results in comparison with those by Ozturk and Erdogan [23].

Table 5
The effect of the Poisson's ratio on the normalized SIFs in a non-homogeneous orthotropic plate with crack face shear loading for the mixed-mode problem ($\kappa_0 = 5.0$, $K_0 = \tau\sqrt{\pi a}$)

Method	βa	δ^4	0.25			10.0		
		ν	0.15	0.3	0.45	0.15	0.3	0.45
Ozturk and Erdogan [23]	0.5	$K_I(a)/K_0$	−0.1980	−0.1971	−0.1963	−0.0366	−0.0365	−0.0365
		$K_{II}(a)/K_0$	0.9898	0.9915	0.9931	0.9956	0.9961	0.9965
	1.0	$K_I(a)/K_0$	−0.3203	−0.3186	−0.3169	−0.0660	−0.0657	−0.0654
		$K_{II}(a)/K_0$	0.9888	0.9921	0.9953	0.9913	0.9925	0.9938
MCC S16, $\Delta a = a/24$	0.5	$K_I(a)/K_0$	−0.1883	−0.1872	−0.1861	−0.0340	−0.0341	−0.0339
		$K_{II}(a)/K_0$	0.9529	0.9546	0.9359	0.9168	0.9173	0.9179
	1.0	$K_I(a)/K_0$	−0.3034	−0.3012	−0.2991	−0.0598	−0.0595	−0.0591
		$K_{II}(a)/K_0$	0.9495	0.9529	0.9557	0.9117	0.9134	0.9145
DCT S16, $\Delta a = a/24$	0.5	$K_I(a)/K_0$	−0.2009	−0.1997	−0.1987	−0.0349	−0.0344	−0.0343
		$K_{II}(a)/K_0$	0.9224	0.9241	0.9252	0.8507	0.8519	0.8919
	1.0	$K_I(a)/K_0$	−0.3264	−0.3242	−0.3221	−0.0638	−0.0620	−0.0615
		$K_{II}(a)/K_0$	0.9185	0.9213	0.9241	0.8445	0.8479	0.8496

Table 6
Effect of crack tip mesh discretization on the SIFs in a non-homogeneous orthotropic plate with crack face pressure loading for the mixed-mode problem considering $\nu = (0.15, 0.30, 0.45)$ and $\beta a = 0.5$, $\kappa_0 = 5.0$, $\delta^4 = 10$, $K_0 = \sigma\sqrt{\pi a}$

Method	ν	0.15	0.30	0.45
Ozturk and Erdogan [23]	$K_I(a)/K_0$	1.0748	1.0776	1.0804
	$K_{II}(a)/K_0$	0.1252	0.1252	0.1251
MCC S8, $\Delta a = a/32$	$K_I(a)/K_0$	1.070	1.074	1.076
	$K_{II}(a)/K_0$	0.1639	0.1646	0.1653
MCC S16, $\Delta a = a/24$	$K_I(a)/K_0$	1.073	1.076	1.079
	$K_{II}(a)/K_0$	0.1297	0.1297	0.1298

6.4. Two interacting offset cracks

This example is based on the experiments for SIFs for interacting cracks conducted by Mehdi-Soozani et al. [39]. They have used polycarbonate PSM1 (Product of Measurements Group, Inc.), which is a photoelastic material free of time-edge effects and exhibits very little creep at room temperature. The configuration that they [39] tested and analyzed is illustrated by Fig. 26(a), which shows two interacting offset cracks of length $2a$ located with angles $\theta = 0°$ and $\theta = 45°$ in a finite two-dimensional plate under uniform remote tension. Fig. 26(b) shows the complete finite element mesh configuration, and Fig. 26(c) shows the mesh detail with 16 sectors (S16) around the four crack tips. The applied load corresponds to $\sigma_{22} = \pm\sigma = \pm1.22$ MPa along the top and bottom edges. The displacement boundary condition is prescribed such that $u_1 = u_2 = 0$ for the node in the middle of the left edge and $u_2 = 0$ for the node in the middle of the right edge. For the isotropic case, we consider the plate as homogeneous. For the orthotropic case, the variations of $E_{11}(x)$, $E_{22}(x)$, and $G_{12}(x)$ are assumed to be exponential and non-proportional functions, while the Poisson's ratio is constant. The mesh has 2621 Q8, 574 T6, and 64 T6qp elements with a total of 3259 elements and 9460 nodes. The following data were used for the FEM analysis:

$$L = 381 \text{ mm}, \quad W = 114.3 \text{ mm}, \quad t = 3.17 \text{ mm}, \quad 2a = 20.32 \text{ mm}$$

For isotropic homogeneous case:

$$E = 2390 \text{ MPa}, \nu = 0.38,$$

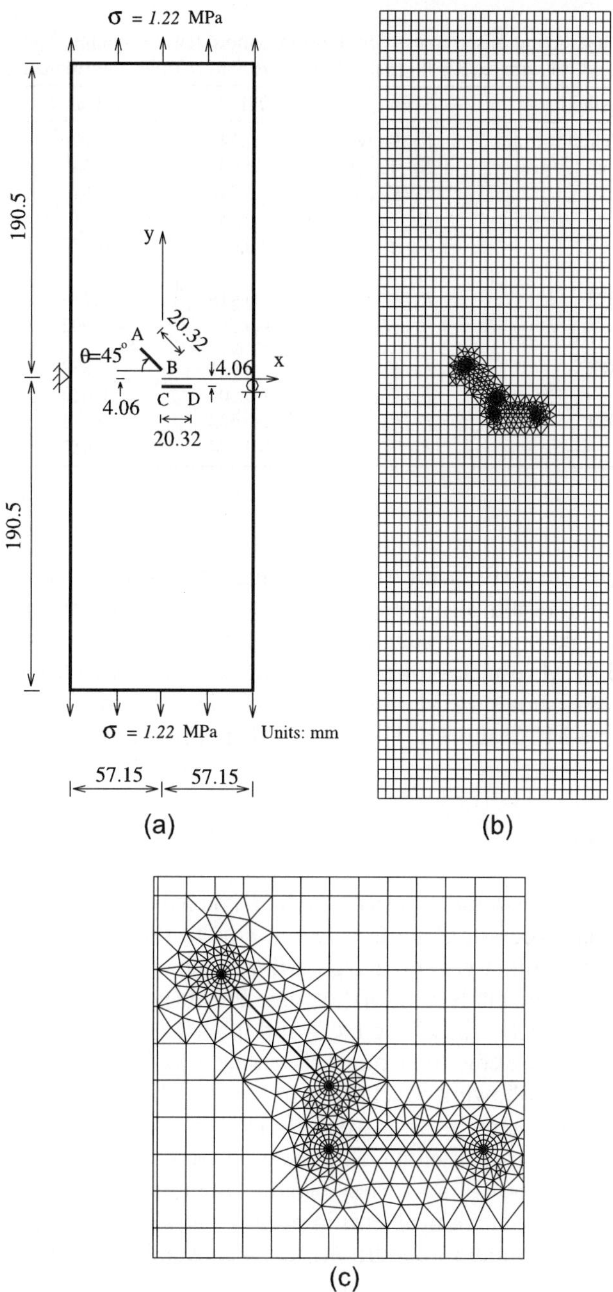

Fig. 26. Two interacting offset cracks in a finite plate; (a) geometry and BCs; (b) complete finite element mesh; (c) mesh detail showing 6 rings (R6) and 16 sectors (S16) around crack tips. (Units: N, mm).

For orthotropic FGM case:

$$E_{11}(x_1) = E_{11}^0 e^{\alpha x_1}, \quad E_{22}(x_1) = E_{22}^0 e^{\beta x_1}, \quad G_{12}(x_1) = G_{12}^0 e^{\gamma x_1},$$

$$E_{11}^0 = 2390 \text{ MPa}, \quad E_{22}^0 = 1195 \text{ MPa}, \quad G_{12}^0 = 688.7 \text{ MPa}, \quad v_{12} = 0.38,$$

$$(\alpha, \beta, \gamma) = (0.01970, 0.01576, 0.01183), \quad \text{generalized plane stress}, \quad 2 \times 2 \text{ Gauss quadrature}.$$

Table 7
The normalized SIFs with two interacting offset cracks for uniform remote tension loading considering a homogeneous isotropic material and also an orthotropic FGM with $(\alpha, \beta, \gamma) = (0.01970, 0.01576, 0.01183)$. The normalizing factor is $K_0 = \sigma\sqrt{\pi a}$

Problem	Method	SIFs	TIP A	TIP B	TIP C	TIP D
Isotropic homogeneous	Mehdi-Soozani et al. [39]	**K_I/K_0 (exper.)**	**0.64**	**0.70**	**1.14**	**1.16**
		K_{II}/K_0 (exper.)	**−0.59**	**−0.58**	**0.14**	**0.04**
		K_I/K_0 (numer.)	0.79	0.64	1.18	1.18
		K_{II}/K_0 (numer.)	−0.65	−0.61	0.14	0.01
	MCC	K_I/K_0	0.613	0.752	1.129	1.124
		K_{II}/K_0	−0.576	−0.614	0.130	0.012
	DCT	K_I/K_0	0.616	0.756	1.131	1.129
		K_{II}/K_0	−0.565	−0.602	0.128	0.005
Orthotropic FGM (proposed)	MCC	K_I/K_0	0.631	0.872	1.216	1.341
		K_{II}/K_0	−0.607	−0.630	0.155	0.021
	DCT	K_I/K_0	0.633	0.875	1.220	1.343
		K_{II}/K_0	−0.594	−0.613	0.152	0.004

Table 7 shows FEM results for SIFs using the MCC and DCT for the isotropic homogeneous case in comparison with experimental results and numerical results using the boundary element method (BEM) by Mehdi-Soozani et al. [39]. The present results for the isotropic homogeneous case show good agreement with the experimental and BEM results. Notice that the MCC results are closer to the experimental results than the BEM. Table 7 also shows FEM results for SIFs in orthotropic FGM with non-proportional material gradation using the MCC and DCT for which there are no available reference solutions.

7. Conclusions and extensions

FGMs feature an intentional material property grading, whereas the property orientation or orthotropy is usually a result of material processing/manufacturing. Due to the need to understand the mechanics of such materials, this paper presents a general purpose finite element formulation and implementation for elasticity and mixed-mode fracture analysis of orthotropic FGMs where cracks are arbitrarily oriented with respect to the principal axes of orthotropy. The DCT and the MCC method are assessed by using carefully designed singular and transition elements around the crack tips in orthotropic FGMs. From extensive numerical results presented, the following conclusions are made. The MCC and DCT provide very accurate SIFs for mixed-mode crack problems in orthotropic FGMs with cracks either aligned with the principal directions of orthotropy or arbitrarily oriented. Moreover, in contrast to homogeneous materials, Poisson's ratio and boundary conditions are found to have significant influence on the energy release rates and, consequently, SIFs in mixed-mode crack problems in orthotropic FGMs. Thus, special attention is required in defining Poisson's ratio and boundary conditions for problems involving such materials.

Acknowledgements

We gratefully acknowledge the support from the National Science Foundation (NSF) under grant no. CMS-0115954 (Mechanics and Materials Program) and from the NASA Ames Research Center (NAG 2-1424) to the University of Illinois at Urbana-Champaign. At NASA, Dr. Tina Panontin serves as the project Technical Monitor. We thank Mr. M.C. Walters for useful suggestions.

References

[1] Hirai T. Functionally gradient materials and nanocomposites. In: Holt JB, Koizumi M, Hirai T, Munir ZA, editors. Proceedings of the Second International Symposium on Functionally Gradient Materials, Ceramic Transactions, vol 34. Westerville, Ohio: The American Ceramic Society; 1993. p. 11–20.

[2] Delale F, Erdogan F. The crack problem for a nonhomogeneous plane. J Appl Mech 1983;50:609–14.

[3] Erdogan F, Wu BH. In: Analysis of FGM specimen for fracture toughness testing. Ceramic Transactions, vol 34. Westerville, Ohio: The American Ceramic Society; 1993. p. 39–46.

[4] Marur PR, Tippur HV. Numerical analysis of crack-tip fields in functionally graded materials with a crack normal to the elastic gradient. Int J Solids Struct 2000;37:5353–70.

[5] Noda N, Jin ZH. Thermal stress intensity factors for a crack in a strip of a functionally graded materials. Int J Solids Struct 1993;30:1039–56.

[6] Jin ZH, Noda N. An internal crack parallel to the boundary of a nonhomogeneous half plane under thermal loading. Int J Eng Sci 1993;31:793–806.

[7] Erdogan F, Wu BH. Crack problems in FGM layers under thermal stresses. J Therm Stresses 1996;19:237–65.

[8] Jin ZH, Batra RC. Stress intensity relaxation at the tip of an edge crack in a functionally graded material subjected to a thermal shock. J Therm Stresses 1996;19:317–39.

[9] Wang BL, Han JC, Du SY. Crack problems for functionally graded materials under transient thermal loading. J Therm Stresses 2000;23(2):143–68.

[10] Jin ZH, Paulino GH. Transient thermal stress analysis of an edge crack in a functionally graded material. Int J Fract 2001;107(1):73–98.

[11] Wang BL, Noda N. Thermally induced fracture of a smart functionally graded composite structure. Theor Appl Fract Mech 2001;35(2):93–109.

[12] Anlas G, Santare MH, Lambros J. Numerical calculation of stress intensity factors in functionally graded materials. Int J Fract 2000;104:131–43.

[13] Chen J, Wu L, Du S. A modified *J* integral for functionally graded materials. Mech Res Commun 2000;27(3):301–6.

[14] Eischen JW. Fracture of nonhomogeneous materials. Int J Fracture 1987;34:3–22.

[15] Kim JH, Paulino GH. Finite element evaluation of mixed mode stress intensity factors in functionally graded materials. Int J Numer Meth Eng 2002;53(8):1903–35.

[16] Gu P, Dao M, Asaro RJ. A simplified method for calculating the crack-tip field of functionally graded materials using the domain integral. ASME J Appl Mech 1999;66(1):101–8.

[17] Bao G, Wang L. Multiple cracking in functionally graded ceramic/metal coatings. Int J Solids Struct 1995;32(19):2853–71.

[18] Bao G, Cai H. Delamination cracking in functionally graded coating/metal substrate systems. Acta Mechanica 1997;45(3):1055–66.

[19] Sampath S, Herman H, Shimoda N, Saito T. Thermal spray processing of FGMs. MRS Bull 1995;20(1):27–31.

[20] Kaysser WA, Ilschner B. FGM research activities in Europe. MRS Bull 1995;20(1):22–6.

[21] Gu P, Asaro RJ. Cracks in functionally graded materials. Int J Solids Struct 1997;34(1):1–17.

[22] Ozturk M, Erdogan F. Mode I crack problem in an inhomogeneous orthotropic medium. Int J Eng Sci 1997;35(9):869–83.

[23] Ozturk M, Erdogan F. The mixed mode crack problem in an inhomogeneous orthotropic medium. Int J Fract 1999;98:243–61.

[24] Lekhnitskii SG. Anisotropic plates. New York: Gordon and Breach Science Publishers; 1968.

[25] Shih CF, de Lorenzi HG, German MD. Crack extension modeling with singular quadratic isoparametric elements. Int J Fract 1976;12:647–51.

[26] Irwin GR. Analysis of stresses and strains near the end of a crack traversing a plate. ASME J Appl Mech 1957;24:354–61.

[27] Rybicki EF, Kanninen MF. A finite element calculation of stress intensity factors by a modified crack closure integral. Eng Fract Mech 1977;9:931–8.

[28] Raju IS. Calculation of strain-energy release rates with high order and singular finite elements. Eng Fract Mech 1987;28(3):251–74.

[29] Sih GC, Paris PC, Irwin GR. On cracks in rectilinearly anisotropic bodies. Int J Fract Mech 1965;1:189–203.

[30] Ramamurthy TS, Krishnamurthy T, Narayana KB, Vijayakumar K, Dataguru B. Modified crack closure integral method with quarter point elements. Mech Res Commun 1986;13(4):179–86.

[31] Kim JH, Paulino GH. Isoparametric graded finite elements for nonhomogeneous isotropic and orthotropic materials. ASME J Appl Mech, in press.

[32] Paulino GH, Kim JH. The weak patch test for nonhomogeneous materials modeled with graded finite elements, submitted for publication.

[33] Wawrzynek PA. Interactive finite element analysis of fracture processes: an integrated approach, MS Thesis, Cornell University, 1987.

[34] Wawrzynek PA, Ingraffea AR. Discrete modeling of crack propagation: theoretical aspects and implementation issues in two and three dimensions. Report 91-5, School of Civil Engineering and Environmental Engineering, Cornell University, 1991.

[35] Kim JH, Paulino GH. Mixed-mode *J*-integral formulation and implementation using graded elements for fracture analysis of nonhomogeneous orthotropic materials, submitted for publication.

[36] Krenk S. On the elastic constants of plane orthotropic elasticity. J Compos Mat 1979;13:108–16.

[37] Atluri SN, Kobayashi AS, Nakagaki M. A finite element program for fracture mechanics analysis of composite material. Fract Mech Compos, ASTM STP 1975;593:86–98.

[38] Wang SS, Yau JF, Corten HT. A mixed mode crack analysis of rectilinear anisotropic solids using conservation laws of elasticity. Int J Fract 1980;16:247–59.

[39] Mehdi-Soozani A, Miskioglu I, Burger CP, Rudolphi TJ. Stress intensity factors for interacting cracks. Eng Fract Mech 1987;27:345–59.

PERGAMON

Engineering Fracture Mechanics 69 (2002) 1587–1606

www.elsevier.com/locate/engfracmech

Engineering Fracture Mechanics

Analysis of spallation mechanism in thermal barrier coatings with graded bond coats using the higher-order theory for FGMs

Marek-Jerzy Pindera [a,*], Jacob Aboudi [b], Steven M. Arnold [c]

[a] *Department of Civil Engineering and Applied Mechanics, University of Virginia, Thornton Hall, Charlottesville, VA 22901, USA*
[b] *Tel-Aviv University, Ramat-Aviv 69978, Israel*
[c] *NASA Glenn Research Center, Cleveland, OH 44135, USA*

Received 21 February 2001; received in revised form 2 May 2001; accepted 11 May 2001

Abstract

A major failure mechanism in plasma-sprayed thermal barrier coatings (TBCs) is spallation of the top coat due to the top/bond coat thermal expansion mismatch concomitant with deposition-induced interfacial roughness, oxide film growth and creep-induced normal stress reversal at the rough interface's crest. Experiments indicate that reduction of the thermal expansion mismatch by embedding alumina particles in the bond coat may increase coating durability. This approach is examined using the higher-order theory for functionally graded materials (FGMs). Specifically, combined effects of a graded bond coat microstructure and oxide film thickness on the crack-tip stress field in the vicinity of a rough top/bond coat interface are investigated during thermal cycling in the presence of a local horizontal delamination situated within the homogeneous top coat at the rough interface's crest. The analysis, which accounts for the high-temperature creep/relaxation effects within the individual constituents, is conducted in two distinct ways which reflect the present dichotomy in the modeling of FGMs. The coupled approach explicitly accounts for the micro–macro-structural interaction due to the particles' presence and is an intrinsic feature of the higher-order theory. The uncoupled approach, which is the prevailing paradigm in the mechanics/materials communities, replaces the actual graded microstructure with a fictitious material having continuously or discretely varying homogenized properties. These homogenized properties are then used in the structural analysis of a TBC without regard for particle–particle or particle–interface interactions. The feasibility of using graded bond coat microstructures to reduce horizontal delamination driving forces is critically examined and the limitations of the homogenization-based approach are highlighted. © 2002 Elsevier Science Ltd. All rights reserved.

1. Introduction

Plasma-sprayed thermal barrier coatings (TBCs) are widely employed on nickel-based structural components in the hot section of aircraft gas turbine engines. They enable the structural component to operate

---

* Corresponding author. Tel.: +1-804-924-1040; fax: +1-804-982-2951.
E-mail addresses: mp3g@virginia.edu, marek@virginia.edu (M.-J. Pindera).

at a higher temperature, thereby increasing engine efficiency, or to operate longer at a given service temperature, thereby increasing the component durability. Zirconia-based coatings are typically used due to the low-conductivity of plasma-sprayed zirconia, whose microstructure is characterized by porosities and microcracks, and its chemical stability at elevated temperatures. To prevent oxidation of the structural component, and improve the thermomechanical compatibility between the coating (top coat) and the substrate, an intermediate layer of a nickel-based alloy (bond coat) is first plasma-sprayed onto the substrate material, followed by the top coat itself. The plasma-spray process results in a rough top/bond coat interface. At elevated temperatures, the bond coat alloy composition produces a thin oxide film at the rough interface which acts as a protective oxidation barrier.

Early experimental investigations into failure mechanisms of plasma-sprayed TBCs conducted by Miller and Lowell [1] identified delamination at (or just above) the top/bond coat interface as the primary failure mode, ultimately leading to spallation of the zirconia coating. The failure mechanism was hypothesized to involve residual stresses induced into the coating during the cooldown stage of a thermal cycle due to the thermal expansion mismatch between the top coat and the bond coat. The failure initiation process was shown to be substantially influenced by the oxide film formation at the top/bond coat interface during exposure to elevated temperatures, and based on the observed location of delamination the interface's irregularity was hypothesized to play a role.

The influence of rough interface, oxide film, and thermomechanical property mismatch on the spallation mechanism has been subsequently studied by a number of investigators. Evans et al. [2] employed elasticity solutions for simplified geometries to demonstrate the presence of thermally induced normal stress at the curved oxide film/top coat interface and discussed the role of compressive inplane stress leading to buckling of the delaminated region and spallation. Chang et al. [3,4] conducted finite-element analyses of coatings on cylindrical substrates with sinusoidally varying interfaces simulating cooldown from a stress-free temperature. Based on these studies, a model was proposed for failure events leading to spallation. Validation of the proposed model was subsequently conducted by Petrus and Ferguson [5] and Freborg et al. [6]. In these studies, cyclic thermal loading that closely simulated actual inservice thermal exposure was employed, oxide film growth was modeled explicitly, and a creep model was used for the inelastic behavior of the TBC constituents to account for the actual time-dependent response at elevated temperatures. The results demonstrated the evolution of a non-zero stress component normal to the interface during cyclic thermal loading. Due to stress relaxation during the hold period of a heatup–hold–cooldown thermal cycle, this stress component becomes tensile at the interface's crest upon cooldown, setting up a delamination initiation site at this location. In the trough region, the normal stress becomes compressive, also due to stress relaxation. The growth of the oxide film at the top coat/bond coat interface subsequently changes the normal stress sign in the trough region, facilitating horizontal delamination growth. This is the presently accepted model that explains initiation and growth of delamination at or just above the top coat/oxide film interface ultimately leading to spallation. Recently, this scenario was confirmed independently by Pindera et al. [7] using an alternate analysis technique.

The above investigations also demonstrated that the top/bond coat thermal expansion mismatch has the greatest effect on the evolution of the normal stress at the rough top/bond coat interface. In order to reduce this mismatch, and thus increase coating durability, Brindley et al. [8] fabricated TBCs with heterogeneous bond coats consisting of a uniform dispersion of submicron alumina particles in the nickel-based alloy bond coats. Durability experiments subsequently demonstrated consistent doubling of TBC life with just a 5% alumina particle content. Bond coats with up to 20% alumina particle content were also fabricated and tested, but showed substantial variability in durability due to oxidation problems which currently are being addressed through improved deposition techniques. It must be noted that attempts to increase TBC durability through the use of graded top coat microstructures consisting of a dispersion of metallic particles in plasma-sprayed top coats failed due to oxidation of the metallic phase at elevated temperatures [1]. The problem arises due to the high diffusivity of oxygen in zirconia. In addition, embedding high-conductivity particles in a low-conductivity matrix increases the overall coating conductivity, requiring a thicker coating for the same temperature drop. Therefore, for gas turbine applications in the aerospace environment this approach to increasing coating durability has been abondoned.

Motivated by the experimental results of Brindley et al. [8], Pindera et al. [9] quantified the reduction of the normal stress at the top/bond coat interface's crest with increasing bond coat alumina particle content. The normal stress reduction at the crest, however, was achieved at the expense of stress magnification in the trough region. This provides the motivation for grading the bond coat microstructure to optimize the normal stress magnitudes at both the crest and trough locations.

In this paper, we investigate the effect of grading the bond coat with alumina particles in the vicinity of the rough interfacial region on the potential reduction of horizontal delamination driving forces. As in our previous investigations [7,9], the analysis of thermally induced stresses in TBCs with graded bond coats during a spatially uniform thermal cycle simulating a furnace heating durability test is conducted using the *higher-order theory for functionally graded materials* (*FGMs*) [10–12]. The analysis, which accounts for the important creep/relaxation effects within the individual TBC constituents, is conducted in two distinct ways which reflect the present dichatomy in the modeling of FGMs. The coupled approach explicitly accounts for the micro–macrostructural interaction arising from the heterogeneous bond coat microstructure and is an intrinsic feature of the higher-order theory. The uncoupled approach, which is the prevailing paradigm in the mechanics/materials communities, replaces the heterogeneous bond coat microstructure by fictitious homogenized properties determined locally using a micromechanics model. The feasibility of using graded bond coat microstructures to reduce horizontal delamination driving forces is critically examined and the limitations of the homogenization-based approach are highlighted.

2. An outline of the higher-order theory for FGMs

The higher-order theory for FGMs is an approximate analytical approach, demonstrated to be sufficiently accurate in comparison with results obtained from more detailed finite-element analyses, for modeling stress and deformation fields in spatially non-uniform heterogeneous materials with different microstructural details. It circumvents the yet-to-be resolved problem of local homogenization of the material's spatially variable heterogeneous microstructure, based on the concept of a representative volume element, by explicitly accounting for the microstructural details and local particle–particle interactions in the course of obtaining approximate analytical solutions for the local field quantities. This is accomplished by discretizing the volume occupied by a FGM into subvolumes in a manner that mimics the material's microstructure. The analytical solution for the local field quantities is subsequently obtained by satisfying the field equations, namely the heat conduction equation

$$\frac{\partial q_1}{\partial x_1} + \frac{\partial q_2}{\partial x_2} + \frac{\partial q_3}{\partial x_3} = 0 \tag{1}$$

and the equilibrium equations

$$\frac{\partial \sigma_{1j}}{\partial x_1} + \frac{\partial \sigma_{2j}}{\partial x_2} + \frac{\partial \sigma_{3j}}{\partial x_3} = 0 \quad (j = 1, 2, 3) \tag{2}$$

in each subvolume of the discretized microstructure in a volumetric sense (0th, 1st, and 2nd moments of these equations), subject to the continuity of temperatures, heat fluxes, displacements and tractions between the individual subvolumes, and the external thermomechanical boundary conditions, imposed in an average sense. Assuming that the material within each subvolume is isotropic, the heat flux components q_i in the above equations are related to the temperature field T through Fourier's law

$$q_i = -k \frac{\partial T}{\partial x_i} \tag{3}$$

where k is the thermal conductivity, while the stress components σ_{ij} are related to the total strain components ε_{ij} through Hooke's law

$$\sigma_{ij} = 2\mu\varepsilon_{ij} + \lambda\varepsilon_{kk}\delta_{ij} - 2\mu\varepsilon_{ij}^{in} - \sigma_{ij}^{T} \tag{4}$$

where λ and μ are the Lamé's constants of the material filling the given subvolume, ε_{ij}^{in} are the inelastic strain components based on the chosen inelastic constitutive theory, and σ_{ij}^{T} are the thermal stress components consisting of the products of the stiffness tensor elements, thermal expansion coefficients and the temperature change. The total strain components in the individual subvolumes are obtained from the strain–displacement relations

$$\varepsilon_{ij} = \frac{1}{2}\left(\frac{\partial u_i}{\partial x_j} + \frac{\partial u_j}{\partial x_i}\right) \tag{5}$$

The solution of Eqs. (1) and (2) in the manner discussed above is accomplished by approximating the temperature and displacement fields within each subvolume using a quadratic expansion of the Legendre type in terms of local coordinates, as discussed in the sequel. This is in sharp contrast with the linear expansion employed in the generalized method of cells (GMC) developed to model the response of homogenized periodic multi-phase materials, in which a similar subvolume discretization was employed [13]. As demonstrated in our previous investigations (cf. Refs. [10–12]), a higher-order representation of the temperature and displacement fields is necessary in the present case in order to accurately capture the local effects created by the thermomechanical field gradients, the material's microstructure and the finite dimensions in the functionally graded directions. Hence *higher-order* is used to describe the present approach in order to differentiate it from GMC.

The next section briefly outlines the manner of volume discretization of a heterogeneous material's microstructure employed in the higher-order theory for FGMs. This discretization provides the basis for the approximate analytical solution of the governing field equations (1) and (2), which is described in the subsequent sections. The description of the solution methodology is sufficiently concise to facilitate the interpretation of the presented results, but avoids the lengthy derivations of the equations that provide solutions for the local field variables within the subvolumes of the discretized microstructure described in detail in Refs. [10–12].

2.1. Volume discretization

The version of the higher-order theory employed in the present investigation is based on a geometric model of a heterogeneous composite that occupies the region bounded by $|x_1| < \infty$, $0 \leqslant x_2 \leqslant H$, and $0 \leqslant x_3 \leqslant L$ (Fig. 1). The loading applied to the composite may involve an arbitrary temperature or heat flux distribution and mechanical effects represented by a combination of surface displacements and/or tractions in the x_2–x_3 plane, and a uniform strain in the x_1 direction. The composite may be reinforced by an arbitrary distribution of continuous fibers or finite-length inclusions in the x_2–x_3 plane, continuously or periodically arranged in the direction of the x_1 axis. The microstructure of the heterogeneous composite in the x_2–x_3 plane is modeled by discretizing the heterogeneous composite's cross-section into N_q and N_r generic cells in the intervals $0 \leqslant x_2 \leqslant H$, $0 \leqslant x_3 \leqslant L$, respectively. The indices q and r, whose ranges are $q = 1, 2, \ldots, N_q$ and $r = 1, 2, \ldots, N_r$, identify the generic cell in the x_2–x_3 plane. The generic cell (q, r) consists of eight subcells designated by the triplet $(\alpha\beta\gamma)$, where each index α, β, γ takes on the values 1 or 2 which indicate the relative position of the given subcell along the x_1, x_2 and x_3 axis, respectively. The dimensions of the generic cell

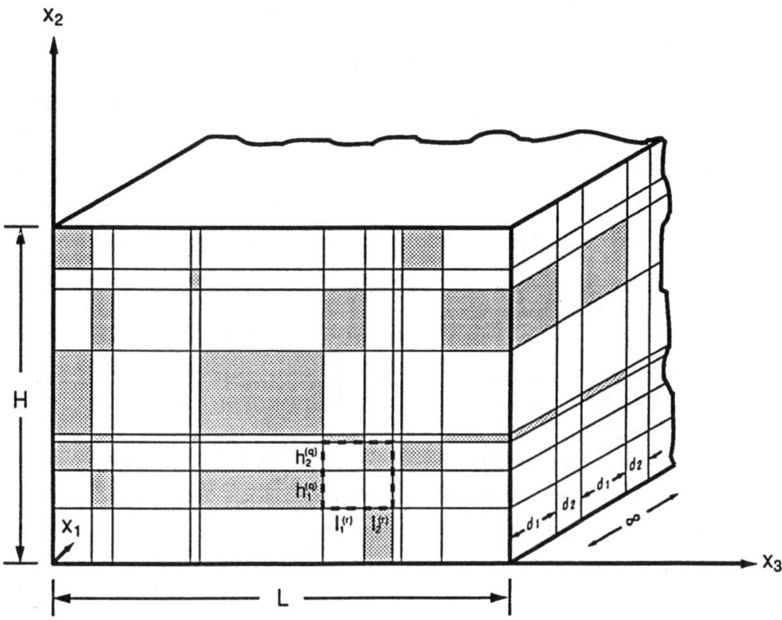

Fig. 1. A geometric model of a material functionally graded in the x_2–x_3 plane with periodicity in the x_1 direction, illustrating the volume discretization employed in the higher-order theory.

along the periodic x_1 direction, d_1 and d_2 are fixed, whereas the dimensions along the functionally graded directions x_2 and x_3, $h_1^{(q)}$, $h_2^{(q)}$, and $l_1^{(r)}$, $l_2^{(r)}$, can vary arbitrarily such that

$$H = \sum_{q=1}^{N_q} \left(h_1^{(q)} + h_2^{(q)} \right), \qquad L = \sum_{r=1}^{N_r} \left(l_1^{(r)} + l_2^{(r)} \right) \tag{6}$$

Given the applied thermomechanical loading, an approximate solution for the temperature and displacement fields within each $(\alpha\beta\gamma)$ subcell, referred to the local coordinate system $\bar{x}_1^{(\alpha)} - \bar{x}_2^{(\beta)} - \bar{x}_3^{(\gamma)}$ placed at the subcell's centroid, is constructed in two steps as described next.

2.2. Thermal analysis

Let the functionally graded cross-section shown in Fig. 1 be subjected to steady-state temperature or heat flux distributions on its bounding surfaces in the x_2–x_3 plane. Under these circumstances, the heat flux field in the material occupying the subcell $(\alpha\beta\gamma)$ of the (q,r)th generic cell, in the region $\left| \bar{x}_1^{(\alpha)} \right| \leqslant \frac{1}{2} d_\alpha$, $\left| \bar{x}_2^{(\beta)} \right| \leqslant \frac{1}{2} h_\beta^{(q)}$, $\left| \bar{x}_3^{(\gamma)} \right| \leqslant \frac{1}{2} l_\gamma^{(r)}$, must satisfy the heat conduction equation, Eq. (1), expressed in the local subcell coordinate system,

$$\frac{\partial q_1^{(\alpha\beta\gamma)}}{\partial \bar{x}_1^{(\alpha)}} + \frac{\partial q_2^{(\alpha\beta\gamma)}}{\partial \bar{x}_2^{(\beta)}} + \frac{\partial q_3^{(\alpha\beta\gamma)}}{\partial \bar{x}_3^{(\gamma)}} = 0 \quad (\alpha, \beta, \gamma = 1, 2) \tag{7}$$

The subcell heat flux components $q_i^{(\alpha\beta\gamma)}$ are obtained from the temperature field using the local version of Eq. (3),

$$q_i^{(\alpha\beta\gamma)} = -k^{(\alpha\beta\gamma)} \frac{\partial T^{(\alpha\beta\gamma)}}{\partial \bar{x}_i^{(\cdot)}} \tag{8}$$

where $k^{(\alpha\beta\gamma)}$ are the coefficients of heat conductivity of the material in the subcell $(\alpha\beta\gamma)$, and no summation is implied by repeated Greek letters in the above and henceforth.

The temperature distribution in the subcell $(\alpha\beta\gamma)$ of the (q,r)th generic cell, measured with respect to a reference temperature T_{ref}, is denoted by $T^{(\alpha\beta\gamma)}$. This temperature field is approximated by a second-order expansion in the local coordinates $\bar{x}_1^{(\alpha)}$, $\bar{x}_2^{(\beta)}$, $\bar{x}_3^{(\gamma)}$ as follows:

$$T^{(\alpha\beta\gamma)} = T_{(000)}^{(\alpha\beta\gamma)} + \bar{x}_2^{(\beta)} T_{(010)}^{(\alpha\beta\gamma)} + \bar{x}_3^{(\gamma)} T_{(001)}^{(\alpha\beta\gamma)} + \frac{1}{2}\left(3\bar{x}_1^{(\alpha)2} - \frac{d_\alpha^2}{4}\right) T_{(200)}^{(\alpha\beta\gamma)}$$

$$+ \frac{1}{2}\left(3\bar{x}_2^{(\beta)2} - \frac{h_\beta^{(q)2}}{4}\right) T_{(020)}^{(\alpha\beta\gamma)} + \frac{1}{2}\left(3\bar{x}_3^{(\gamma)2} - \frac{l_\gamma^{(r)2}}{4}\right) T_{(002)}^{(\alpha\beta\gamma)} \tag{9}$$

where the unknown volume-averaged temperature $T_{(000)}^{(\alpha\beta\gamma)}$, and the higher-order coefficients $T_{(lmn)}^{(\alpha\beta\gamma)}$ ($l, m, n = 0, 1,$ or 2 with $l + m + n \leqslant 2$), are determined in the manner described below.

Given the six unknown quantities associated with each subcell $\left(\text{i.e., } T_{(000)}^{(\alpha\beta\gamma)}, \ldots, T_{(002)}^{(\alpha\beta\gamma)}\right)$ and eight subcells within each generic cell, $48N_qN_r$ unknown quantities must be determined for a composite with N_q, and N_r rows and columns of cells containing arbitrarily specified materials. These quantities are determined by first satisfying the local heat conduction equation (7), as well as the first and second moment of this equation in each subcell in a volumetric sense in view of the employed temperature field approximation. Subsequently, continuity of heat flux and temperature is imposed in an average sense at the interfaces separating adjacent subcells, as well as neighboring cells. Fulfillment of these field equations and continuity conditions, together with the imposed thermal boundary conditions on the bounding surfaces of the composite, provides the necessary $48N_qN_r$ equations for the $48N_qN_r$ unknown coefficients $T_{(lmn)}^{(\alpha\beta\gamma)}$ in the temperature field expansion in each $(\alpha\beta\gamma)$ subcell. The final form of this system of equations can be symbolically written as

$$\boldsymbol{\kappa} \mathbf{T} = \mathbf{t} \tag{10}$$

where the structural thermal conductivity matrix $\boldsymbol{\kappa}$ contains information on the geometry and thermal conductivities of the individual subcells $(\alpha\beta\gamma)$ in the N_qN_r cells spanning the x_2 and x_3 directions, the thermal coefficient vector $\mathbf{T}$ contains the unknown coefficients that describe the temperature field in each subcell, i.e.,

$$\mathbf{T} = [T_{11}^{(111)}, \ldots, T_{N_qN_r}^{(222)}]$$

where

$$\mathbf{T}_{qr}^{(\alpha\beta\gamma)} = [T_{(000)}, T_{(010)}, T_{(001)}, T_{(200)}, T_{(020)}, T_{(002)}]_{qr}^{(\alpha\beta\gamma)}$$

and the thermal force vector $\mathbf{t}$ contains information on the boundary conditions.

2.3. Mechanical analysis

Given the temperature field generated by the applied surface temperatures and/or heat fluxes obtained in the preceding section, the resulting displacement and stress fields are then determined. This is carried out for arbitrary mechanical loading, consistent with global equilibrium requirements, applied to the surfaces of the composite.

The stress field in the subcell $(\alpha\beta\gamma)$ of the (q,r)th generic cell generated by the given temperature field must satisfy the equilibrium equations, Eq. (2), expressed in the local coordinate system

$$\frac{\partial \sigma_{1j}^{(\alpha\beta\gamma)}}{\partial \bar{x}_1^{(\alpha)}} + \frac{\partial \sigma_{2j}^{(\alpha\beta\gamma)}}{\partial \bar{x}_2^{(\beta)}} + \frac{\partial \sigma_{3j}^{(\alpha\beta\gamma)}}{\partial \bar{x}_3^{(\gamma)}} = 0 \quad (j = 1, 2, 3) \tag{11}$$

The subcell stress components $\sigma_{ij}^{(\alpha\beta\gamma)}$ are obtained from the local version of Eq. (4)

$$\sigma_{ij}^{(\alpha\beta\gamma)} = 2\mu^{(\alpha\beta\gamma)}\varepsilon_{ij}^{(\alpha\beta\gamma)} + \lambda^{(\alpha\beta\gamma)}\varepsilon_{kk}^{(\alpha\beta\gamma)}\delta_{ij} - 2\mu^{(\alpha\beta\gamma)}\varepsilon_{ij}^{\text{in}(\alpha\beta\gamma)} - \sigma_{ij}^{\text{T}(\alpha\beta\gamma)} \tag{12}$$

where $\lambda^{(\alpha\beta\gamma)}$ and $\mu^{(\alpha\beta\gamma)}$ are the Lamé's constants of the material filling the given subcell $(\alpha\beta\gamma)$, $\varepsilon_{ij}^{\text{in}(\alpha\beta\gamma)}$ are the inelastic strain components, and $\sigma_{ij}^{\text{T}(\alpha\beta\gamma)}$ is the thermal stress consisting of the products of the stiffness tensor components, thermal expansion coefficients and the temperature change. The subcell strain components are then obtained from the strain–displacement relations (5) expressed in the local coordinate system.

The displacement field in the subcell $(\alpha\beta\gamma)$ of the (q, r)th generic cell is approximated by a second-order expansion in the local coordinates $\bar{x}_1^{(\alpha)}$, $\bar{x}_2^{(\beta)}$, and $\bar{x}_3^{(\gamma)}$ as follows:

$$u_1^{(\alpha\beta\gamma)} = \bar{x}_1^{(\alpha)} W_{1(100)}^{(\alpha\beta\gamma)} \tag{13}$$

$$u_2^{(\alpha\beta\gamma)} = W_{2(000)}^{(\alpha\beta\gamma)} + \bar{x}_2^{(\beta)} W_{2(010)}^{(\alpha\beta\gamma)} + \bar{x}_3^{(\gamma)} W_{2(001)}^{(\alpha\beta\gamma)} + \frac{1}{2}\left(3\bar{x}_1^{(\alpha)2} - \frac{d_\alpha^2}{4}\right) W_{2(200)}^{(\alpha\beta\gamma)}$$

$$+ \frac{1}{2}\left(3\bar{x}_2^{(\beta)2} - \frac{h_\beta^{(q)2}}{4}\right) W_{2(020)}^{(\alpha\beta\gamma)} + \frac{1}{2}\left(3\bar{x}_3^{(\gamma)2} - \frac{l_\gamma^{(r)2}}{4}\right) W_{2(002)}^{(\alpha\beta\gamma)} \tag{14}$$

$$u_3^{(\alpha\beta\gamma)} = W_{3(000)}^{(\alpha\beta\gamma)} + \bar{x}_2^{(\beta)} W_{3(010)}^{(\alpha\beta\gamma)} + \bar{x}_3^{(\gamma)} W_{3(001)}^{(\alpha\beta\gamma)} + \frac{1}{2}\left(3\bar{x}_1^{(\alpha)2} - \frac{d_\alpha^2}{4}\right) W_{3(200)}^{(\alpha\beta\gamma)}$$

$$+ \frac{1}{2}\left(3\bar{x}_2^{(\beta)2} - \frac{h_\beta^{(q)2}}{4}\right) W_{3(020)}^{(\alpha\beta\gamma)} + \frac{1}{2}\left(3\bar{x}_3^{(\gamma)2} - \frac{l_\gamma^{(r)2}}{4}\right) W_{3(002)}^{(\alpha\beta\gamma)} \tag{15}$$

where $W_{i(000)}^{(\alpha\beta\gamma)}$, which are the volume-averaged displacements, and the higher-order terms $W_{i(lmn)}^{(\alpha\beta\gamma)}$ $(i = 1, 2, 3)$ must be determined from conditions similar to those employed in the thermal problem. In this case, there are 104 unknown coefficients $W_{i(lmn)}^{(\alpha\beta\gamma)}$ in a generic cell (q, r), accounting for the periodicity in the x_1 direction using a homogenization procedure (cf. [12]). The determination of these coefficients parallels that of the thermal problem. Here, the heat conduction equation is replaced by the equilibrium equation (11), and the continuity of tractions and displacements at the various interfaces replaces the continuity of heat fluxes and temperature. Finally, the boundary conditions involve the appropriate mechanical quantities.

Application of the above equations and conditions in a volumetric and average sense, respectively, produces a system of $104N_qN_r$ algebraic equations in the unknown coefficients $W_{i(lmn)}^{(\alpha\beta\gamma)}$. The final form of this system of equations is symbolically represented by

$$\mathbf{KU} = \mathbf{f} + \mathbf{g} \tag{16}$$

where the structural stiffness matrix $\mathbf{K}$ contains information on the geometry and thermomechanical properties of the individual subcells $(\alpha\beta\gamma)$ within the cells comprising the functionally graded composite, the displacement coefficient vector $\mathbf{U}$ contains the unknown coefficients that describe the displacement field in each subcell, i.e.,

$$\mathbf{U} = [\mathbf{U}_{11}^{(111)}, \ldots, \mathbf{U}_{N_qN_r}^{(222)}]$$

where

$$\mathbf{U}_{qr}^{(\alpha\beta\gamma)} = [W_{1(100)}, \ldots, W_{3(002)}]_{qr}^{(\alpha\beta\gamma)}$$

and the mechanical force vector **f** contains information on the boundary conditions and the thermal loading effects generated by the applied temperature. In addition, the inelastic force vector **g** appearing on the right hand side of Eq. (16) contains inelastic effects given in terms of the integrals of the inelastic strain distributions that are represented by the coefficients $R^{(\alpha\beta\gamma)}_{ij(l,m,n)}$ defined below

$$R^{(\alpha\beta\gamma)}_{ij(l,m,n)} = \mu^{(\alpha\beta\gamma)} A_{(lmn)} \int_{-1}^{+1} \int_{-1}^{+1} \int_{-1}^{+1} \varepsilon^{\text{in}(\alpha\beta\gamma)}_{ij} P_l(\zeta_1^{(\alpha)}) P_m(\zeta_2^{(\beta)}) P_n(\zeta_3^{(\gamma)}) \, \mathrm{d}\zeta_1^{(\alpha)} \mathrm{d}\zeta_2^{(\beta)} \mathrm{d}\zeta_3^{(\gamma)} \tag{17}$$

where

$$A_{(lmn)} = \tfrac{1}{4}\sqrt{(1+2l)(1+2m)(1+2n)}$$

and the Legendre polynomials $P_.(\zeta_i^{(\cdot)})$ are functions of the non-dimensionalized variables $\zeta_i^{(\cdot)}$'s, defined in the interval $-1 \leqslant \zeta_i^{(\cdot)} \leqslant +1$, which are given in terms of the local subcell coordinates as follows: $\zeta_1^{(\alpha)} = \bar{x}_1^{(\alpha)}/(d_\alpha/2)$, $\zeta_2^{(\beta)} = \bar{x}_2^{(\beta)}/(h_\beta^{(q)}/2)$, and $\zeta_3^{(\gamma)} = \bar{x}_3^{(\gamma)}/(l_\gamma^{(r)}/2)$. These integrals depend implicitly on the elements of the displacement coefficient vector **U**, requiring an incremental solution of Eq. (16) at each point along the loading path.

Similarities and differences between the higher-order theory and standard displacement-based finite-element formulations have been described elsewhere [14]. An important advantage of the present approach is the simultaneous satisfaction of displacement and traction continuity conditions between the different subvolumes of the spatially variable microstructure. This facilitates the determination of thermally induced stresses across the top/bond coat interface in the present application which cannot be easily captured due to very localized and very steep gradients in the thickness direction.

3. Problem definition

3.1. TBC system geometry

We consider a flat TBC system shown in Fig. 2 that remains in a generalized plane strain state under the imposed spatially uniform thermal loading which simulates furnace thermal cycling used for durability testing, Fig. 3. As is customary, the rough interface between the top and bond coats is modeled by a periodically varying form with the wavelength of 25 μm and total (crest-to-trough) amplitude of 10 μm at the mean elevation of 630 μm from the substrate's bottom. An oxide film with two different thicknesses (1 and 2 μm) is introduced into the bond coat at the top/bond coat interface to simulate progressive bond coat

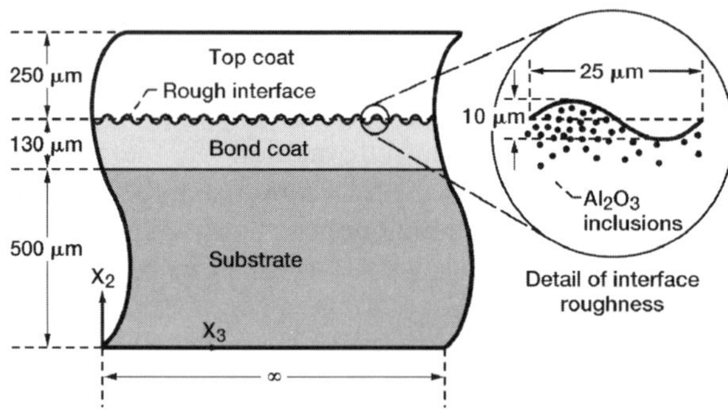

Fig. 2. A TBC system with a sinusoidally varying top coat/bond coat interface and a functionally graded bond coat in the interfacial region.

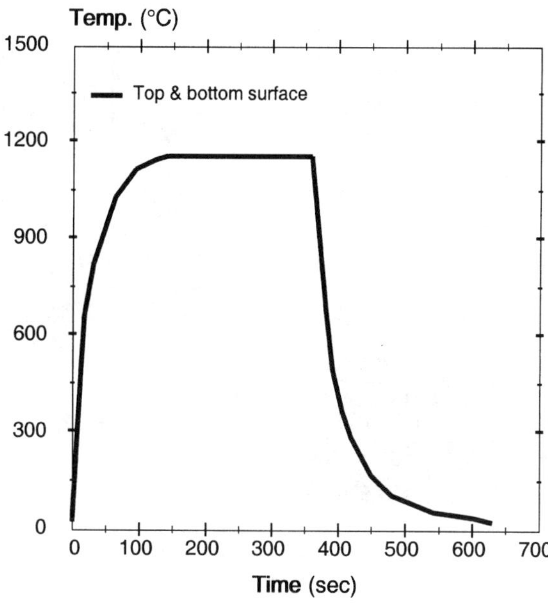

Fig. 3. Thermal loading history.

oxidation. In the present analysis, the actual oxide film growth is not modeled as it occurs primarily at the elevated temperature during the hold portion of the heating cycle where relaxation rates are much greater than the film growth itself.

Due to the periodic nature of the modeled TBC's geometry, the analysis of thermally induced stress fields is performed on the repeating cross-section element bounded by two vertical planes passing through the crest and trough of the sinusoidally varying interface. Symmetry boundary conditions on the deformation of these planes are imposed consistent with the modeled geometry and the generalized plane strain state (namely, $\bar{\sigma}_{11} = \bar{\sigma}_{33} = 0$, where the bars denote average values of the normal stresses in the x_2–x_3 and x_1–x_2 planes). The repeating cross-section element of the TBC system is discretized into generic cells and subcells in order to approximate the geometry of the sinusiodally varying interface with sufficient accuracy. Fig. 4 illustrates the employed volume discretization in the region containing the wavy interface for the investigated TBCs without an oxide film. The same discretization was employed for TBCs with different oxide film thicknesses. The discretized region for the pure TBC system (consisting of the top coat, bond coat and substrate only) is shown in Fig. 4(a), while the TBC system with the actual graded bond coat microstructure is shown in Fig. 4(b). The grading is accomplished by gradually decreasing the alumina particle content from 40% at the wavy interface's crest to 0% below the trough. The homogenized microstructure of the graded bond coat is shown in Fig. 4(c). It consists of eight layers, with progressively decreasing uniform alumina particle content in increments of 5%, whose homogenized or macroscopic properties have been calculated using the GMC. Fig. 5 illustrates the reduction of the top/bond coat thermal expansion coefficient mismatch as a function of temperature for different alumina particle content in the homogenized bond coat layers.

In addition, a horizontal crack of total length 4.25 μm is introduced at a distance of 0.5 μm above the interface's crest within the top coat, with traction-free boundary conditions imposed on the subcell faces separated by the crack. The crack arises upon cooldown due to the high normal stress in the crest region relative to the low-transverse strength of the plasma-sprayed top coat. In order to capture the large stress gradients in the vicinity of the crack tip, the mesh in this region is further refined as shown in Fig. 4(a) for

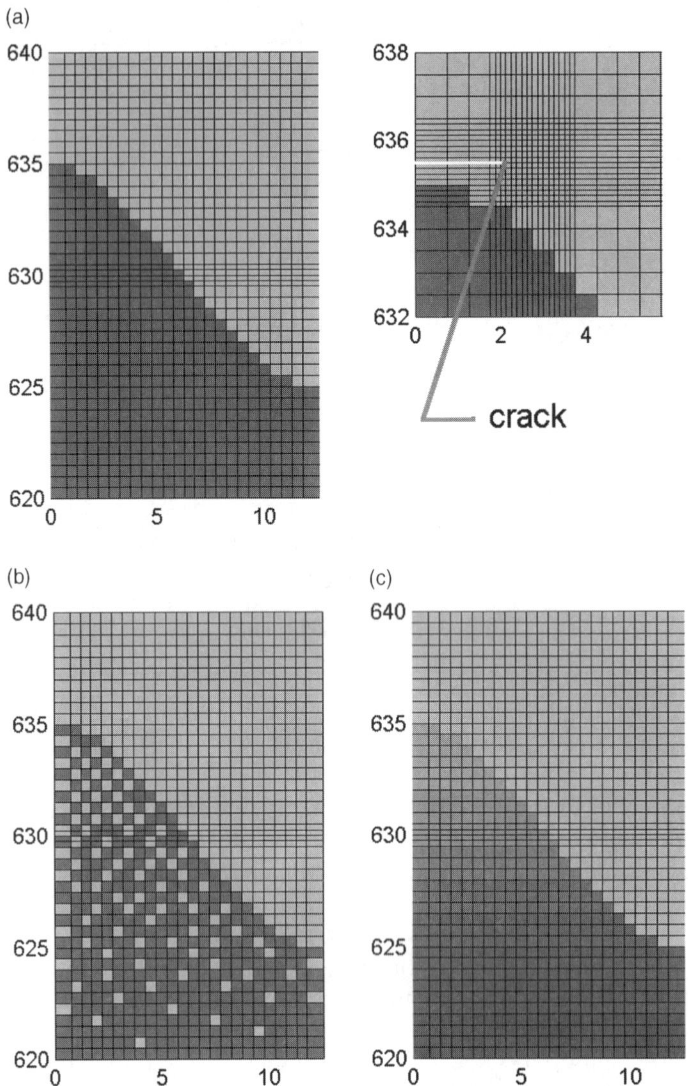

Fig. 4. Volume discretization of the sinusoidally varying interface region: (a) pure TBC (left) and a magnified region (right) showing refined mesh around the horizontal crack in the top coat above the crest of the top/bond coat interface; (b) TBC with actual graded bond coat microstructure and (c) TBC with homogenized graded bond coat microstructure.

the pure TBC. The same mesh refinement was employed for the graded bond coat TBCs. We note that the higher-order theory cannot capture the singular behavior at the crack tip as it is an approximate analysis based on volume-averaging of the field equations within discretized subvolumes of the material. However, by discretizing the region in the vicinity of the crack tip sufficiently well, the higher-order theory is capable of capturing the high stress gradients, as well as high magnitudes of the stress field, with sufficient accuracy. This has been demonstrated previously in the context of the free-edge problem in the case of a laminated composite plate [10,11]. Our objective in this paper is to investigate the relative effect of grading on the stress gradients and stress magnitudes in the presence of a horizontal crack in order to determine the feasibility of using this approach to reduce the crack-driving forces responsible for top coat delamination

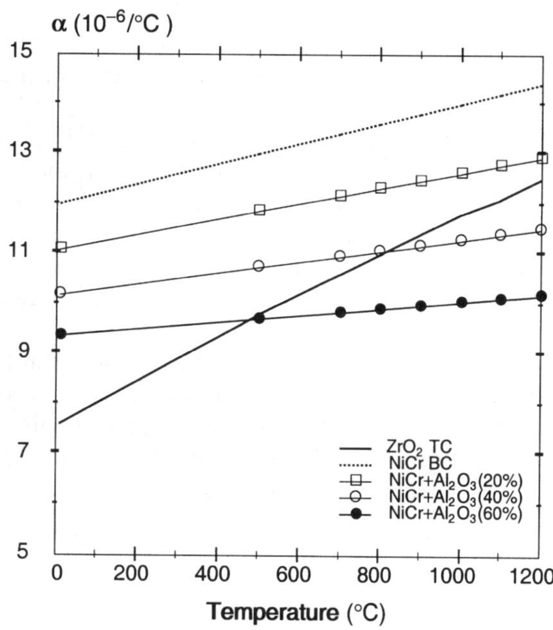

Fig. 5. Thermal expansion coefficient vs temperature of pure zirconia, and pure and heterogeneous NiCr-based bond coat material with different alumina particle content.

and ultimately spallation, rather than to calculate the critical stress intensity factors and energy release rates. These parameters are yet to be fully characterized for plasma-sprayed zirconia coatings.

3.2. Constituents' material response

The material thermoelastic and viscoplastic parameters that govern the response of the individual TBC constituents are given in Table 1. These are the same as the parameters used in our previous investigation dealing with the effects of interfacial roughness and oxide film thickness on the inelastic response of TBCs with homogeneous bond coats [7], and are based on the properties employed by Freborg et al. [6]. The viscoplastic response of the zirconia top coat, NiCr-based bond coat, and Ni-based substrate is modeled by a power-law creep equation, generalized to multi-axial loading situations as follows,

$$\dot{\varepsilon}_{ij}^{c} = \frac{3F(\sigma_e, T)}{2\sigma_e} \sigma'_{ij} \tag{18}$$

where σ'_{ij} are the components of the stress deviator, $\sigma_e = \sqrt{3/2\sigma'_{ij}\sigma'_{ij}}$, and

Table 1
Thermoelastic and creep material parameters of the TBC constituents (see Petrus and Ferguson [5] for actual values at different temperatures)

Material	E (GPa)	v	α (10^{-6}/°C) $10° \rightarrow 1200$ °C	A (MPa^{-n}/s) $10° \rightarrow 1200$ °C	n $10° \rightarrow 1200$ °C
Zr-based top coat	25.4	0.33	$7.6 \rightarrow 12.5$	$2.01 \times 10^{-30} \rightarrow 1.85 \times 10^{-7}$	$1.0 \rightarrow 3.0$
Al₃O₂ film/particles	380	0.26	8.6	–	–
NiCr-based bond coat	156	0.27	$12.0 \rightarrow 14.4$	$4.39 \times 10^{-40} \rightarrow 7.40 \times 10^{-6}$	$1.0 \rightarrow 3.0$
Ni-based substrate	156	0.27	$12.0 \rightarrow 19.3$	$4.85 \times 10^{-36} \rightarrow 2.25 \times 10^{-9}$	$1.0 \rightarrow 3.0$

$$F(\sigma_\mathrm{e}, T) = A(T)\sigma_\mathrm{e}^{n(\mathrm{T})} \tag{19}$$

where the temperature-dependent parameters A, n are listed in the table for the range 10–1200 °C, together with the thermal expansion coefficients.

The alumina particles and film are assumed to be elastic since their creep response is small for the employed thermal loading compared with the other materials. Further, the pronounced inelastic response of plasma-sprayed zirconia compared to fully densified zirconia is due to the local deformation mechanisms activated by porosities and microcracks introduced during the plasma-sprayed process, not modeled in this investigation, as discussed by DiMassi-Marcin et al. [15]. The power-law creep model for the inelastic response of zirconia has been shown to model the overall effect of these mechanisms with sufficient accuracy.

The functional form of the multi-axial creep model for the homogenized layers with different alumina particle content determined from the GMC remains the same as for the pure constituents, with appropriate modifications of the creep parameters A and n which depend on the alumina particle content at different temperatures [9]. This is consistent with the results reported by Crossman et al. [16] and Crossman and Karlak [17] in the context of metal matrix composites.

4. Results

Here, we focus on the normal stress σ_{22} and the inplane shear stress σ_{23} in the top coat itself in the vicinity of the horizontal crack above the rough top/bond coat interface's crest after one thermal cycle ($t = 630$ s, see Fig. 3). The influence of the normal stress on top coat delamination leading to spallation has been extensively investigated for pure TBCs with evolving oxide film (albeit without explicitly incorporating the crack's presence). The role that the inplane shear stress plays in the delamination process, on the other hand, has not been extensively discussed. Due to the pronounced stress relaxation during the hold period of the thermal cycle, these stress components do not change significantly at the end of subsequent thermal cycles for the chosen spatially uniform loading, and therefore one thermal cycle is sufficient to delineate the trends. As noted previously, the horizontal crack arises due to the relatively large normal stress σ_{22} that develops at the crest of the top/bond coat interface during the cooldown portion of the thermal cycle in the absence of an oxide film. We note that during the first thermal cycle, this stress component is initially compressive at the crest of the interface during heating, but relaxes to nearly zero during the hold period. Hence, stress relaxation produces a sign reversal of the normal stress at the crest location upon cooldown. The same effect changes the sign of the normal stress in the trough region, from tensile during heating to compressive upon cooldown. During subsequent cycling, the residual stresses decrease to nearly zero during the heating and hold portions of the thermal cycle, assuming the same values upon cooldown as after the first cycle. The above illustrates the importance of the inelastic constituent behavior, thereby precluding strictly elastic analysis of the top coat spallation mechanism.

The normal and shear stress fields illustrating the differences that arise due to oxide film thickness and grading have been generated in the same region as shown in Fig. 4. The color scale in the contour plots that follow has been adjusted to clearly delineate the impact of these two effects on the crack-tip stress field in the top coat itself. Since the top coat properties are smaller than those of the bond coat and oxide film/particles, the top coat stress field generally is significantly smaller in magnitude than in the bond coat and oxide film regions. In some areas of these regions, therefore, the stress fields fall outside of the employed color scale.

Fig. 6 presents comparison of the σ_{22} stress distributions in pure and graded TBCs in the absence of an oxide film. The large stress concentration at the crack tip in the pure TBC, Fig. 6(a) is substantially reduced by grading the bond coat according to the analysis based on the graded bond coat's homogenized

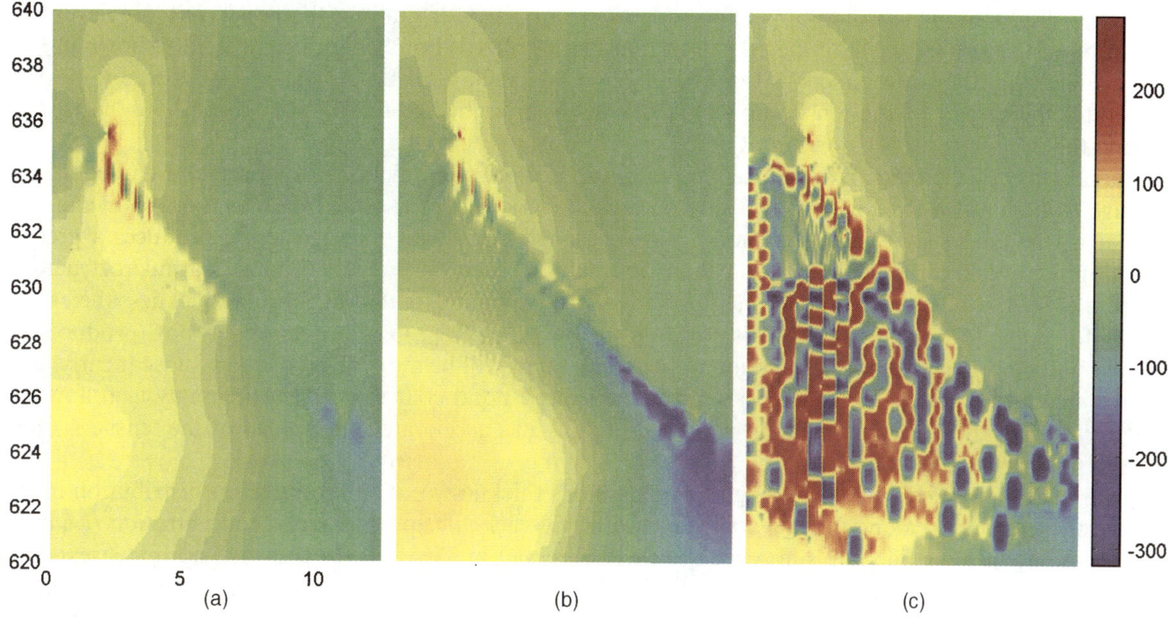

Fig. 6. Distribution of σ_{22} in the interfacial region of the investigated TBC system without an oxide film after cooldown ($t = 630$ s): (a) pure TBC; (b) graded TBC with homogenized bond coat microstructure and (c) graded TBC with actual bond coat microstructure (colorbar scale in MPa).

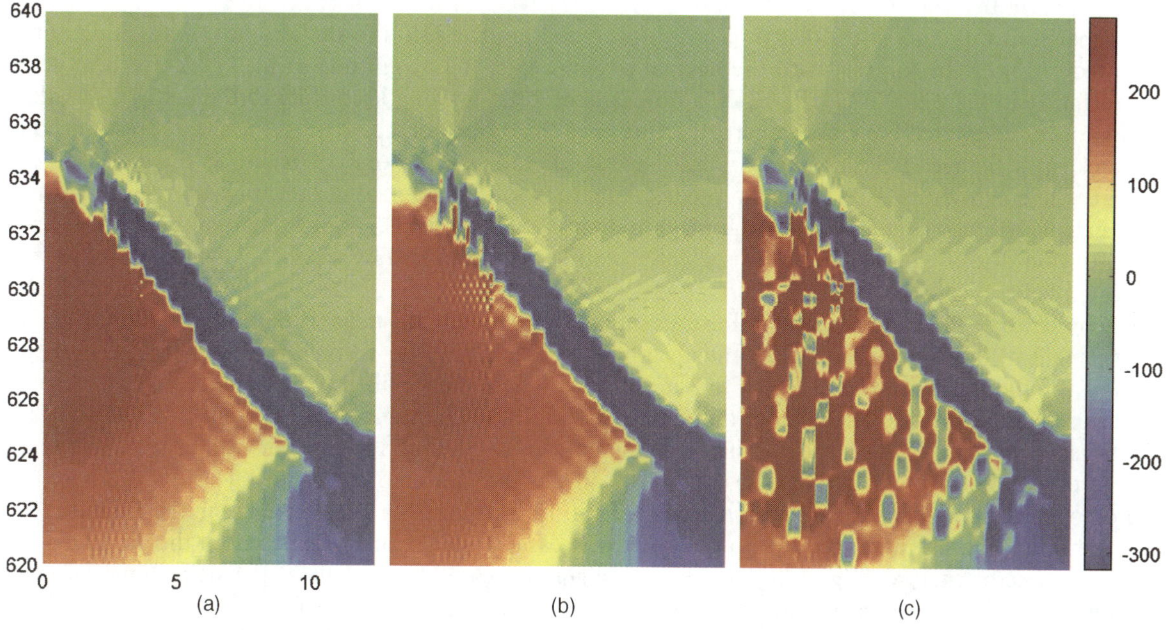

Fig. 7. Distribution of σ_{22} in the interfacial region of the investigated TBC system with a 2 μm oxide film after cooldown ($t = 630$ s): (a) pure TBC; (b) graded TBC with homogenized bond coat microstructure and (c) graded TBC with actual bond coat microstructure (colorbar scale in MPa).

properties, Fig. 6(b). This would appear to make a strong case for grading the bond coat. However, examination of the results based on the actual (or non-homogenized) bond coat microstructure shown in Fig. 6(c) indicates that the stress concentration reduction at the crack tip is not as substantial. Further, grading accelerates the spread of the tensile normal stress in the trough region which has been hypothesized to provide the driving force for horizontal crack growth during thermal cycling.

The presence of the oxide film reduces the crack-tip normal stress field for the pure and graded TBCs. Relative reductions for the TBCs with 1 μm film (not shown) follow the same trends as those shown in Fig. 6. That is, the analysis based on the homogenized graded bond coat microstructure produces a greater reduction relative to the pure TBC result than the analysis based on the actual bond coat microstructure. The results for the TBCs with 2 μm oxide film shown in Fig. 7 indicate that at this film thickness the crack-driving force due to σ_{22} has virtually disappeared for the pure TBC. Grading, however, now produces an increase in the normal stress concentration at the crack tip as will be more clearly shown subsequently. The relative increases predicted by the actual and homogenized bond coat microstructure analyses follow the same trend as in the preceding case. In addition, grading has accelerated the spread of the tensile normal stress in the trough region.

Fig. 8 quantifies the effects of grading and oxide film's thickness on the normal stress distribution in the vicinity of the crack tip along the line coincident with the crack. The normal stress concentration *reduction* due to grading in the oxide film's absence is clearly observed in Fig. 8(a). In contrast, normal stress concentration *increase* in the presence of a 2 μm oxide film is seen in Fig. 8(b). Further, the dramatic effect of oxide film thickness, which overshadows the grading effect, is clearly observed upon comparison of Fig. 8(a) and (b). These results support the hypothesis that the normal stress by itself is not sufficient to provide the necessary crack-driving force with increasing oxide film thickness during the initial crack extension stage. It is possible that TBC curvature, taken into account in previous investigations, increases the size of the tensile normal stress zone in the trough region, providing the necessary crack-driving force. However, examination of the inplane shear stress σ_{23} reveals an interesting, rarely discussed influence on the spallation mechanism as illustrated in the sequel.

Comparison of the σ_{23} stress distributions in pure and graded TBCs in the absence of an oxide film is presented in Fig. 9. In contrast with the normal stress results, the magnitude of the crack-tip shear stress concentration in the pure TBC, Fig. 9(a), is now enhanced by grading according to the analysis based on the homogenized graded bond coat microstructure, Fig. 9(b). Further, the sign of the shear stress ahead of the crack tip is changed from positive to negative due to grading. The analysis based on the actual graded bond coat microstructure, Fig. 9(c), also predicts shear stress sign reversal in the crack-tip region, but to a lesser extent. Ahead of the crack tip, the magnitude of the shear stress is quite small, although the stress concentration directly at the crack tip is greater relative to the pure TBC result, as will be more clearly illustrated below.

The presence of the 2 μm oxide film illustrates more dramatically the deleterious effect of grading on the enhancement of the shear stress field zone ahead of the crack tip. Fig. 10 presents comparison of the σ_{23} stress distributions in pure and graded TBCs in this case. The presence of the oxide film changes the sign of the shear stress field at the crack tip in the pure TBC, Fig. 10(a) (compare with Fig. 9(a)). The size of the shear stress zone ahead of the crack tip is further magnified by grading. However, as will be seen below, the shear stress concentration directly at the crack tip is only slightly affected by grading. The analysis based on the homogenized graded bond coat microstructure, Fig. 10(b), predicts a small increase in the magnitude of the shear stress concentration relative to the pure TBC case, while the analysis based on the actual graded bond coat microstructure, Fig. 10(c), predicts a slight decrease. Nevertheless, the deleterious effect of grading on the size of the shear stress field zone ahead of the crack tip is clear.

Fig. 11 quantifies the effects of grading and oxide film's presence on the shear stress distribution in the vicinity of the crack tip along the line coincident with the crack. The shear stress concentration *increase* directly at the crack tip due to grading is clearly observed in Fig. 11(a) in the oxide film's absence. The effect

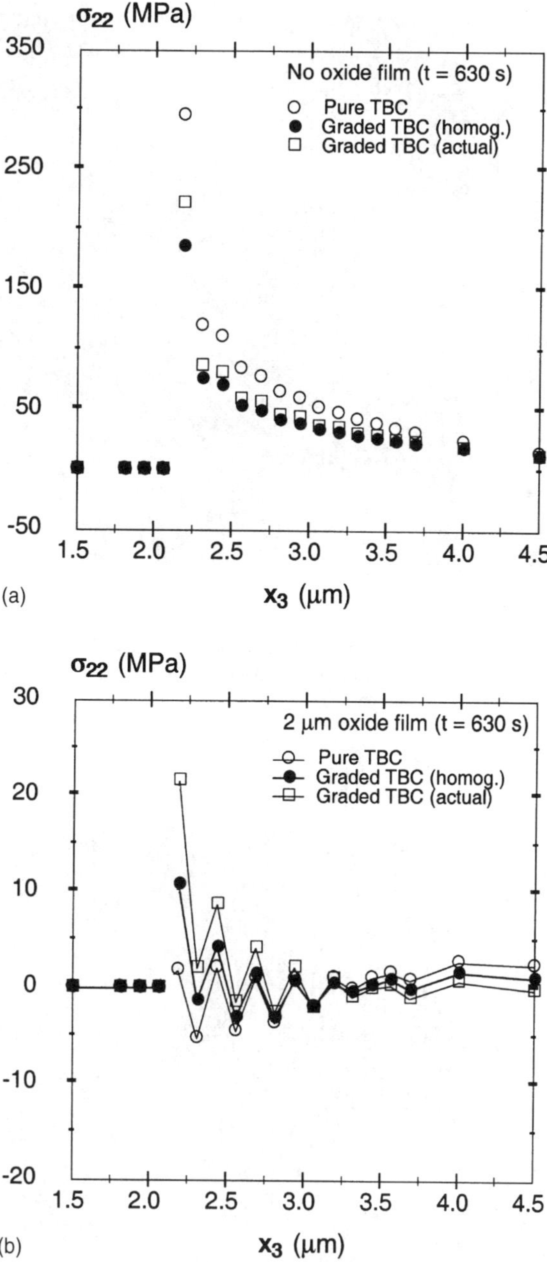

Fig. 8. Comparison of crack tip σ_{22} distributions along the line coincident with the crack in pure and graded bond coat TBCs after cooldown ($t = 630$ s): (a) no oxide film and (b) 2 μm oxide film.

of grading on the crack-tip shear stress concentration virtually disappears with increasing oxide film thickness as seen in Fig. 11(b) for the 2 μm oxide film, in contrast with the enhancement of the shear stress-affected zone size observed in Fig. 10. These results support the hypothesis that the shear stress is an important component of the crack-driving force with increasing oxide film thickness during the initial crack

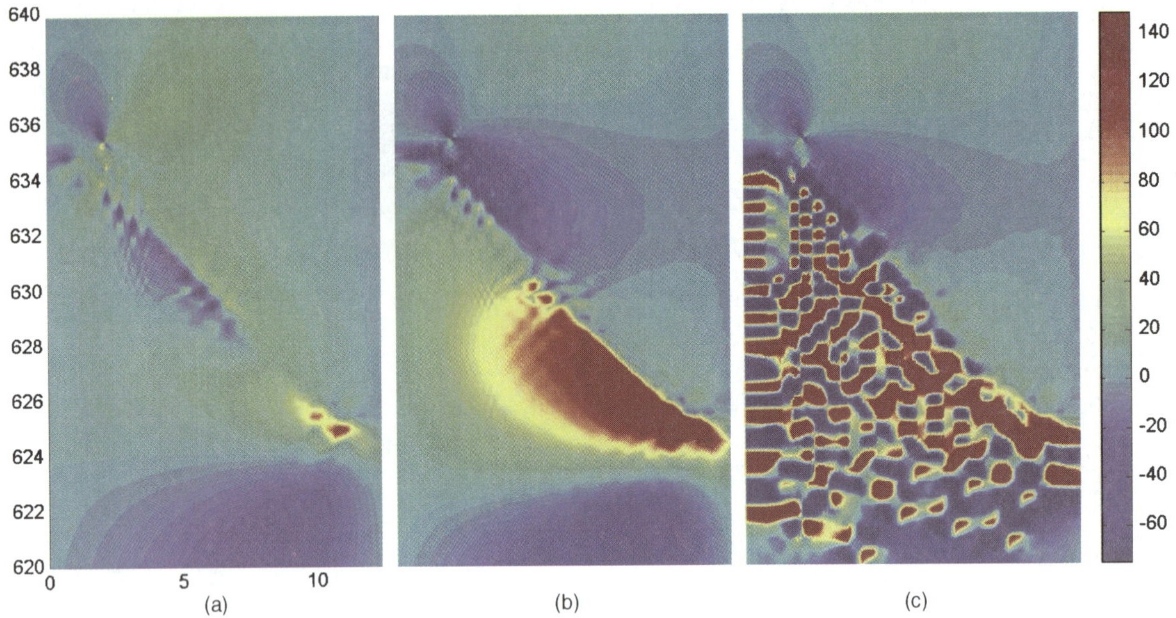

Fig. 9. Distribution of σ_{23} in the interfacial region of the investigated TBC system without an oxide film after cooldown ($t = 630$ s): (a) pure TBC; (b) graded TBC with homogenized bond coat microstructure and (c) graded TBC with actual bond coat microstructure (colorbar scale in MPa).

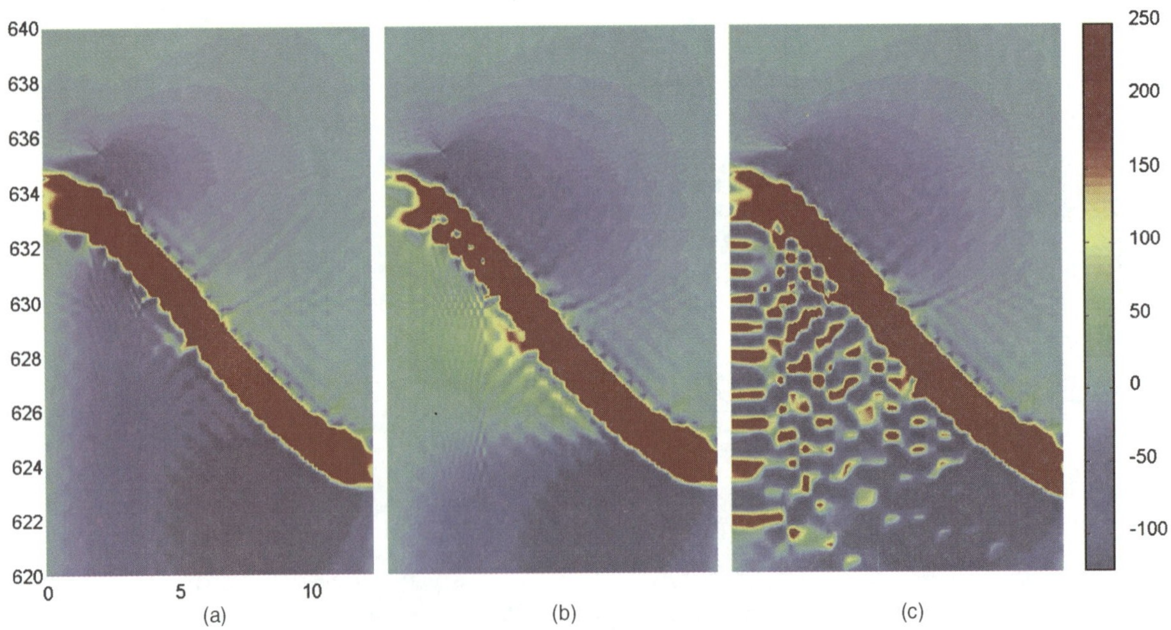

Fig. 10. Distribution of σ_{23} in the interfacial region of the investigated TBC system with a 2 μm oxide film after cooldown ($t = 630$ s): (a) pure TBC; (b) graded TBC with homogenized bond coat microstructure and (c) graded TBC with actual bond coat microstructure (colorbar scale in MPa).

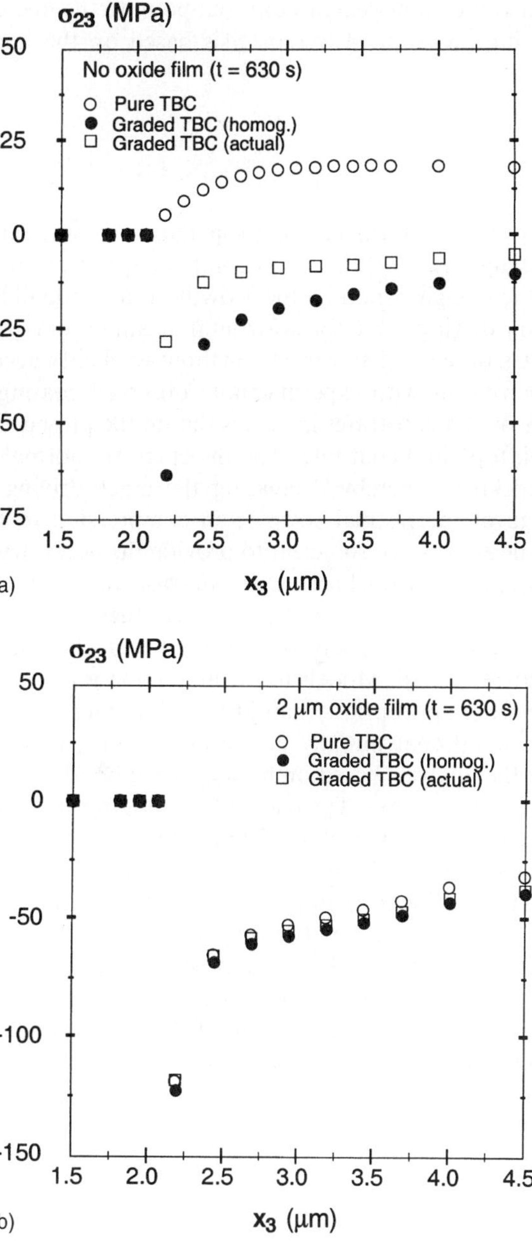

Fig. 11. Comparison of crack tip σ_{23} distributions along the line coincident with the crack in pure and graded bond coat TBCs after cooldown ($t = 630$ s): (a) no oxide film and (b) 2 μm oxide film.

extension stage. They also indicate that the grading-induced enhancement of the shear stress zone size ahead of the crack tip provides additional energy for crack extension.

It is important to note that the differences predicted by the homogenized and non-homogenized higher-order theory analyses occur in the top coat which has been modeled as *homogeneous*. The two approaches are expected to produce different results in the graded region, as readily observed in the preceding contour plots. The differences observed in the homogeneous top coat point to the micro–macrostructural interaction which is not explicitly taken into account in the analysis based on the homogenized bond coat microstructure.

5. Discussion and summary

The mechanism for top coat delamination leading to spallation in pure plasma-sprayed TBCs proposed in the literature [3–6] is primarily based on the stress component normal to the rough top/bond coat interface, which is modified in the trough region by the growth of an oxide film and stress relaxation effects, thereby providing the necessary driving force for horizontal delamination growth. This model is based on finite-element analyses of TBCs on curved substrates (without explicitly accounting for the presence of a local delamination), and is consistent with experimentally observed coating failure by spallation. Alternatively, our analysis of TBCs on flat substrates indicates that in the presence of a horizontal delamination just above the crest of a rough top/bond coat interface, the crack-tip normal stress field actually decreases with increasing oxide film thickness, thereby decreasing the crack-driving force. Increasing oxide film thickness does change the sign of the normal stress from compression to tension in the trough region, however the extent of the tensile zone is not sufficient to provide the necessary crack-extension force for the investigated crack length and oxide film thicknesses. In contrast, the shear stress field at the crack tip increases with increasing oxide film thickness. It is hypothesized, therefore, that this stress component provides the necessary driving force for crack extension during at least the initial delamination growth when the oxide film thickness is relatively small. Models for delamination growth based on mixed-mode fracture mechanics may, therefore, be more appropriate than just mode I considerations.

Further, while grading the bond coat initially decreases the normal stress field at the crack tip, this reduction is overestimated by the analysis based on homogenized bond coat microstructure relative to the analysis based on the actual microstructure. The effect of grading on the crack-tip normal stress field becomes less important with increasing oxide film thickness (at least up to 2 μm). In addition, grading promotes the growth of the tensile normal stress zone in the rough interface's trough region, and has a similar effect as increasing oxide film thickness in pure TBCs. Perhaps more significantly, grading increases the shear stress field at the crack tip, *which is further magnified by the presence of the oxide film.* This magnification is overestimated by the analysis based on homogenized bond coat microstructure relative to the actual microstructure. However, the differences become smaller with increasing film thickness. Therefore, under conditions of a spatially uniform temperature field which simulates furnace heating durability test, grading is likely to accelerate delamination growth based on our hypothesis regarding the importance of the shear stress component. The assessment of the desirability of grading the bond coat under conditions of gradient heating will be addressed separately.

6. Conclusions

The presented results generated using the higher-order theory for FGMs indicate that, under spatially uniform cyclic heating, grading the bond coat to reduce the top/bond coat thermal expansion mismatch is likely to accelerate the growth of horizontal cracks in the top coat at, or just above, the rough interface's crests. These cracks are thought to be initiated by the relatively large normal stress at the crests of the rough top/bond coat interface which becomes tensile upon cooldown due to stress relaxation at elevated temperatures. According to the presently accepted model, the growth of these cracks, which is driven by an evolving oxide film in the bond coat at the interface, ultimately leads to top coat spallation. Grading the bond coat, while *initially reducing* the normal stress field at the tip of a horizontal crack above the rough interface's crest, *increases* the shear stress field at the crack tip, thereby providing the necessary crack-driving force. This is further amplified by the oxide film's presence. Further, grading the bond coat accelerates the evolution of the tensile normal stress in the trough region of the rough interface, thereby potentially providing additional synergistic mechanism for crack extension when the oxide film thickness becomes sufficiently large. The overall effect of grading is similar to increasing the oxide film thickness.

In addition, the results indicate that micro–macrostructural coupling which explicitly accounts for particle–particle and particle–interface interactions, and which is an intrinsic feature of the higher-order theory, is important in the analysis of plasma-sprayed TBCs with graded bond coat microstructures. In particular, the extent of the crack-tip normal stress field reduction was overestimated by the analysis based on the homogenized properties of the graded bond coat microstructure relative to the actual microstructure. The extent of the shear stress magnification was also overestimated by the analysis based on the homogenized properties. The micro–macrostructural coupling effects propagate sufficiently far into the top coat, modeled here as homogeneous, to affect the crack-tip stress fields in the immediate vicinity of the top/graded bond coat interface. These results suggest that the porous microstructure of plasma-sprayed coatings should be explicitly taken into account in order to gain a better understanding of the horizontal delamination growth. This will be addressed in future studies using the higher-order theory.

Acknowledgements

The first two authors gratefully acknowledge the support provided by the NASA-Glenn Research Center through the NASA grant NAG3-2359.

References

[1] Miller RA, Lowell CE. Failure mechanisms of thermal barrier coatings exposed to elevated temperatures. Thin Solid Films 1982;95:265–73.

[2] Evans AG, Crumley GB, Demaray RR. On the mechanical behavior of brittle coatings and layers. Oxid Met 1983;20(5/6):193–216.

[3] Chang GC, Phucharoen W, Miller RA. Behavior of thermal barrier coatings for advanced gas turbine blades. Surf Coat Technol 1987;30:13–28.

[4] Chang GC, Phucharoen W, Miller RA. Thermal expansion mismatch and plasticity in thermal barrier coatings. Turbine Engine Hot Section Technology, NASA CP 1987;2493:357–68.

[5] Petrus GJ, Ferguson BL. A software tool to design thermal barrier coatings: a technical note. J Therm Spray Technol 1997;6(1):29–34 (See also: NASA Phase I Final Report, Project No. 93-1-04-23-8477).

[6] Freborg AM, Ferguson BL, Brindley WJ, Petrus GJ. Modeling oxidation induced stresses in thermal barrier coatings. Mater Sci Eng, A 1998;245:182–90.

[7] Pindera M-J, Aboudi J, Arnold SM. The effect of interface roughness and oxide film thickness on the inelastic response of thermal barrier coatings to thermal cycling. Mater Sci Eng, A 2000;284:158–75.

[8] Brindley WJ, Miller RA, Aikin BJ. Improved bond-coat layers for thermal-barrier coatings. NASA Tech Briefs, 1998. 63–5.

[9] Pindera, M-J., Arnold, S. M., Aboudi, J. Analysis of TBCs with homogeneous and heterogeneous bond coats under spatially uniform cyclic thermal loading. NASA Technical Memorandum 2001-210803.

[10] Aboudi J, Pindera M-J, Arnold SM. Thermoelastic theory for the response of materials functionally graded in two directions. Int J Solids Struct 1996;33(7):931–66.

[11] Aboudi J, Pindera M-J, Arnold SM. Thermoplasticity theory for bidirectionally functionally graded materials. J Therm Stresses 1996;19:809–61.

[12] Aboudi J, Pindera M-J, Arnold SM. Higher-order theory for functionally graded materials. Composites Part B: Engineering 1999;30(8):777–832.

[13] Aboudi J. Micromechanical analysis of thermo-inelastic multiphase short-fiber composites. Composites Engineering 1995;5(7):839–50.

[14] Pindera M-J, Aboudi J, Arnold SM. Thermomechanical analysis of functionally graded thermal barrier coatings with different microstructural scales. J Am Ceram Soc 1998;81(6):1525–36.

[15] DeMassi-Marcin JT, Sheffler KD, Bose S. Mechanisms of degradation and failure in a plasma-deposited thermal barrier coating. J Eng Gas Turb Power 1990;112:521–6.

[16] Crossman FW, Karlak RK, Barnett DM. Creep of B/Al composites as influenced by residual stress, bond strength, and fiber packing geometry. In: Failure Modes in Composites II. New York: TMS of AIME Publication; 1974. p. 8–31.

[17] Crossman FW, Karlak RK. Multiaxial creep of metal matrix fiber reinforced composites. In: Failure Modes in Composites III. New York: TMS of AIME Publication; 1976. p. 260–87.

PERGAMON

Engineering Fracture Mechanics 69 (2002) 1607–1634

Engineering Fracture Mechanics

www.elsevier.com/locate/engfracmech

On the use of effective properties for the fracture analysis of microstructured materials

J.E. Dolbow, J.C. Nadeau *

Department of Civil and Environmental Engineering, Duke University, 121 Hudson Hall, Box 90287, Durham, NC 27708-0287, USA

Received 2 February 2001; received in revised form 25 May 2001; accepted 10 July 2001

Abstract

This paper addresses some fundamental theoretical and numerical issues concerning the application of effective properties for the failure analysis of microstructured materials, with a focus on functionally graded materials. An edge-cracked square region is considered which is taken to be on the size of a representative volume element in a larger structural system. The region is made functionally graded by utilizing a non-homogeneous probability density function (PDF) governing the spatial distribution of the constituents. The non-homogeneity of the PDF exists only in the direction of the crack plane. Based upon the type of microstructure considered in the region, an accurate homogenization method is employed to obtain effective properties. For functional forms of the PDF, energy release rates and stress intensity factors (SIFs) are then calculated using both effective properties based on pointwise homogenization and realizations of specific microstructures generated from the PDF. For all calculations, we employ the eXtended finite element method, alleviating the need to remesh the domain between different microstructural realizations. Enrichment strategies are employed by the method to capture the singular stress fields near the crack tip as well as the discontinuous strain fields at fiber–matrix interfaces, even with relatively coarse meshes. SIFs are calculated using a domain form of the *J*-integral that does not exhibit any domain dependence. The results show excellent agreement between the implicit and the mean of the explicit trials for the energy release rates and strain energy densities. The results also seem to indicate that the implicit SIFs predict the mean SIF of the explicit trials for an arbitrary initial crack geometry. Furthermore, the results indicate that a good approximation of the SIF at specific crack tips may be estimated by reinterpreting the implicit results according to knowledge of the distinct crack-tip location in the microstructure. Quasi-static crack growth studies are performed to investigate the probability of finding crack tips within either of the composite phases.
© 2002 Elsevier Science Ltd. All rights reserved.

Keywords: FGM; Effective properties; SIFs; Microstructure; X-FEM

1. Introduction

The use of functionally graded materials (FGMs) provides the possibility of optimizing the response of safety-critical structures to highly adverse operating conditions. For example, the outer skin of modern

---

* Corresponding author. Tel.: +1-919-660-5216; fax: +1-919-660-5219.
E-mail address: nadeau@duke.edu (J.C. Nadeau).

aircraft fuselages may reach temperatures as high as 175 °C, and advanced material systems are deemed necessary to withstand the severe thermo-mechanical cycling and associated wear. In these applications, FGMs formed from optimized microstructures can result in structural systems with substantially better performance than those using conventional coating technology [34] even with respect to significant variability in material properties, loading, and overall geometry [35]. By effectively smoothing the transition in material properties exhibited at bimaterial interfaces, FGM coatings also appear to be more resistant to failure due to thermal shocks [45]. It bears emphasis, however, that many FGMs do exhibit sharp discontinuities in material properties at the microstructural level, and a rigorous failure analysis must account for any multi-scale effects. In this paper, we seek to examine the limitations and implications of using effective properties to determine energy release rates and stress intensity factors (SIFs) in microstructured materials, with a focus on FGMs.

Effective properties of materials (see, e.g., [7,23,36]) are a great resource when they may be employed. When a representative volume element (RVE) of a material is large with respect to the microstructure, while at the same time being small relative to the length scale of the system, then effective properties may be defined so as to capture average response. For related discussions, the reader is referred to Ostoja-Starzewski [37], Huet [25] and Auriault [2]. It is important to note that effective elastic properties can be defined either by equating energies in RVEs or relating average stress to the average strain within the RVEs, and that these definitions are equivalent when the Hill condition is satisfied [25]. Adopting a homogenization method based on a microstructural idealization ensures that all of the effective properties are consistently determined. A well known result of Levin [30], nicely elaborated on by Rosen and Hashin [43], is that the effective compliance and the effective coefficient of thermal expansion are related. Thus, neither homogenization methods nor functional forms of effective properties should be arbitrarily employed in a given analysis. With the appropriate application of effective properties, however, tremendous computational efficiencies can be achieved while still capturing the response quantity of interest, provided that that quantity does not depend on the fluctuations of the solution fields on the scale of the microstructure.

The use of homogenization in the vicinity of a crack tip is problematic as the geometry possesses a zero length scale, seemingly implying that a local RVE cannot be defined. By the same token, in elastic materials the SIFs characterizing the strength of the crack-tip singularity are related to the energy release rate of the global structure, suggesting that they contain more than just local information. Several researchers have examined the limitations of homogenization methods to examine local quantities, including Pindera et al. [39] and Reiter et al. [40] for FGMs. Unfortunately, much of the analysis does not easily generalize to examining the behavior near a crack tip, and fairly little research has focused on the evaluation of SIFs using effective properties. Jha and Charalambides [26] investigated the effect of microstructure on SIFs in layered composites, and proposed alternative homogenization models that accounted for microscale variations. An excellent article on related issues is provided by Dagan [9]. Clearly the subject area is a rich one, and further analysis is required to provide a clear picture of how effective properties may be employed in determining fracture related quantities such as energy release rates and SIFs.

One of the primary motivations for adopting homogenization methods is the severe computational costs associated with the explicit simulation of microstructural details. In addition, FGMs implicitly possess non-uniform, non-periodic microstructures, which complicate the development of purely analytical models. Although finite element models that explicitly model fiber distributions have been adopted by some researchers, these are rarely used to model entire macroscopic specimens. An evolving crack discontinuity complicates the situation by effectively creating a moving boundary value problem. To a degree, these concerns have been mitigated by recent advances in computational technology. In particular, the eXtended finite element method (X-FEM) [10] provides a means to capture local, evolving solutions at a fraction of the cost of alternative numerical methods. The method is based on the partition of unity framework developed by Melenk and Babuška [31], wherein local non-polynomial enrichment spaces are embedded within a standard Galerkin approximation. The X-FEM makes the best use of this framework to model geometric features that are independent of the element boundaries, allowing for accurate simulations of crack growth [32] and evolving material interfaces [12] without remeshing. The method is therefore ideally suited for the fracture analysis of FGMs, and in particular for examining the detailed effects of composite microstructures near a crack tip.

Motivated by the above considerations, in this paper we compare energy release rates and SIFs predicted using effective properties to those obtained from distinct realizations of the corresponding microstructure. In particular, we focus our investigation to cases where an accurate homogenization method is expected to provide a good estimation for energy-based quantities within the RVE. In this investigation we employ a generalized self-consistent model which is subsequently confirmed to yield accurate homogenization results for the adopted microstructure. For the sake of brevity, we refer to the use of effective properties as an "implicit" method and to the direct analysis of microstructural features as an "explicit" method. For the explicit analyses presented herein, the functionally graded microstructures are randomly generated using a probability density function (PDF). We study edge-cracked specimens which are assumed to be representative of regions at crack tips within a larger structural system. Explicit and implicit results are compared for energy release rates and SIFs for a range of gradients in the fiber volume fraction. We also examine the effect of the ratio of fiber to matrix properties and explore crack growth simulations.

This paper is organized as follows. In the next section, we provide the details in the approach, methodology, and formulation for the present failure analysis of FGMs. In particular, we describe the explicit and implicit approaches and the X-FEM strategies employed for both. After verifying the computational approach with some benchmark problems, comparisons of energy release rates and SIFs for the explicit and implicit approaches are provided in Section 3. Finally, Section 4 provides a summary and some concluding remarks.

2. Methodology and formulation

2.1. Approach

Our objective is to investigate the validity of non-homogeneous fracture mechanics for the fracture analysis of material systems with microstructure, such as FGMs. Consider the body shown in Fig. 1, corresponding to a fractured FGM whose microstructure is composed of distinct fiber and matrix phases.

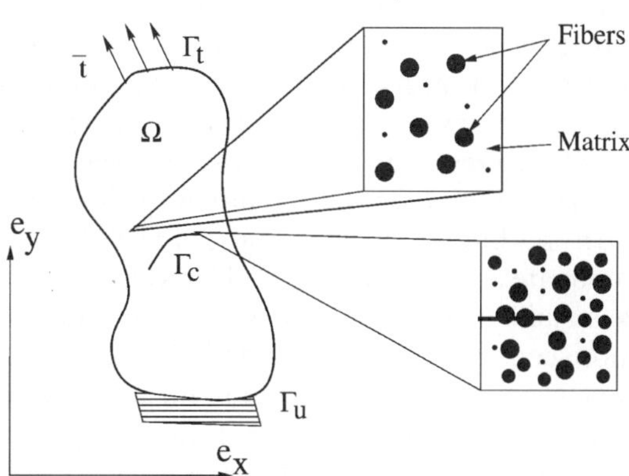

Fig. 1. An FGM with a crack subjected to loads. A zoom of the microstructure is shown at two points, indicating the variation in fiber volume fraction.

Under the assumption of a smooth variation in material properties, the following expression for the asymptotic near-tip stress fields in non-homogeneous materials has been derived [27]:

$$\boldsymbol{\sigma} = \frac{K_{\mathrm{I}}}{\sqrt{2\pi r}}\tilde{\boldsymbol{\sigma}}^{\mathrm{I}}(\theta) + \frac{K_{\mathrm{II}}}{\sqrt{2\pi r}}\tilde{\boldsymbol{\sigma}}^{\mathrm{II}}(\theta) \tag{1}$$

where $\boldsymbol{\sigma}$ is the Cauchy stress tensor, (r,θ) are the local angular coordinates at the crack tip, and $\tilde{\boldsymbol{\sigma}}(\theta)$ are the well known angular functions for homogeneous materials. The SIFs $(K_{\mathrm{I}}, K_{\mathrm{II}})$ are functions of the geometry, loading, and gradient in material properties. Of additional interest is the energy release rate, which can be written in terms of the SIFs as

$$G = \frac{1 - v_0^2}{E_0}(K_{\mathrm{I}}^2 + K_{\mathrm{II}}^2) \tag{2}$$

for plane strain conditions, where (E_0, v_0) are Young's modulus and Poisson's ratio at the crack tip. The quantities are of critical interest to the failure analysis of FGMs, as fracture criterion are often expressed in terms of the above in relation to fracture toughness. For example, in pure mode-I loading

$$K_{\mathrm{I}} \leqslant K_{\mathrm{Ic}}, \quad G \leqslant G_{\mathrm{Ic}}$$

where both K_{Ic} and G_{Ic} are referred to as the fracture toughness. Presumably, the fracture toughness is not homogeneous for an FGM.

A common approach in the fracture analysis of FGMs is to employ homogenization to obtain effective properties for use in the above relationships. This approach is obviously an approximation, and several issues come to the forefront upon examining the above equations in more detail. With reference to (1), at the microscale the variation in material properties is no longer smooth and we expect this to affect the local stress fields, but the influence on the SIFs is not clear. The tip of a mathematically sharp crack has a length scale of zero and thus it it not possible to define a RVE in the vicinity of the crack tip. Thus, the definition of effective material properties at the crack tip for use in (2) seems problematic. Nevertheless, in so much as effective properties are suitable for determining the average energetic response of the structure, we would like to employ them to the full extent possible. In essence, we wish to determine precisely how applicable non-homogeneous fracture mechanics is to material systems for which the non-homogeneous material properties represent effective material properties.

To address this question, we will adopt the approach of investigating the difference between an explicit and implicit modeling of the microstructure for determining the SIFs and energy release rate. By explicit modeling we mean that an actual realization of the microstructure within the system of interest is analyzed whereas by implicit we mean that the use of effective properties are used based on the PDFs used to characterize the explicit microstructures. It bears emphasis that this approach is hardly a new one; explicit calculations are often employed to both improve homogenization models and/or to disregard them altogether [18,26,33]. The distinction of the present effort concerns the focus on energy release rates and SIFs in FGMs, the choice of numerical method, and the interpretation of the results.

The specific system we consider is a bi-unit, plane strain square with a crack passing from the center of the left edge to the center of the square as shown in Fig. 2. This system is meant to be representative of the region near a crack tip, and we assume that a generalization of our findings for this simple system will be indicative of the behavior in more complicated arrangements. In order to pursue our objective, it is necessary to adopt a microstructure and homogenization method for which we know that the implicit model is appropriate for the explicit microstructure, exclusive of considering the scale of the crack tip. It is well known that homogenization methods have limitations (see e.g., [4,6,17,38]). However, we endeavor to implement a microstructure for which homogenization is expected to be appropriate (again, exclusive of considering the scale of the crack tip). Any difference in the character of the results between the implicit and

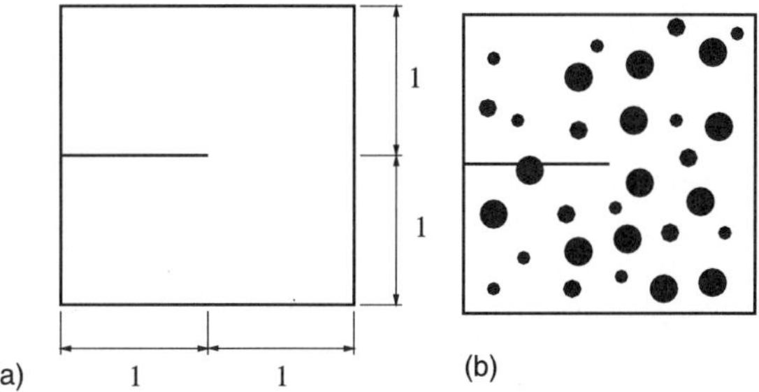

Fig. 2. (a) Implicit and (b) explicit models of the bi-unit domain with a crack.

explicit models can therefore be attributed to the difference between the explicit and implicit crack tips, and not to the inaccuracy of the homogenization method.

We adopt a space filling microstructure in the form of a continuous matrix with circular cylindrical fibers. Subsequently, we adopt a generalized self-consistent approach based on a circular cylinder for the transversely isotropic effective properties (see Appendix B). Christensen [6] has shown that the results of the generalized self-consistent model are physically realistic for explicit microstructures which admit full packing. More specifically, the form of four of the five results are equivalent to the exact results of composites cylinder assemblage [24] which is one form of a space filling arrangement.

We restrict ourselves to small fiber volume fractions, primarily for numerical efficiency but also to limit the extreme relative sizes of fibers that would be required at larger volume fractions in order to achieve full packing. We recognize that many FGM applications (e.g., thermal barrier coatings) and analyses involve a transition from 100% of one material to 100% of another. Though this situation is not specifically addressed herein, it is within the scope of the approach in so far as Fig. 2 is a subregion within the structure of Fig. 1.

2.2. Problem statement

We consider the body shown in Fig. 1, corresponding to a fractured FGM. We assume plane strain conditions, allowing for a two dimensional representation of the domain $\Omega \subset \mathscr{R}^2$, bounded by Γ. The boundary Γ consists of the disjoint sets Γ_u, Γ_t, and Γ_c, such that $\Gamma = \Gamma_u \cup \Gamma_t \cup \Gamma_c$. Displacements are prescribed on Γ_u, while tractions are imposed on Γ_t. The displacement field is allowed to be discontinuous across the crack surface Γ_c.

In the present investigation, we make the following additional assumptions:

- the variation in fiber volume fraction at the macroscale is a function of x only.
- the crack faces are traction free;
- the strains and displacements are small;
- the matrix and fiber phases are perfectly bonded;
- the fibers are cylindrical with uniform radii.

Given these conditions, we proceed in this section to present the governing equations and the corresponding weak formulation.

2.2.1. Governing equations
The equilibrium equations and boundary conditions are

$$\nabla \cdot \boldsymbol{\sigma} + \boldsymbol{b} = \boldsymbol{0} \quad \text{in } \Omega \tag{3}$$

$$\boldsymbol{\sigma} \cdot \boldsymbol{n} = \begin{cases} \bar{\boldsymbol{t}} \text{ on } \Gamma_t \\ \boldsymbol{0} \text{ on } \Gamma_{c^+} \\ \boldsymbol{0} \text{ on } \Gamma_{c^-} \end{cases} \tag{4}$$

where $\boldsymbol{n}$ is the unit outward normal. In the above, $\boldsymbol{\sigma}$ is the Cauchy stress, and $\boldsymbol{b}$ is the body force per unit volume. Γ_{c^+} and Γ_{c^-} indicate the top and bottom crack faces, respectively.

In the present investigation, we consider small strains and displacements. The kinematics equations therefore consist of the strain–displacement relation

$$\boldsymbol{\epsilon} = \boldsymbol{\epsilon}(\boldsymbol{u}) = \nabla_s \boldsymbol{u} \tag{5}$$

where ∇_s is the symmetric part of the gradient operator, and the boundary conditions

$$\boldsymbol{u} = \bar{\boldsymbol{u}} \text{ on } \Gamma_u \tag{6}$$

The constitutive relation is given by Hooke's law:

$$\boldsymbol{\sigma} = \boldsymbol{C} : \boldsymbol{\epsilon} \tag{7}$$

where $\boldsymbol{C}$ is the fourth order elasticity tensor, which varies spatially for an FGM. For a transversely isotropic material the components of $\boldsymbol{C}$ are given by

$$[\boldsymbol{C}] = \begin{bmatrix} C_{1111} & C_{1122} & C_{1133} \\ C_{1122} & C_{1111} & C_{1133} \\ C_{1133} & C_{1133} & C_{3333} \\ & & & C_{2323} \\ & & & & C_{2323} \\ & & & & & \frac{1}{2}(C_{1111} - C_{1122}) \end{bmatrix} \tag{8}$$

2.2.2. Weak form

We now describe the variational formulation of Eqs. (3)–(7). The space of admissible displacement fields is defined by

$$\mathcal{U} = \{\boldsymbol{v} \in \mathcal{V} : \boldsymbol{v} = \bar{\boldsymbol{u}} \text{ on } \Gamma_u\} \tag{9}$$

where the space $\mathcal{V}$ is related to the regularity of the solution. The details on this matter when the domain contains an internal boundary or re-entrant corner may be found in Babuška and Rosenzweig [3] and Grisvard [20]. We note that the space $\mathcal{V}$ allows for discontinuous functions across the crack line. The test function space is defined similarly as

$$\mathcal{U}_0 = \{\boldsymbol{v} \in \mathcal{V} : \boldsymbol{v} = 0 \text{ on } \Gamma_u\} \tag{10}$$

Using the constitutive relation and the kinematics constraints in the weak form, the problem is to find $\boldsymbol{u} \in \mathcal{U}$ such that

$$\int_{\Omega} \boldsymbol{\epsilon}(\boldsymbol{u}) : \boldsymbol{C} : \boldsymbol{\epsilon}(\boldsymbol{v}) \, \mathrm{d}\Omega = \int_{\Omega} \boldsymbol{b} \cdot \boldsymbol{v} \, \mathrm{d}\Omega + \int_{\Gamma_t} \bar{\boldsymbol{t}} \cdot \boldsymbol{v} \, \mathrm{d}\Gamma \quad \forall \boldsymbol{v} \in \mathcal{U}_0 \tag{11}$$

The discretization of the above weak form depends on the choice of modeling approach in describing the influence of the fiber and matrix phases and their respective variation in volume fraction over the domain. In the following subsection, we describe two general approaches: *implicit* and *explicit*. In the implicit approach, the components of $\boldsymbol{C}$ vary continuously in the domain, according to homogenized models of the microstructure. In the explicit approach, the fiber and matrix phases are modeled discretely, and the components of $\boldsymbol{C}$ therefore exhibit discontinuities across fiber–matrix boundaries.

2.3. Implicit models of FGM fracture

For the implicit model of FGM fracture, we require an appropriate homogenization method for the adopted microstructural characterization. The microstructure adopted is that of circular cylindrical fibers in a well defined matrix. From a computational perspective, the primary concern is then the representation of the singular fields near the crack tip. These issues are presented in detail in the following subsections.

2.3.1. Homogenization techniques

Since the constitutive relation C for the explicit cases to be considered is independent of the out-of-plane coordinate z, it is appropriate to adopt a homogenization method for aligned, continuous fibers. More specifically, fibers of circular cross section will be adopted in the explicit trials, therefore a homogenization method for an aligned circular cylindrical fiber composite is desired. The effective properties of such a composite are transversely isotropic with five elastic material properties: E_{zz}^*, v_{zx}^*, κ_{xy}^*, μ_{zx}^*, and μ_{xy}^* where the axes of the fibers are parallel to the z-axis. The relations between these five engineering material constants noted above and the five components presented in Eq. (8) are well known and they are presented in Appendix B.

The homogenization method that we adopt herein is a generalized self-consistent model which is specifically addressed in Appendix B. The generalized self-consistent expressions for E_{zz}^*, v_{zx}^*, κ_{xy}^*, and μ_{zx}^* are the same form as the exact results of Hashin and Rosen [24] for the composite cylinders assemblage model which admits a full packing capability through a continuous distribution of fiber sizes. Thus, if the specific explicit microstructure which is adopted is that of the composite cylinders assemblage then the implemented effective properties (for these four properties) are exact; the fifth constant μ_{xy}^* remains approximate. It will be shown in Section 3 that in terms of strain energy the implicit case is generally within 1% of the mean of the explicit trials for the adopted microstructure and fiber volume fraction. Thus, the implemented homogenization method is satisfactory for the adopted microstructure.

We initially considered a distribution in fiber diameters so as to approximate a volume filling microstructural characterization. As will be seen below, however, sufficiently accurate homogenization results were obtained using constant diameter fibers at the volume fraction of fibers considered.

2.3.2. Discretization with the X-FEM

We now describe the extended finite element approximation to (11) for implicit models of FGM fracture. The present application is similar to that for modeling fracture in homogeneous materials, and so we do not present many of the details provided in Dolbow [10] and Moës et al. [32]. It is important to note, however, that many features of the approximation will be similar for the explicit models described in Section 2.4.

The Galerkin approximation to the weak form (11) begins by considering a finite dimensional subspace $\mathcal{U}_0^h$ of $\mathcal{U}_0$ spanned by N linearly independent functions in $\mathcal{U}_0$. We then pose the weak form in $\mathcal{U}_0^h$ as follows:

$$\int_\Omega \epsilon(\boldsymbol{u}) : C : \epsilon(\boldsymbol{v}) \, \mathrm{d}\Omega = \int_\Omega \boldsymbol{b} \cdot \boldsymbol{v} \, \mathrm{d}\Omega + \int_{\Gamma_t} \bar{\boldsymbol{t}} \cdot \boldsymbol{v} \, \mathrm{d}\Gamma \quad \forall \boldsymbol{v} \in \mathcal{U}_0^h \tag{12}$$

For the sake of concreteness, we now consider a rectangular domain Ω and a regular finite element triangulation $\mathcal{T}^h = \mathcal{U}p_{e=1}^{n_{el}} \mathcal{T}_e$ such that $\mathcal{T}^h = \Omega$ as shown in Fig. 3. This figure also depicts a crack whose geometry Γ_c is taken to be independent of the mesh. A standard finite element basis for $\mathcal{U}_0^h$ is constructed from the space of complete polynomials $P^k(\mathcal{T}_e^h)$ of order $\leqslant k$ over each element:

$$\mathcal{U}_0^h = \mathrm{span}\{\phi_i\}_{i=1}^N \quad \text{where } \{\phi_i \in [C^0(\mathcal{T}^h)]^2 : \phi_i|_{\mathcal{T}_e^h} \in [P^k(\mathcal{T}_e^h)]^2 \text{ and } \phi_i|_{\Gamma_u} = 0\} \tag{13}$$

where the functions $\phi_i(\boldsymbol{x})$ are typically the nodal shape functions. Any linear combination of these functions results in a continuous interpolation for the displacement field, and furthermore possesses poor

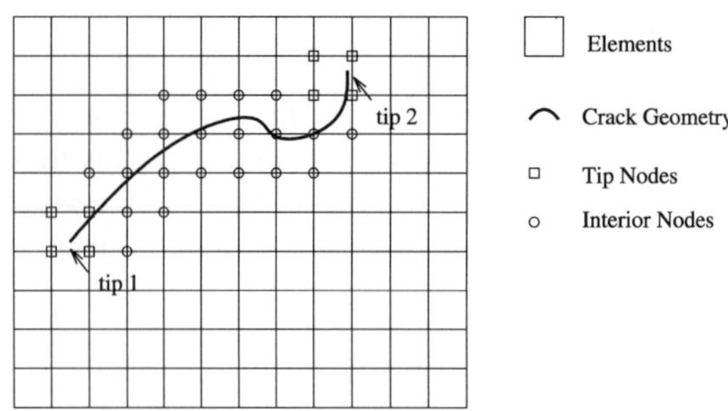

Fig. 3. An arbitrary crack placed on a mesh. The circles indicate nodes whose shape functions are enriched with the function $H(x)$, while the squares indicate those that are enriched with the sets of near-tip functions.

approximation properties for representing the singular stress fields near crack tips. Standard finite element approaches therefore construct "broken" meshes that conform to the crack geometry in order to represent the discontinuity, and employ significant mesh refinement or special singular elements near crack tips.

The X-FEM takes an alternative approach by *extending* the standard finite element approximation. We consider the set of overlapping subdomains $\{\omega_i\}$ defining the support of each nodal shape function and sets of enrichment functions $\{\mathscr{E}_i^k\}$ that possess desirable approximation properties over each subdomain. The method follows the *partition-of-unity* framework [31] through multiplying the enrichment functions by the nodal shape functions ϕ_i in order to ensure a conforming approximation. A general X-FEM basis for $\mathscr{U}_0^h$ is therefore

$$\mathscr{U}_0^h = \operatorname{span}\left\{ \{\phi_i\} \cup \{\phi_i \mathscr{E}_i^k\}_{k=1}^{n_i^{\mathrm{E}}} \right\}_{i=1}^{n} \tag{14}$$

where n is the number of standard nodal shape functions and n_i^{E} is the number of enrichment functions for node i. The approximation to the field $u(x)$ is then written as

$$u^h(x) = \sum_i u_i \phi_i(x) + \sum_i \phi_i(x) \left(\sum_k a_i^k \mathscr{E}_i^k(x) \right) \tag{15}$$

where the u_i and $a_i^{(k)}$ are the constant (vector-valued) degrees of freedom. It bears emphasis that as a consequence of the above construction, the degrees of freedom u_i *are not* equal to the value of the displacement vector $u^h(x_i)$ at the nodes.

The above construct is the most general form of the X-FEM approximation where every nodal shape function may be enriched with an arbitrary number of additional functions $\{\mathscr{E}_i^k\}$. In practice, only those functions whose supports are in the vicinity of a feature of interest are enriched, giving the approximation a local character. For example, the X-FEM approximation for a crack is given by

$$u^h(x) = \sum_{i \in I} u_i \phi_i(x) + \sum_{j \in J} b_j \phi_j(x) H(x) + \sum_{k \in K_1} \phi_k(x) \left(\sum_{l=1}^{4} c_k^{l1} F_1^l(r, \theta) \right) + \sum_{k \in K_2} \phi_k(x) \left(\sum_{l=1}^{4} c_k^{l2} F_2^l(r, \theta) \right) \tag{16}$$

where I is the set of all nodes in the mesh, J is the set of nodes enriched with the generalized Heaviside function $H(x)$, and (K_1, K_2) are the sets of nodes enriched with the sets of near tip functions $F_1^l(r, \theta)$ and $F_2^l(r, \theta)$ respectively. The set J is circled in Fig. 3 while the sets (K_1, K_2) are squared. As shown in the figure, only those nodes in the vicinity of the crack are enriched; the X-FEM extends a standard FE approximation locally. A detailed definition of these sets and the criteria used to determine them numerically is provided in Dolbow et al. [13].

The generalized Heaviside function $H(x)$ is simple to construct; the function is constant and changes sign across the crack front. Details for its construction can be found in Moës et al. [33]. The set of near-tip functions $F^l(r, \theta)$ are taken to be those functions which span the exact asymptotic crack-tip fields for linear elasticity:

$$\{F^l(r, \theta)\} \equiv \left\{ \sqrt{r} \sin\left(\frac{\theta}{2}\right), \sqrt{r}\cos\left(\frac{\theta}{2}\right), \sqrt{r}\sin\left(\frac{\theta}{2}\right)\sin(\theta), \sqrt{r}\cos\left(\frac{\theta}{2}\right)\sin(\theta) \right\} \tag{17}$$

where (r, θ) are the local polar coordinates at the crack tip. These functions are appropriate for the fracture analysis of FGMs, as the asymptotic near-tip displacement fields have been shown to be the same as those for homogeneous materials [21].

Substitution of the approximation (16) into (12) and invoking the arbitrariness of the test functions results in a linear algebraic equation for these degrees of freedom:

$$\boldsymbol{Kd} = \boldsymbol{f} \tag{18}$$

where $\boldsymbol{K}$ is the elastic stiffness matrix and $\boldsymbol{f}$ is the vector of nodal forces. The vector of nodal unknowns $\boldsymbol{d}$ gathers the degrees of freedom $\{\boldsymbol{u}_i, \boldsymbol{b}_j, \boldsymbol{c}_k^{l1}, \boldsymbol{c}_k^{l2}\}$.

As the geometry of the crack Γ_c is independent of the mesh with the X-FEM, it is necessary to modify the quadrature routines used to assemble the volume integrals in (12). As the crack is allowed to be arbitrarily oriented in an element, the use of standard Gauss quadrature may not adequately integrate the discontinuous field. The matrix system of Eq. (18) arising from the discrete weak form is normally constructed with a loop over all elements, as the domain is approximated by

$$\overline{\Omega} = \bigcup_{e=1}^{m} \overline{\Omega}_e \tag{19}$$

where m is the number of elements, and Ω_e is the element subdomain. For elements "cut" by a crack, we define the element subdomain to be a union of a set of subpolygons whose boundaries align with the crack geometry:

$$\overline{\Omega}_e = \bigcup_{e=1}^{m_{es}} \overline{\Omega}_{es} \tag{20}$$

where m_{es} denotes the number of subpolygons for the element.

In earlier investigations [13], each side of a cut element was triangulated to form a set of subtriangles. We adopt a slightly different approach for this study, choosing instead to partition cut elements into subquadrilaterals as shown in Fig. 4. This technique was proposed in Dolbow [10] and results in a more efficient integration scheme, particularly for the curved fiber boundaries considered in Section 2.4.2. While the subquadrilaterals do not necessarily conform to the local geometry, the integration error can be sufficiently bounded by choosing enough divisions. For the present investigation, we use five divisions in each direction. These subpolygons are also used to determine which nodes are enriched with the discontinuous function $H(x)$, by calculating the percentage of a node's support above and below the crack.

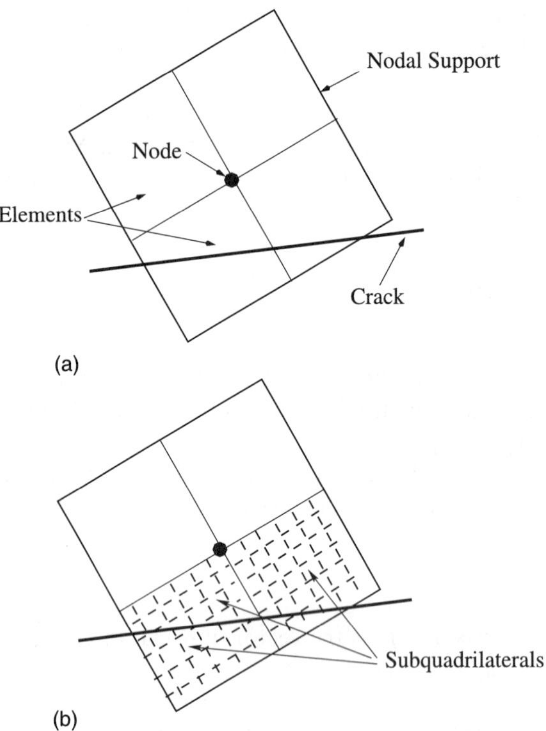

Fig. 4. (a) Nodal support cut by a crack. (b) The subquadrilaterals associated with elements cut by the crack for node selection and the integration of element stiffness matrices.

2.3.3. Implicit benchmark problems

We present numerical results for some FGM faracture problems with analytical solutions. These problems serve to validate the domain form of the J-integral (see Appendix A) and illustrate the advantages of the near-tip enrichment in the X-FEM approximation.

We consider a rectangular plate of width W and height h with an edge crack of length a subjected to a far-field stress σ° as shown in Fig. 5. The coordinate system is taken to coincide with the left hand side of the plate and the crack axis as shown. We take Poisson's ratio as constant and Young's modulus as

$$E(x) = C_1 e^{C_2 x} \tag{21}$$

where

$$C_1 = E_1, \quad C_2 = \ln(E_2/E_1)/W$$

with E_1 and E_2 denoting the Young's moduli for two different materials. The problem therefore models an FGM whose properties transition from material 1 to material 2 in the x-direction. Analytical solutions for the SIF K_1 for this problem as $h \rightarrow \infty$ have been derived [16]. In order to approximate an infinite boundary, we set $h/W = 10$. In the following, all results are reported for a uniform mesh of 160×16 quadrilateral elements, and symmetry conditions are not employed.

SIFs were calculated for crack lengths of $a/W = 0.2$ and 0.4 and various ratios (E_2/E_1). The results are summarized in Table 1. The table indicates a good comparison between the numerical and analytical re-

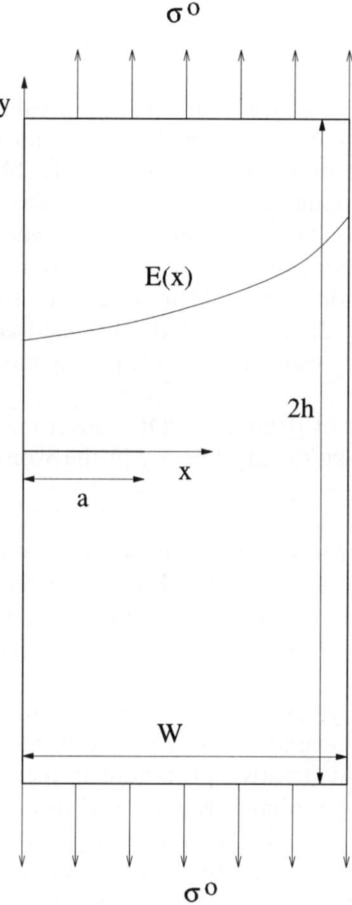

Fig. 5. Edge-cracked plate for the benchmark problems.

Table 1
Normalized K_{I} values for various crack lengths and material properties

E_2/E_1	Crack length					
	$a/W = 0.2$			$a/W = 0.4$		
	$K_{\mathrm{I}}/(\sigma^\circ\sqrt{\pi a})$	Analytical	% error	$K_{\mathrm{I}}/(\sigma^\circ\sqrt{\pi a})$	Analytical	% error
0.1	1.268	1.2965	2.2	2.492	2.570	3.0
0.2	1.370	1.396	1.8	2.387	2.443	2.3
1.0	1.353	1.373	1.5	2.083	2.107	1.1
5.0	1.124	1.132	0.7	1.732	1.748	0.9
10.0	0.996	1.024	2.6	1.575	1.626	3.1

sults. The error ranges from 0.7% to 3.0% over all trials. Results are reported using a domain size generated from $r_{\mathrm{d}}/h_{\mathrm{local}} = 2.0$ (see Appendix A), and changing this ratio did not result in a significant domain dependence. We note, however, that the error generally increases as the Young's moduli for the two materials diverge. These observations are consistent with the results reported in Anlas et al. [1].

2.4. Explicit models of FGM fracture

2.4.1. Microstructure generation for an FGM

In this section, we present the method by which explicit microstructure realizations are generated. For the sake of concreteness, we consider a domain given by the bi-unit square $\Omega = [-1, 1] \times [-1, 1]$. Our task here is to place an appropriate number of non-overlapping circular fibers of constant radius r_f such that the total volume fraction within Ω is approximately c. Furthermore, we desire to have a variable distribution of the fibers in the x-direction thereby achieving a functionally graded distribution of fibers. In the development of the microstructure generation, we initially considered a distribution in fiber diameters so as to implement a microstructural realization that could more closely approach a volume filling arrangement than that afforded by a constant fiber diameter. For the volume fraction of fibers considered, however, a constant fiber diameter was sufficiently general to be adequately modeled by the adopted homogenization method.

We begin the generation procedure by defining a PDF f governing the spatial distribution of the centers x_c^f of fiber cross sections which we take to vary linearly in the x-direction:

$$f(x,y) = f(x) = mx + k \tag{22}$$

where m and k are constants. In order to avoid a boundary or "wall effect", we define the PDF f over the domain $\widetilde{\Omega} = [-(1 + r_f), 1 + r_f] \times [-(1 + r_f), 1 + r_f]$. Because of the finite radius r_f of the fibers, a fiber whose center falls outside of Ω, but within a distance r_f of it, will have a portion of that fiber within Ω. The constant k is determined such that $\int_{\widetilde{\Omega}} f \, \mathrm{d}\Omega = 1$ and limits are placed on m such that f, on $\widetilde{\Omega}$, is strictly positive.

A fiber center is placed within $\widetilde{\Omega}$ by first generating its position along the x-axis. This is accomplished by using a uniform random number generator and the CDF of the marginal PDF, $f_x = 2f(x)$. Given the x-coordinate of the fiber's center, an iterative procedure is then used to determine the y-coordinate, where $y \in [-1 - r_f, 1 + r_f]$. A trial y-coordinate is generated using a uniform random number generator and the CDF of the marginal PDF, f_y, which is uniform by construction. If the generated y-coordinate results in the current fiber overlapping any existing fibers, then the y-coordinate is regenerated until the fiber can be placed without overlapping any of the existing fibers. This procedure yields an explicit microstructural realization consistent with the PDF. It is important to generate the explicit microstructures in the manner described above. If, for example, fiber centers (x_c^f, y_c^f) are generated, and excluded due to overlap, the resulting explicit microstructure will have an artificially smaller effective m than the actual PDF.

The implemented termination requirement is as follows. Let A_f^n be the area of fiber material within Ω after n fibers have been placed. The corresponding fiber volume fraction within Ω is thus $A_f^n/4$. With the addition of another fiber, the maximum, and mostly likely (for "small" $r_f/2$), increment in the change of volume fraction is $\pi r_f^2/4$. As such, when $A_f^n > c - \pi r_f^2/8$ the addition of fibers will be terminated. A typical example of a generated explicit microstructure is shown in Fig. 6. The number of fibers required to satisfy the termination requirement is denoted by n_f.

2.4.2. X-FEM strategies for matrix–fiber interaction

We focus attention on a typical mesh of Ω as shown in Fig. 7, containing n_f fibers. We denote the set of matrix–fiber interfaces by $\{\Gamma_1^f\}_{f=1}^{n_f}$, and the fiber domains by $\{\Omega_f^f\}_{f=1}^{n_f}$. It is necessary to make a few modifications to the formulation presented in Section 2.3.2 in order to model the fibers explicitly. In particular, as the matrix–fiber interfaces represent surfaces of discontinuity in normal strain, we enrich the displacement approximation with special interface functions that capture this effect. We argue that this approach is much more convenient than constructing a mesh that conforms to each matrix–fiber interface, especially for the statistical studies which follow in Section 3.

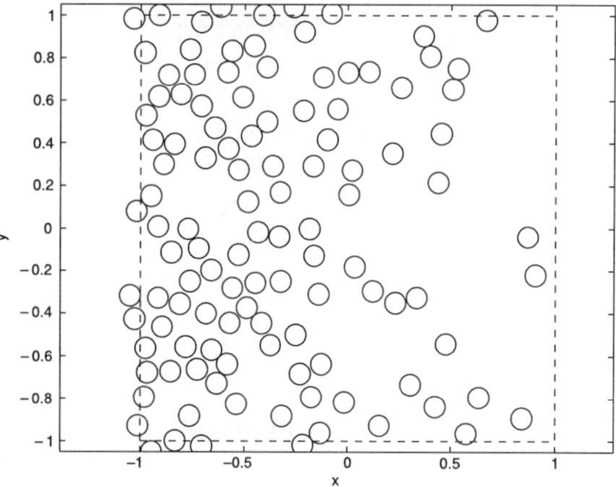

Fig. 6. Example of an extreme explicit FGM microstructure: $r_f = 0.05$, $m = -0.21$, $c = 0.2$. The dashed line denotes the boundary Γ of Ω. The total volume fraction of fiber material strictly within Ω is 0.2006.

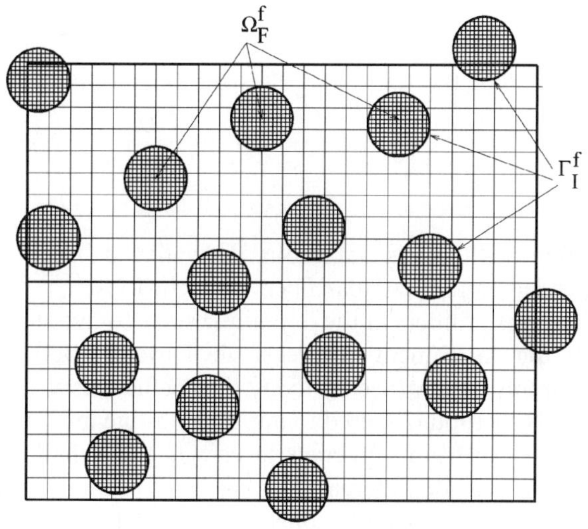

Fig. 7. A regular mesh with arbitrarily located fibers.

The X-FEM approximation to the displacement field takes the form:

$$\boldsymbol{u}^h(\boldsymbol{x}) = \sum_{i\in I} \boldsymbol{u}_i\phi_i(\boldsymbol{x}) + \sum_{j\in J} \boldsymbol{b}_j\phi_j(\boldsymbol{x})H(\boldsymbol{x}) + \sum_{k\in K_1}\phi_k(\boldsymbol{x})\left(\sum_{l=1}^{4} \boldsymbol{c}_k^{l1}F_1^l(r,\theta)\right) + \sum_{f=1}^{n_f}\left(\sum_{m\in G_f}\boldsymbol{g}_m^f\phi_m(\boldsymbol{x})I_f(\boldsymbol{x})\right)$$

$$(23)$$

where $\boldsymbol{g}_m^f$ are the additional degrees of freedom corresponding to the interface function $I_f(\boldsymbol{x})$ defined below. In the above, the set of nodes $m \in G_f$ enriched with this function is given by

$$G_f = \{m \in I : \omega_m \cap \Gamma_I^f \neq \emptyset\} \tag{24}$$

The interface functions $I_f(\boldsymbol{x})$ are constructed to have discontinuous derivatives across each matrix–fiber interface, allowing for the discontinuity in normal strain to be represented with (23). This idea was first

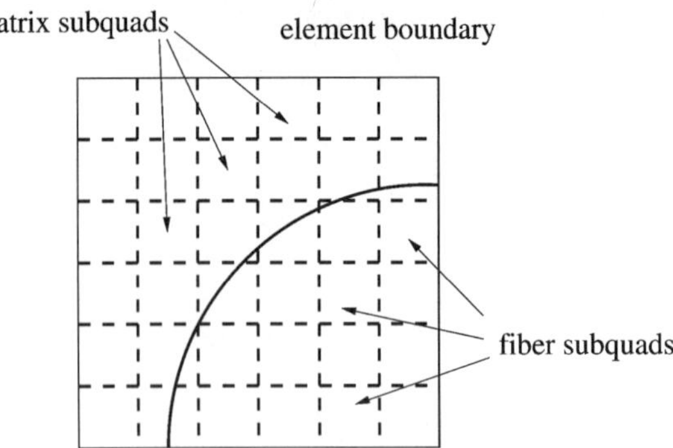

Fig. 8. Integration strategy for fibers. Fiber properties are used in those subquads whose centroids lie in the fiber domain, while matrix properties are employed elsewhere.

employed in mesh free methods by Krongauz and Belytschko [28], and it leads to optimal rates of convergence. In the present investigation, we construct a signed distance function from each fiber:

$$d_f(\mathbf{x}) = \|\mathbf{x} - \mathbf{x}_c^f\| - r_f \tag{25}$$

where $\mathbf{x}_c^f \in \Omega_F^f$ is the center point of the fiber, and r_f is the fiber radius. The interface function is then constructed as

$$I_f(\mathbf{x}) = \begin{cases} d_f(\mathbf{x}) & \text{if } d_f \geqslant 0 \\ 0 & \text{if } d_f < 0 \end{cases} \tag{26}$$

yielding a continuous function with a discontinuous derivative normal to the interface. The above framework can easily be extended to non-circular fibers or inclusions.

As we will consider distributions of fibers in the domain Ω so as to avoid a wall effect, certain fibers may intersect the boundary Γ. If the enrichment intersects the Dirichlet boundary, i.e., $\Gamma_u \cap \{\Gamma_1^f\} \neq \emptyset$, the boundary conditions cannot simply be imposed by fixing the coefficients $\mathbf{u}_i$ and $\mathbf{g}_m^f$. We therefore adopt the iterative technique of the LATIN method [29] used to impose non-linear constitutive laws. This method was used to enforce frictional contact constraints on crack faces with the X-FEM in Dolbow et al. [14], and the details of the implementation are provided therein. For the case of purely Neumann boundary conditions investigated in Section 3, the method sufficiently constrains the system of equations and achieves convergence in one iteration.

Concerning the integration of the terms in the matrix equations, we employ the same scheme detailed in Section 2.3.2 of dividing those elements cut by a fiber boundary into subquadrilaterals as shown in Fig. 8. If the center of an element or subquadrilateral is covered by Ω_F^f, then the constitutive properties of the fiber material is used in (7). We note that this approach with the approximation (23) can model any number of fibers, and no modifications are necessary when the crack intersects a fiber domain or when two fiber boundaries intersect.

3. Numerical examples

To investigate the applicability of effective properties in the fracture analysis of microstructured materials we present the results of a number of numerical studies in this section involving energy release rates and SIFs in FGMs.

Table 2
Numerical data of mechanical properties of Al and SiC

Properties	Constituents	
	Al	SiC
Young's modulus (GPa)	69.7	419.2
Poisson's ratio	0.34	0.19

We consider an FGM whose microstructure is in the form of an aluminum matrix with circular, cylindrical, silicon carbide fibers. Both constituent materials are assumed to be isotropic, and their material properties are given in Table 2. Unless otherwise noted, the overall volume fraction of fibers in the domain Ω is set to $c = 0.2$. In addition, all numerical calculations employ a single, uniform mesh of 70×70 quadrilateral elements. This discretization was found to yield results within 1% of the converged solution in a number of preliminary trials.

We begin our investigation by first ascertaining whether or not the homogenization method described in Section 2.3.1 (also see Appendix B) is adequately capturing the mean energetic response of the explicit trials, which is the basis by which the effective elastic properties are determined. We desire that the diameter of the fibers be small relative to the dimension of the edge-cracked domain detailed in Section 2.4.1. The ratio of fiber diameter to the edge-length of the domain was varied and in the end $2r_f/2 = 0.05$ was found to be small enough and was thus adopted. Fig. 9 depicts strain energy density versus the gradient m of the PDF governing the fiber distribution within the continuous matrix. The dashed line corresponds to the implicit results for the homogenized domain using the generalized self-consistent method. The circles with error bars correspond to the mean and plus/minus one standard deviation, respectively, for 10 realizations of the explicit microstructure. It is seen that the implicit method utilizing the generalized self-consistent model is within approximately 1% of the means of the explicit trials and always within one standard deviation of the mean. These results not only support the use of the generalized self-consistent model but also supports the adoption of pointwise homogenization in cases where $m \neq 0$. This last result is consistent with the findings of Reiter et al. [41].

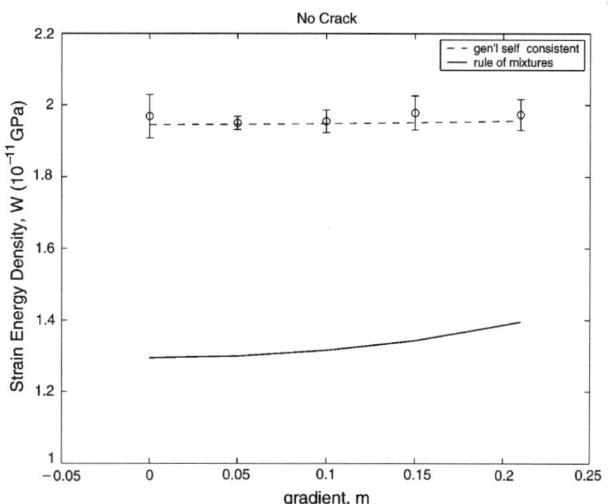

Fig. 9. Comparison of implicit and explicit strain energies versus gradient m of the PDF for Ω with no crack. The circles denote the mean of the 10 explicit trials and the error bars denote plus/minus one standard deviation.

As a means of comparison and to further promote confidence in the adopted homogenization method, we also present implicit results using the rule of mixtures. The results obtained using the rule of mixtures are shown as a solid line in Fig. 9. The implicit strain energies are seen to exhibit significant error, in marked contrast to those obtained with the generalized self-consistent method. We emphasize that the rule of mixtures is only employed here for the sake of comparison, to illustrate the importance of using an accurate homogenization method. All remaining implicit calculations employ the generalized self-consistent method described in Section 2.3.1.

We now compare implicit and explicit results for the edge-cracked plate illustrated in Fig. 2. For the explicit results we generated 10 microstructures, and report means and standard deviations in all cases. Fig. 10 presents the results for the energy release rate as a function of the gradient m of the PDF characterizing the spatial distribution of fibers parallel to the direction of the crack. The energy release rates were calculated using suitable forms of domain and interaction integrals described in Appendix A. The dashed line corresponds to the implicit results using the homogenized material properties. The circles denote the mean of 10 explicit trials and the error bars correspond to plus/minus one standard deviation. From Fig. 10 it is seen that the mean energy release rate of the explicit trials are within 2% of the energy release rate calculated using effective properties. This is an inspiring result indicating that the use of effective properties may yield energy release rates in very close agreement with a mean explicit energy release rate.

In a similar manner, Fig. 11 presents the results for the normalized K_I SIF, as a function of the gradient m of the PDF. From Fig. 11 it is seen that K_I varies slightly with m for both the implicit and explicit results, however, the implicit results predict a consistently larger K_I than the explicit results. This is a notable difference considering the previous comparison of energy release rates. The strain energies for both the implicit and explicit cases with the crack also agree quite well (see Fig. 12). Plots of the percent difference between the implicit and explicit results for K_I and W are presented in Fig. 13. We observe that on average, the strain energies are about 2–3% different while the mode-I SIFs are 12–13% different.

It bears emphasis that the above results are all for the case where the crack tip x_c is surrounded by a small ball $B(x_c, r)$ of matrix material. This measure was necessary to appropriately define the auxiliary fields used in the domain form of the interaction integral (see Appendix A). Therefore, we have not yet allowed for the possibility of a crack tip located on a matrix–fiber boundary, nor entirely within a fiber. In order to investigate the potential implications of having invoked such a region, an explicit trial was considered for a

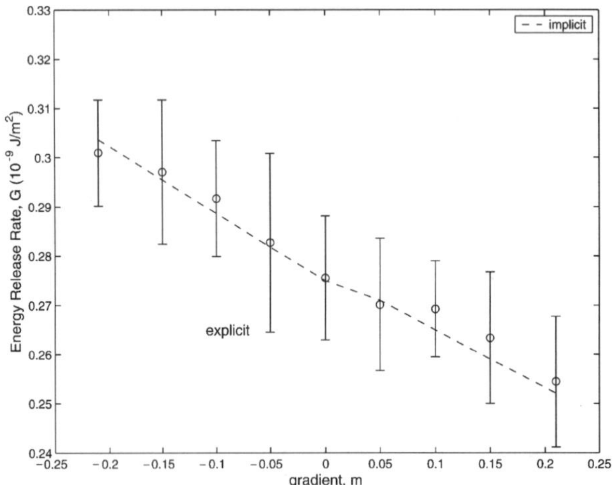

Fig. 10. Explicit and implicit results for the energy release rate. The circles denote the mean of the 10 explicit trials and the error bars denote plus/minus one standard deviation.

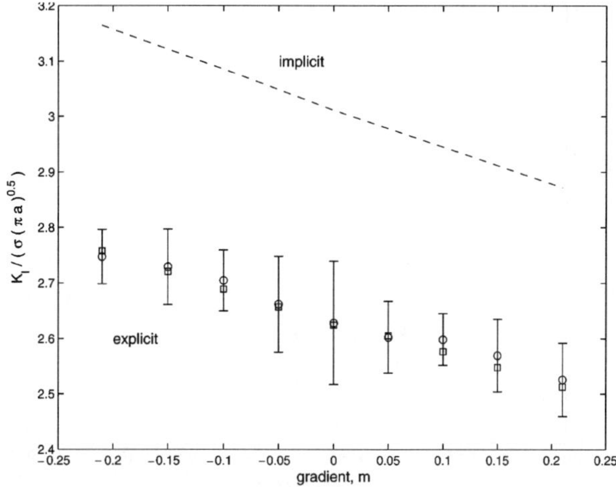

Fig. 11. Explicit and implicit results for K_I. The circles denote the mean of the 10 explicit trials and the error bars denote plus/minus one standard deviation. The squares denote the calculated values of K_I utilizing the implicit energy release rate as discussed later in this section.

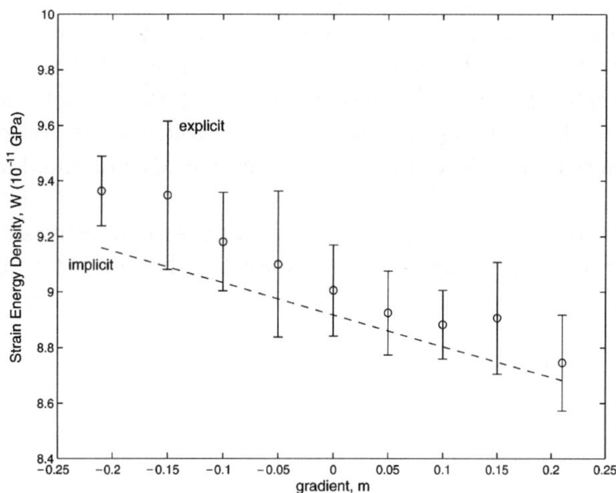

Fig. 12. Explicit and implicit results for strain energy density W. The circles denote the mean of the 10 explicit trials and the error bars denote plus/minus one standard deviation.

uniform distribution of fibers (i.e., $m = 0$) but with the crack tip located at the center of a fiber. This trial yielded $K_I/(\sigma\sqrt{\pi a}) = 5.92$. From Fig. 11 it is seen that the mean $K_I/(\sigma\sqrt{\pi a})$ for $m = 0$ with an exclusion region is 2.63. Since the fiber volume fraction is 0.2 it would be expected that in the case without an exclusion region that the crack tip would be in matrix material 80% of the explicit trials and in the fiber material 20%. Thus, one can estimate that without the exclusion region the mean of the explicit trials to be approximately, $0.8(2.63) + 0.2(5.92) = 3.2$ which compares quite closely (within approximately 6%) with the implicit result of 3.01. Thus, it appears that the implicit SIFs predict fairly well the average explicit SIFs assuming that the probability of a crack tip being within a given material is equal to the volume fraction of that material. In physically realized material systems, however, this is not likely to be the case.

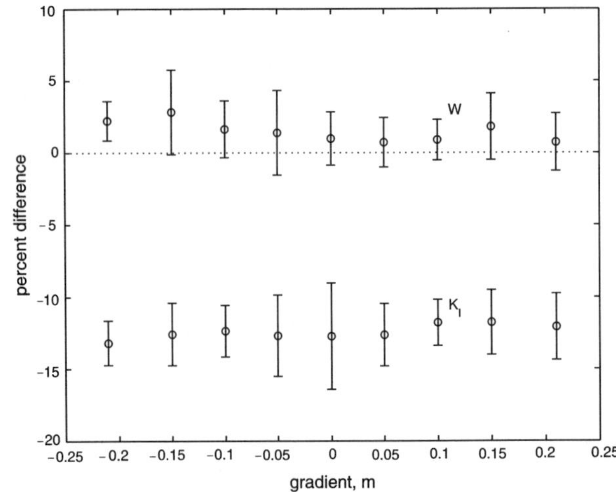

Fig. 13. Percent error plots for explicit and implicit results. The circles denote the error of the mean results and the error bars denote error for plus/minus one standard deviation in the results.

In order to shed further light on the present results, the importance of degree of mismatch between matrix and fiber properties was explored by varying the Young's modulus of the fibers. We considered a single microstructure generated from a PDF with $m = -0.21$ whose SIF was representative of the mean in Fig. 11. Fig. 14 depicts the ratio in the explicit mode-I SIF K_I^e to the implicit value K_I^i as a function of the ratio E_f/E_m. The results show that as this ratio decreases, the explicit value becomes larger than the implicit value. These findings are consistent with results from a similar study presented in Jha and Charalambibdes [26].

The above results give rise to an interesting question. Is it possible to estimate the mean K_I SIF using the implicit energy release rate and knowledge of which material the crack tip lies within? To address this

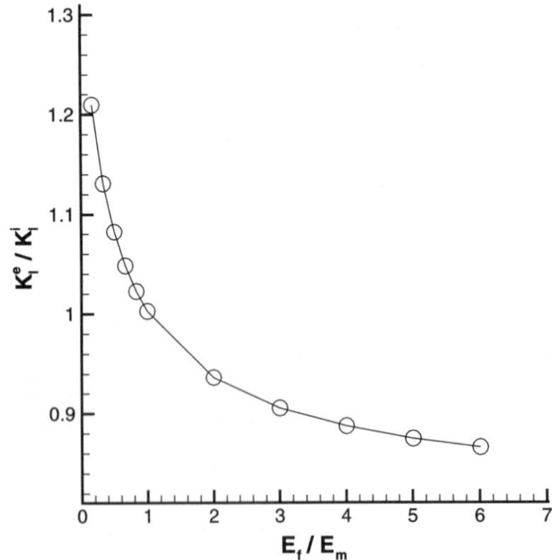

Fig. 14. Comparison of explicit and implicit trials as fiber properties vary. For SiC fibers, and the Al matrix, the ratio of E_f/E_m is 6.01.

question we utilize the implicit energy release rate and the knowledge that for the results presented in Fig. 11 the crack tip lies in the Al matrix material. Assuming that $K_{II} = 0$ we can solve Eq. (2) for K_I where we now take E_0 and v_0 to be the Young's modulus and Poisson's ratio of the aluminum matrix. The resulting K_I SIFs are plotted as squares in Fig. 11. The K_I calculated in this manner very closely approximates the mean of the explicit trials. Using the energy release rate for $m = 0$, we can alternatively estimate the explicit mode-I SIF for the problem considered in the previous paragraph where the crack tip terminates at the center of a silicon carbide fiber. Calculating the SIF using the implicit energy release rate and the silicon carbide material properties yields a normalized K_I of 6.17 which compares quite well with the explicit result of 5.92. Thus it appears that there is promise in utilizing the implicit energy release rate together with knowledge of the location of the crack tip to accurately predict the mean explicit SIFs.

The degree to which SIFs obtained from implicit calculations can be effectively used to determine crack growth remains unclear. This is due in large part because even for the pure mode-I loading examined in this investigation, the asymmetry of the explicit microstructures with respect to the x-axis gives rise to non-zero K_{II} SIFs. The results are shown in Fig. 15 as a function of m; the mode-II SIFs have means between -0.02 and 0.02 with standard deviations as large (in comparison to the mean) as 0.06. The local fields around crack tips due to neighboring fibers will therefore give rise to crack deflection. To some extent, the tendency for cracks to grow into fibers or avoid fibers is a function of the constituent properties as well as the matrix–fiber interface [19].

We now present the results of some limited quasi-static crack growth studies. These studies serve to illustrate the complex behavior of crack trajectories at the microstructural level and to examine the probability of a crack tip propagating within a given material. We assume that the cracks propagate according to the maximum hoop stress criterion [15]. Using this criterion and considering general mixed-mode loadings, the angle of crack propagation is written in terms of K_I and K_{II} as

$$\theta_c = 2 \arctan\frac{1}{4}\left(K_I/K_{II} \pm \sqrt{(K_I/K_{II})^2 + 8} \right) \tag{27}$$

We seed two cracks in the domain, and Fig. 16(a) shows their orientation and initial von Mises stress σ^M contours, normalized by the applied stress σ^0. We note that both fiber and crack boundaries are independent of the underlying mesh. In these studies, we investigated slightly larger fiber radii but maintained

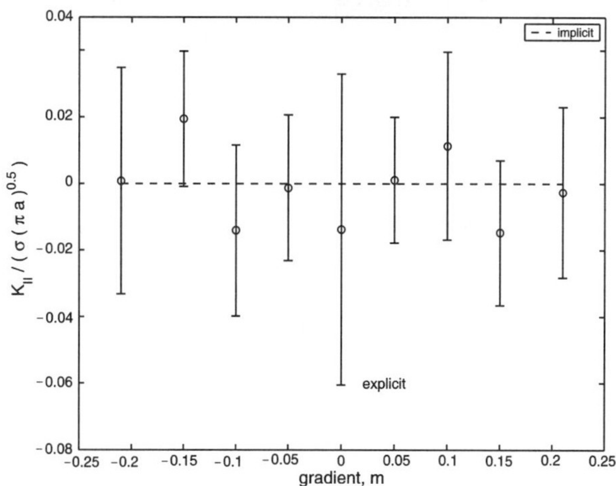

Fig. 15. Explicit and implicit results for K_{II}. The circles denote the error of the mean results and the error bars denote error for plus/minus one standard deviation in the results.

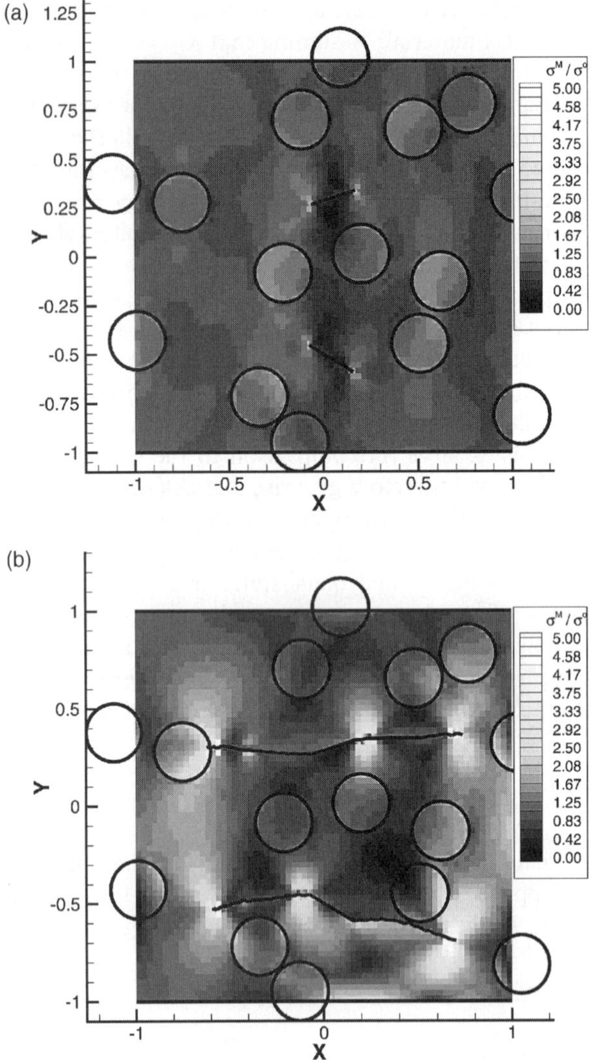

Fig. 16. Crack growth simulation using $c = 0.2$ and $r_f = 0.15$. (a) Initial crack configuration and von Mises stress contours. (b) Final crack configuration and von Mises stress contours.

an overall volume fraction of $c = 0.2$. Crack growth is simulated for a total of 19 steps with $\Delta a = 0.04$, and the final configuration is shown in Fig. 16(b). Although we did not conduct an exhaustive number of simulations, the results shown here are consistent with other cases, where the cracks exhibit an equal tendency to grow toward and deflect away from local fibers.

4. Summary and concluding remarks

This paper has presented an investigation that sheds light onto the applicability of the use of effective properties, or equivalently non-homogeneous properties, in fracture mechanics problems. In the process the development of the X-FEM method to the application of fracture of FGMs was achieved and energy release rates and SIFs were computed for an edge-cracked domain where the microstructure was treated in both an implicit and explicit manner. The implicit results were obtained by pointwise homogenization within a domain utilizing the PDF governing the spatial distribution of the fiber centers. The fiber diameters were sufficiently small that generalized self-consistent effective properties were appropriate, disregarding the scale of the crack tip. The explicit results presented are statistical measures of 10 trials from modeling the actual microstructural realizations generated in a manner consistent with the underlying PDF.

The microstructural characterization of the domain of Fig. 2 was specific as it regards actual computations. That is, the domain considered had a relatively small volume fraction of circular, silicon carbide fibers embedded within a well defined aluminum matrix. The motivation for these choices has been set forth in the body of the paper. Furthermore, the domain was subjected to pure mode-I loading. Thus, the specific results of this paper are limited to this case. However, as is stated in the following conclusions, we believe that the results may be indicative of more general configurations which should be investigated further.

Implicit and explicit strain energies for both the uncracked and cracked domain compared very well for all values of m considered. These results indicate that for the system considered, approximating the effective properties by the generalized self-consistent model and utilizing pointwise homogenization were acceptable. Perhaps most importantly, the implicit energy release rates also compared very well with the mean of the energy release rate from the explicit trials. It is therefore proposed that if an RVE can be defined—exclusive of the scale of the crack tip—and an appropriate homogenization method is employed, then the use of effective properties will permit the accurate estimation of the energy release rate at the crack tip. A simple re-interpretation of the results can then yield implicit SIFs that compare very favorably with the mean SIFs of the explicit trials. The re-interpretation is limited by the knowledge of the tip location within the microstructure.

It appears that the SIFs computed using effective properties are representative of the mean SIFs of the explicit trials if the probability of finding a crack tip within a given material is proportional to the volume fraction of that material. In physically realized material systems, it is not likely that this relationship will hold true. We suspect that as with fiber-reinforced composites, matrix–fiber debonding may be a more predominant failure mode than fiber cracking. Future work will examine these effects, as well as develop improved techniques for the determination of SIFs in FGMs utilizing effective properties. Modification of such approaches used herein are required when local fields are particularly relevant, as demonstrated by the non-zero explicit K_{II} SIFs of Fig. 15. These non-zero quantities give rise to crack deflection, and the importance of this phenomenon merits further study.

Acknowledgements

The authors are grateful to Professor Gideon Dagan, Tel Aviv University, for an early preprint of his paper. We also acknowledge Professor Michael Gosz, Illinois Institute of Technology, for many helpful comments concerning the derivation of the domain integral. Finally, the helpful comments of the reviewers are gratefully acknowledged.

Appendix A. Domain integrals for stress intensity factors

We provide the derivation for the domain integrals used to extract the pure mode-I SIF for the implicit models of FGM fracture as well as those used to extract the mixed-mode SIFs for the explicit studies. In the following, we assume that the crack faces are traction free and consider an arbitrary orientation of the crack with respect to the local variation in material properties.

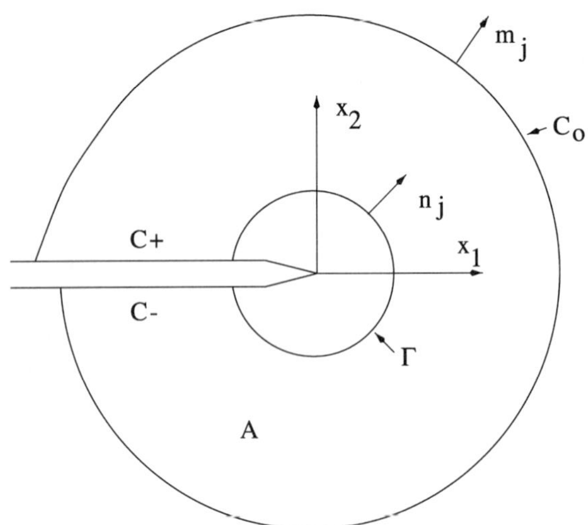

Fig. 17. Conventions at the crack tip. Domain A is enclosed by Γ, C_+, C_-, and C_0. Unit normal $m_j = n_j$ on C_+, C_-, and C_0 and $m_j = -n_j$ on Γ.

We focus attention on a crack tip in an FGM under plane strain conditions, and use the local crack tip coordinates with the x_1 axis parallel to the crack faces as shown in Fig. 17. For the contour Γ that begins at a point on the lower crack face and ends at a point on the upper crack face, the J-integral [42] is given by

$$J = \lim_{\Gamma \to 0} \int_{\Gamma} (W\delta_{1j} - \sigma_{ij}u_{i,1})n_j \, \mathrm{d}\Gamma \tag{A.1}$$

where W is the strain energy density and n_j are the components of the unit outward normal to Γ.

A.1. The implicit case

With the exception of particular crack orientations, the integrand in (A.1) is not divergence free for FGMs due to the local variation in material properties. As a result, an evaluation of the integral on finite contours will exhibit path dependence. It is therefore only equivalent to the energy release rate in the limit as the contour Γ is shrunk to the crack tip, a situation which is not very convenient to numerical implementation.

Gu et al. [22] approach this problem by using an extremely fine mesh near the crack tip and consider a contour that is sufficiently small such that the effect of material property gradients is negligible. An alternative that is more conducive for modeling crack growth was recently suggested by both Anlas et al. [1] and Chen et al. [5], who propose a modified path–domain J-integral of the form

$$J^* = \int_{\Gamma} (W\delta_{1j} - \sigma_{ij}u_{i,1})n_j \, \mathrm{d}\Gamma - \int_{S} \frac{1}{2}\epsilon_{ij}C_{ijkl,1}\epsilon_{kl} \, \mathrm{d}\Omega \tag{A.2}$$

where C is the fourth order elasticity tensor and S is the domain surrounded by Γ, as shown in Fig. 17. In effect, the surface integral removes the contributions due to material property gradients in the standard J-integral. Chen et al. [5] showed that the above integral is domain independent, and argued that it reduces to the standard J-integral in the limit as $\Gamma \to 0$. The J^*-integral is therefore equal to the energy release rate, regardless of the domain size.

Both of the above references recast the modified integral (A.2) into a domain form, as the above integral is not in a form particularly suitable for evaluation with the finite element method. We note, however, that when domain integrals are employed no modification whatsoever to the standard J-integral is required. We demonstrate this in the following derivation.

We begin recasting the standard J-integral (A.1) into an equivalent domain form by considering a sufficiently smooth weighting function $q(x)$ which takes a value of unity on an open set containing the crack tip and vanishes on an outer prescribed contour C_0. Then for each contour Γ as in Fig. 17, assuming the crack faces are traction free and straight in the region A bounded by the contour C_0, we define the following contour integral

$$H = \int_C \left[W\delta_{1j} - \sigma_{ij}\frac{\partial u_i}{\partial x_1} \right] q m_j \, \mathrm{d}C$$

where the contour $C = \Gamma + C_+ + C_- + C_0$ and $\vec{m}$ is the unit outward normal to the contour C.

With this choice of weight function (and noting that $m_j = -n_j$ on Γ), we obtain the following expression in the limit as the contour is shrunk onto the crack tip

$$\lim_{\Gamma \to 0} H = \int_{C_0 \cup C^+ \cup C^-} \left[W\delta_{1j} - \sigma_{ij}\frac{\partial u_i}{\partial x_1} \right] q m_j \, \mathrm{d}C - J$$

The first integral above vanishes when the crack faces are traction free, and so an alternative expression for the J-integral can be written as

$$J = -\lim_{\Gamma \to 0} H$$

We now use the divergence theorem on the closed integral over the contour C and pass to the limit as the contour Γ is shrunk to the crack tip. The condition that the weighting function q is unity on an open set containing the crack tip is easily relaxed to be valid just at the crack tip. We arrive at the following equation for the J-integral in domain form:

$$J = \int_A \left[\sigma_{ij}\frac{\partial u_i}{\partial x_1} - W\delta_{1j} \right] \frac{\partial q}{\partial x_j} \, \mathrm{d}A - \int_A (\epsilon_{ij} C_{ijkl,1}\, \epsilon_{kl}) q \, \mathrm{d}A \tag{A.3}$$

The second surface integral above arises due to the gradient in material properties. We see that the above is equivalent to the domain form presented in both Anlas et al. [1] and Chen et al. [5], but was derived without modification to the original J-integral.

A.2. The explicit case

We assume that the region A consists entirely of the isotropic matrix material, such that the integrand in the J-integral (A.1) is divergence free. As the presence of fibers outside of this region will generally trigger mode-mixity, we present the domain form of the interaction integral [44,46].

For general mixed-mode problems we have the following relationship between the value of the J-integral and the SIFs (through equivalence with the energy release rate):

$$J = \frac{K_{\mathrm{I}}^2}{E_m^*} + \frac{K_{\mathrm{II}}^2}{E_m^*} \tag{A.4}$$

where E_m^* is defined in terms of material parameters E_m (Young's modulus of the matrix) and v_m (Poisson's ratio for the matrix) as

$$E_m^* = \begin{cases} E_m & \text{plane stress} \\ \dfrac{E_m}{1 - v_m^2} & \text{plane strain} \end{cases} \tag{A.5}$$

Two states of a cracked body are considered. State 1, $(\sigma_{ij}^{(1)}, \epsilon_{ij}^{(1)}, u_i^{(1)})$, corresponds to the present state and state 2, $(\sigma_{ij}^{(2)}, \epsilon_{ij}^{(2)}, u_i^{(2)})$, is an auxiliary state which will be chosen as the asymptotic fields for mode-I or mode-II. The J-integral for the sum of the two states is

$$J^{(1+2)} = \int_\Gamma \left[\frac{1}{2}(\sigma_{ij}^{(1)} + \sigma_{ij}^{(2)})(\epsilon_{ij}^{(1)} + \epsilon_{ij}^{(2)})\delta_{1j} - (\sigma_{ij}^{(1)} + \sigma_{ij}^{(2)})\frac{\partial(u_i^{(1)} + u_i^{(2)})}{\partial x_1} \right] n_j \, d\Gamma \tag{A.6}$$

Expanding and rearranging terms gives

$$J^{(1+2)} = J^{(1)} + J^{(2)} + I^{(1,2)} \tag{A.7}$$

where $I^{(1,2)}$ is called the interaction integral for states 1 and 2

$$I^{(1,2)} = \int_\Gamma \left[W^{(1,2)}\delta_{1j} - \sigma_{ij}^{(1)}\frac{\partial u_i^{(2)}}{\partial x_1} - \sigma_{ij}^{(2)}\frac{\partial u_i^{(1)}}{\partial x_1} \right] n_j \, d\Gamma \tag{A.8}$$

where $W^{(1,2)}$ is the interaction strain energy

$$W^{(1,2)} = \sigma_{ij}^{(1)}\epsilon_{ij}^{(2)} = \sigma_{ij}^{(2)}\epsilon_{ij}^{(1)} \tag{A.9}$$

Writing Eq. (A.4) for the combined states gives after rearranging terms

$$J^{(1+2)} = J^{(1)} + J^{(2)} + \frac{2}{E^*}\left(K_I^{(1)}K_I^{(2)} + K_{II}^{(1)}K_{II}^{(2)} \right) \tag{A.10}$$

Equating (A.7) with (A.10) leads to the following relationship:

$$I^{(1,2)} = \frac{2}{E^*}\left(K_I^{(1)}K_I^{(2)} + K_{II}^{(1)}K_{II}^{(2)} \right) \tag{A.11}$$

Making the judicious choice of state 2 as the pure mode-I asymptotic fields with $K_I^{(2)} = 1$ gives the mode-I SIF for state 1 in terms of the interaction integral

$$K_I^{(1)} = \frac{2}{E_m^*} I^{(1,\text{mode-I})} \tag{A.12}$$

The mode-II SIF can be determined in a similar fashion.

Analogous to the development of the domain integral for the implicit case, the above contour integral (A.8) is not in a form best suited for finite element calculations. We therefore repeat the procedure of multiplying the integrand by the weighting function $q(x)$, which after a few manipulations is equivalent to

$$I^{(1,2)} = \int_C \left[W^{(1,2)}\delta_{1j} - \sigma_{ij}^{(1)}\frac{\partial u_i^{(2)}}{\partial x_1} - \sigma_{ij}^{(2)}\frac{\partial u_i^{(1)}}{\partial x_1} \right] q m_j \, d\Gamma \tag{A.13}$$

Using the divergence theorem and passing to the limit as the contour Γ is shrunk to the crack tip, we obtain the following equation for the interaction integral in domain form

$$I^{(1,2)} = \int_A \left[\sigma_{ij}^{(1)}\frac{\partial u_i^2}{\partial x_1} + \sigma_{ij}^{(2)}\frac{\partial u_i^1}{\partial x_1} - W^{(1,2)}\delta_{1j} \right] \frac{\partial q}{\partial x_j} \, dA \tag{A.14}$$

We note that a similar process can be used to derive an interaction integral for the implicit case for FGMs. However, special care must be taken when to consider the construction of the auxiliary fields when applying the divergence theorem. These fields may not satisfy all of the pertinent equations (equilibrium, strain–displacement, Hooke's law) over the domain A. These issues have been explored by Dolbow and Gosz [11].

For the numerical evaluation of the above domain integrals (both implicit and explicit), the domain A is set from the collection of elements about the crack tip. In this paper, we first determine the characteristic length of an element touched by the crack tip and designate this quantity as h_{local}. For two dimensional analysis, this quantity is calculated as the square root of the element area. The domain A is then set to be all elements which have a node within a ball of radius r_{d} about the crack tip. Fig. 18 shows a typical set of elements for the domain A with the domain radius r_{d} taken to be twice the length h_{local}. Fig. 19 shows the contour plot of the weight function q for these elements. The weight function q is taken to have a value of unity for all nodes within the ball r_{d}, and zero on the outer contour. The function is then easily interpolated within the elements using the nodal shape functions.

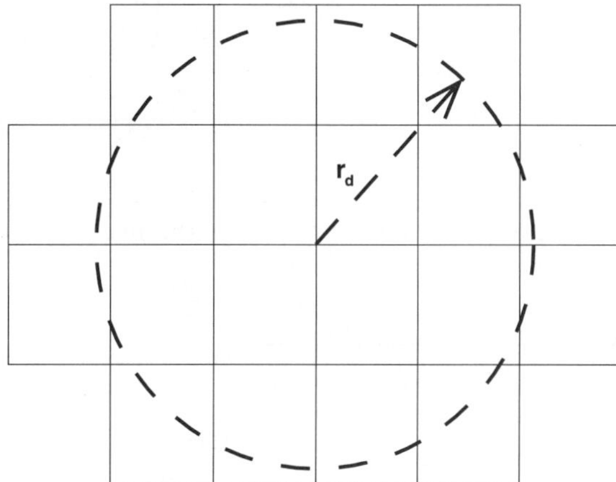

Fig. 18. Elements selected about the crack tip for the interaction integral.

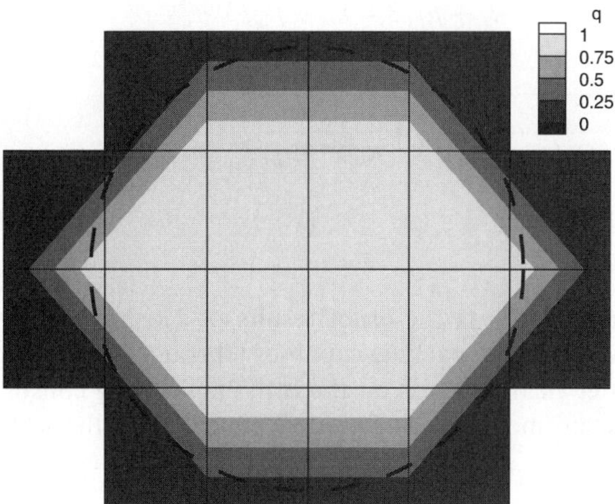

Fig. 19. Weight function q on the elements.

Appendix B. Effective property relationships

In this appendix the effective properties utilized in the body of the paper are presented. The composite is in the form of a continuous matrix with embedded unidirectional, infinitely long, circular cylindrical fibers. Given the fixed global coordinate system $x_1x_2x_3$ we take the direction of the cylindrical fibers to be parallel to the x_3-direction. Let E_m, v_m, κ_m, μ_m and E_f, v_f, κ_f, μ_f denote the Young's modulus, Poisson's ratio, bulk modulus, and shear modulus of the matrix and fiber materials, respectively. The volume fraction of fibers is denoted by c.

The effective properties of the composite are transversely isotropic and therefore there are five independent elastic constants. The engineering constants are E_{33}^*, v_{31}^*, κ_{12}^*, μ_{31}^*, and μ_{12}^*. The relationship between these engineering constants and the components of the effective elasticity tensor $\boldsymbol{C}^*$ are as follows:

$$C_{3333}^* = E_{33}^* + 4\left(v_{31}^*\right)^2\kappa_{12}^* \tag{B.1}$$

$$C_{3311}^* = 2\kappa_{12}^*v_{31}^* \tag{B.2}$$

$$C_{1111}^* = \mu_{12}^* + \kappa_{12}^* \tag{B.3}$$

$$C_{1122}^* = -\mu_{12}^* + \kappa_{12}^* \tag{B.4}$$

$$C_{3131}^* = \mu_{31}^* \tag{B.5}$$

For the generalized self-consistent model the first four of the five engineering material constants are given by [6]

$$E_{33}^* = cE_f + (1-c)E_m + \frac{4c(1-c)(v_f - v_m)^2\mu_m}{\dfrac{(1-c)\mu_m}{\kappa_f + \mu_f/3} + \dfrac{c\mu_m}{\kappa_m + \mu_m/3} + 1} \tag{B.6}$$

$$v_{31}^* = (1-c)v_m + cv_f + \frac{c(1-c)(v_f - v_m)\left[\dfrac{\mu_m}{\kappa_m + \mu_m/3} - \dfrac{\mu_m}{\kappa_f + \mu_f/3}\right]}{\dfrac{(1-c)\mu_m}{\kappa_f + \mu_f/3} + \dfrac{c\mu_m}{\kappa_m + \mu_m/3} + 1} \tag{B.7}$$

$$\kappa_{12}^* = \kappa_m + \frac{\mu_m}{3} + \frac{c}{\dfrac{1}{\kappa_f - \kappa_m + (\mu_f - \mu_m)/3} + \dfrac{1-c}{\kappa_m + 4\mu_m/3}} \tag{B.8}$$

$$\mu_{31}^* = \frac{(1+c)\mu_f + (1-c)\mu_m}{(1-c)\mu_f + (1+c)\mu_m}\mu_m \tag{B.9}$$

These expressions are the same form as the exact results of Hashin and Rosen [24] for the composite cylinders assemblage which admits a full packing capability through a continuous distribution of fiber sizes. The composites cylinder model yields bounds on the fifth engineering constant, μ_{12}^*.

The generalized self-consistent model result for μ_{12}^* [8] is given by the solution to

$$A\left(\frac{\mu_{12}^*}{\mu_m}\right)^2 + 2B\left(\frac{\mu_{12}^*}{\mu_m}\right) + C = 0 \tag{B.10}$$

where

$$A = 3c(1-c)^2\left(\frac{\mu_f}{\mu_m}-1\right)\left(\frac{\mu_f}{\mu_m}-\eta_f\right) + \left[\frac{\mu_f}{\mu_m}\eta_m + \eta_f\eta_m - \left(\frac{\mu_f}{\mu_m}\eta_m - \eta_f\right)c^3\right]$$
$$\times \left[c\eta_m\left(\frac{\mu_f}{\mu_m}-1\right) - \left(\frac{\mu_f}{\mu_m}\eta_m + 1\right)\right] \tag{B.11}$$

$$B = -3c(1-c)^2\left(\frac{\mu_f}{\mu_m}-1\right)\left(\frac{\mu_f}{\mu_m}-\eta_f\right) + \frac{1}{2}\left[\frac{\mu_f}{\mu_m}\eta_m + \left(\frac{\mu_f}{\mu_m}-1\right)c + 1\right]\left[(\eta_m-1)\left(\frac{\mu_f}{\mu_m}-\eta_f\right)\right.$$
$$\left. -2\left(\frac{\mu_f}{\mu_m}\eta_m - \eta_f\right)c^3\right] + \frac{c}{2}(\eta_m+1)\left(\frac{\mu_f}{\mu_m}-1\right)\left[\frac{\mu_f}{\mu_m}+\eta_f + \left(\frac{\mu_f}{\mu_m}\eta_m - \eta_f\right)c^3\right] \tag{B.12}$$

$$C = 3c(1-c)^2\left(\frac{\mu_f}{\mu_m}-1\right)\left(\frac{\mu_f}{\mu_m}-\eta_f\right) + \left[\frac{\mu_f}{\mu_m}\eta_m + \left(\frac{\mu_f}{\mu_m}-1\right)c + 1\right]\left[\frac{\mu_f}{\mu_m}+\eta_f + \left(\frac{\mu_f}{\mu_m}\eta_m + \eta_f\right)c^3\right]$$
$$\tag{B.13}$$

and where $\eta_m = 3 - 4v_m$ and $\eta_f = 3 - 4v_f$.

References

[1] Anlas G, Santare M, Lambros J. Numerical calculation of stress intensity factors in functionally graded materials. Int J Fract 2000;104:131–43.
[2] Auriault JL. Heterogeneous medium. Is an equivalent macroscopic description possible? Int J Eng Sci 1991;29(7):785–95.
[3] Babuška I, Rosenzweig M. A finite element scheme for domains with corners. Numer Math 1972;20:1–21.
[4] Benveniste Y, Dvorak GJ, Chen T. On diagonal and elastic symmetry of the approximate effective stiffness tensor of heterogeneous media. J Mech Phys Solids 1991;39(7):927–46.
[5] Chen J, Wu L, Di S. A modified *j*-integral for functionally graded materials. Mech Res Commun 2000;27(3):301–6.
[6] Christensen RM. A critical evaluation for a class of micromechanics models. J Mech Phys Solids 1990;38(3):379–404.
[7] Christensen RM. Mechanics of composite materials. Malabar, FL: Krieger; 1991.
[8] Christensen RM, Lo KH. Solutions for effective shear properties in three phase sphere and cylinder models. J Mech Phys Solids 1979;27(4):315–30, Erratum 34(6):639.
[9] Dagan G. Effective, equivalent, and apparent properties of heterogeneous media. In: Aref H, Phillips JW, editors. Mechanics for a new millenium. Dordrecht: Kluwer; 2001.
[10] Dolbow J. An extended finite element method with discontinuous enrichment for applied mechanics. PhD thesis, Northwestern University, 1999.
[11] Dolbow JE, Gosz M. On the computation of mixed-mode stress intensity factors in functionally graded materials. Int J Solids Struct 2002;39(9):2557–74.
[12] Dolbow J, Merle R. Modeling dendritic solidification with the extended finite element method. In: Proceedings of the First MIT Conference on Computational Fluid and Solid Mechanics, Boston, 2001.
[13] Dolbow J, Moës N, Belytschko T. Discontinuous enrichment in finite elements with a partition of unity method. Finite Elem Anal Des 2000;36:235–60.
[14] Dolbow J, Moës N, Belytschko T. An extended finite element method for modeling crack growth with frictional contact. Comput Methods Appl Mech Eng 2001;190(51–52):6825–46.
[15] Erdogan F, Sih G. On the crack extension in plates under plane loading and transverse shear. J Basic Eng 1963;85:519–27.
[16] Erdogan F, Wu B. The surface crack problem for a plate with functionally graded properties. ASME J Appl Mech 1997;61:449–56.
[17] Ferrari M. Asymmetry and the high concentration limit of the Mori–Tanaka effective medium theory. Mech Mater 1991;11:251–6.
[18] Foster G, Ibnabdeljalil M, Curtin W. Tensile strength of titanium matrix composites: direct numerical simulations and analytical models. Int J Solids Struct 1998;35(19):2523–36.
[19] Gosz M, Moran B, Achenbach J. On the role of interphases in the transverse failure of fiber composites. Int J Damage Mech 1994;3:357–77.
[20] Grisvard P. Elliptic problems in nonsmooth domains. Boston: Pitman Publishing, Inc; 1985.

[21] Gu P, Asaro R. Cracks in functionally graded materials. Int J Solids Struct 1997;34:1–17.

[22] Gu P, Dao M, Asaro R. A simplified method for calculating the crack-tip field of functionally graded materials using the domain integral. J Appl Mech 1999;66:101–8.

[23] Hashin Z. Analysis of composite materials—a survey. J Appl Mech 1983;50:481–505.

[24] Hashin Z, Rosen BW. The elastic moduli of fiber-reinforced materials. J Appl Mech 1964;31:223–32, see, errata, March 1965, p. 219.

[25] Huet C. Applications of variational concepts to size effects in elastic heterogeneous bodies. J Mech Phys Solids 1990;38(6):813–41.

[26] Jha M, Charalambides PG. Crack-tip micro mechanical fields in layered elastic composites: crack parallel to the interfaces. Int J Solids Struct 1998;35(1–2):149–79.

[27] Jin Z, Noda N. Crack tip singular fields in nonhomogeneous materials. J Appl Mech 1994;61:738–40.

[28] Krongauz Y, Belytschko T. EFG approximation with discontinuous derivatives. Int J Numer Methods Eng 1998;41(7):1215–33.

[29] Ladevèze P. Nonlinear computational structural mechanics. New York: Springer-Verlag; 1998.

[30] Levin VM. Thermal expansion coefficients of heterogeneous materials. Mekhanika Tverdogo Tela 1967;2(1):88–94, Mech Solids 2;58–61 [in English].

[31] Melenk JM, Babuška I. The partition of unity finite element method: Basic theory and applications. Comput Methods Appl Mech Eng 1996;39:289–314.

[32] Moës N, Dolbow J, Belytschko T. A finite element method for crack growth without remeshing. Int J Numer Methods Eng 1999;46:131–50.

[33] Moës N, Oden J, Vemaganti K, Remacle J. Simplified methods and a posteriori error estimation for the homogenization of representative volume elements (RVE). Comput Methods Appl Mech Eng 1999;176(1):265–78.

[34] Nadeau JC, Ferrari M. Microstructural optimization of a functionally graded transversely isotropic layer. Mech Mater 1999;31:637–51.

[35] Nadeau JC, Meng XN. On the response sensitivity of an optimally designed functionally graded layer. Compos: Part B, Eng 2000;31:285–97.

[36] Nemat-Nasser S, Hori M. Micromechanics: Overall properties of heterogeneous materials. In: North-Holland series in applied mathematics and mechanics, vol. 37. Amsterdam: North-Holland; 1993.

[37] Ostoja-Starzewski M. Microstructural disorder, mesoscale finite elements and macroscopic response. Proc R Soc, A 1999; 455:3189–99.

[38] Pindera M-J, Aboudi J, Arnold SM. Limitations of the uncoupled, RVE-based micromechanical approach in the analysis of functionally graded composites. Mech Mater 1995;20(1):77–94.

[39] Pindera M-J, Aboudi J, Arnold SM. Thermomechanical analysis of functionally graded thermal barrier coatings with different microstructural scales. J Am Ceram Soc 1998;81(6):1525–36.

[40] Reiter T, Dvorak G, Tvergaard V. Micromechanical models for graded composite materials. J Mech Phys Solids 1997;45(8):1281–302.

[41] Reiter T, Dvorak GJ, Tvergaard V. Micromechanical models for graded composite materials. J Mech Phys Solids 1997; 45(8):1281–302.

[42] Rice J. A path independent integral and the approximate analysis of strain concentration by notches and cracks. ASME J Appl Mech 1968;35:379–86.

[43] Rosen BW, Hashin Z. Effective thermal expansion coefficients and specific heats of composite materials. Int J Eng Sci 1970;8:157–73.

[44] Shih C, Asaro R. Elastic–plastic analysis of cracks on bimaterial interfaces: part i—small scale yielding. J Appl Mech 1988;55:299–316.

[45] Takahashi H, Ishikawa T, Okugawa D, Hashida T. Laser and plasma-arc thermal shock/fatigue fracture evaluation procedure for functionally gradient materials. In: Schneider G, Petzow G, editors. Thermal shock and thermal fatigue behavior of advanced ceramics. Dordrecht: Kluwer Academic Publishers; 1993. p. 543–54.

[46] Yau J, Wang S, Corten H. A mixed-mode crack analysis of isotropic solids using conservation laws of elasticity. J Appl Mech 1980;47:335–41.

PERGAMON

Engineering Fracture Mechanics 69 (2002) 1635–1645

**Engineering
Fracture
Mechanics**

www.elsevier.com/locate/engfracmech

Theoretical investigation of the effect of plasticity on crack growth along a functionally graded region between dissimilar elastic–plastic solids

Viggo Tvergaard *

*Department of Mechanical Engineering, Solid Mechanics, Technical University of Denmark,
Building 404, DK-2800 Kgs. Lyngby, Denmark*

Received 1 February 2001; received in revised form 23 August 2001; accepted 27 August 2001

Abstract

The influence of a functionally graded layer joining dissimilar elastic–plastic solids is studied in relation to interface crack growth. Conditions of small scale yielding are considered, and the boundary conditions applied on the outer edge of the region analysed are displacements for the elastic oscillating stress singularity fields corresponding to a sharp interface. A cohesive zone model is used to represent the fracture process, where the work of separation per unit area and the peak stress are basic parameters. Only crack growth on the initial crack plane parallel to the graded layer is analysed, but different locations of the crack plane relative to the layer are considered to obtain a parametric understanding. Crack growth resistance curves are illustrated, and the dependence of the steady-state fracture toughness on mode mixity is presented for several combinations of material parameters.
© 2002 Elsevier Science Ltd. All rights reserved.

Keywords: Cohesive zone modelling; Crack growth; Plasticity; *R*-curves

1. Introduction

Crack growth along an interface joining an elastic–plastic solid to a rigid solid has been analysed by Tvergaard and Hutchinson [1], and recently this work has been extended to consider crack growth along an interface between two elastic–plastic solids [2]. In these computations the fracture process has been modelled by a traction–separation law along the crack plane with a specified work of separation per unit area, while the material has been modelled as elastic–plastic. The analyses have shown that the plastic work during crack growth in metals contributes significantly to the fracture toughness, so that the macroscopic work of fracture is much larger than the work absorbed by the local fracture process required to separate the crack surfaces. These results have also shown a strong dependence on the mode mixity, such that the

---

* Tel.: +45-4525-4273; fax: +45-4593-1475.

E-mail address: viggo@mek.dtu.dk (V. Tvergaard).

predicted fracture toughness is much higher in cases where mode II loading dominates at the crack tip than in cases where mode I loading dominates (in good agreement with experimental observations [3–5]).

The previous investigations [1,2] have focussed on crack growth along an interface with a sharp transition between the properties of the two adjacent materials. However, in a number of cases the properties vary gradually over a certain interval, as in the graded composite materials studied by Reiter et al. [6]. A broad overview of functionally graded metals and metal–ceramic composites has been given by Mortensen and Suresh [7,8], where the first part discusses processing of the functionally graded materials, while the second part is devoted to the thermomechanical behaviour of such materials. Cracks in an elastic functionally graded material and the effect of the functional grading on fracture toughness has been analysed in [9], and for debonding of a ceramic coating from a metal substrate it has been shown that the toughness of the interface can vary a great deal between different graded profiles in the metal ceramic composition through the interface layer. For a graded plasticity mismatch between two material layers the effect on crack tip shielding has been analysed in [10] for a crack perpendicularly approaching the interface between the layers.

In the present paper the procedure based on describing the fracture process in terms of a cohesive zone model is used to analyse crack growth in a functionally graded layer of material between dissimilar elastic–plastic solids. As the material around the growing crack is described as elastic–plastic, with spatially varying material properties, the analyses will still represent the contribution of plastic work to the fracture toughness, which often results in a toughness much larger than that corresponding to the work of fracture in the process region. The analyses here are based on the assumption that crack growth remains on the initial crack plane, which is parallel to the functionally graded interface layer. This gives a parametric understanding of the behaviour, but it is noted that in this type of situation crack growth out of the initial crack plane would be quite possible, in some cases leading to a crack growing on another plane parallel to the interface.

2. Problem formulation and numerical procedure

The previous analyses of crack growth along an interface under mixed mode loading have focussed on a sharp interface between an elastic–plastic solid and a rigid solid [1] or between two different elastic–plastic solids [2]. Here, the situation is investigated, where there is a gradual change of the material properties in a material layer between the two solids. As illustrated in Fig. 1 material No. 1 is mostly in the range of positive x^2-coordinates, while material No. 2 is mostly in the range of negative x^2-coordinates. Both materials are taken to be elastic–plastic with the true stress–logarithmic strain curve in uniaxial tension specified by

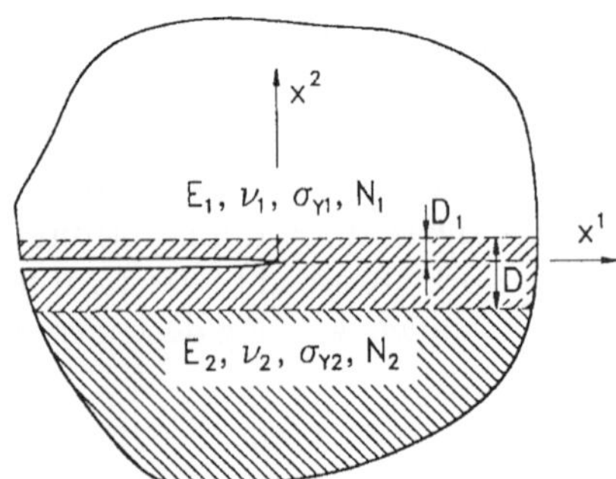

Fig. 1. Crack in a functionally graded layer of thickness D between dissimilar elastic–plastic solids.

$$\epsilon = \begin{cases} \sigma/E & \text{for } \sigma \leqslant \sigma_Y \\ (\sigma_Y/E)(\sigma/\sigma_Y)^{1/N} & \text{for } \sigma \geqslant \sigma_Y \end{cases} \tag{2.1}$$

Here, σ_Y is the initial yield stress and N is the power hardening exponent, while E and v are Young's modulus and Poisson's ratio, respectively. The tensile behaviour is generalized to multiaxial stress states assuming isotropic hardening and using the Mises yield surface. For material No. 1 the parameters are denoted E_1, v_1, σ_{Y1} and N_1, while for material No. 2 they are E_2, v_2, σ_{Y2} and N_2. In the functionally graded material layer of initial width D the same material description is assumed, but the values of the four material parameters inside the layer are taken to be given by linear interpolation between the two neighboring solids. Obviously, this linear variation of the properties through the graded layers is an approximation, and real graded layers will usually show some non-linear variation that may be different for each material parameter.

The crack growth analysed here is assumed to be parallel with the functionally graded layer, at distance D_1, below the upper edge of the layer (Fig. 1). In such a material configuration it is quite possible that the crack will grow out of the initial crack plane, but here the growth is restricted to remain on the initial crack plane, to get a parametric understanding of the effect of D_1/D on the fracture toughness. As for crack growth along a sharp interface between dissimilar solids under mixed mode loading [1,2], the straight ahead crack growth for all load combinations will only occur if the crack grows more easily in this plane than out of the plane.

In a small scale yielding formulation of the crack problem it is assumed that the layer width D is small relative to the outer radius A_0 of the region analysed so that specifying the elastic crack tip singularity field far away from the crack tip will be sufficiently accurate. The elastic interface crack problem was solved long ago by many authors (e.g. [11]). Following the formulation of Rice [12] (see also [1]), the crack has tractions acting on the interface, which are given in terms of the two stress intensity factor components, K_I and K_{II}, by

$$\sigma_{22} + i\sigma_{12} = (K_I + iK_{II})(2\pi r)^{-1/2}r^{i\epsilon} \tag{2.2}$$

Here, r is the distance from the tip, $i = \sqrt{-1}$, ϵ is the oscillation index

$$\epsilon = \frac{1}{2\pi}\ln\left(\frac{1-\beta}{1+\beta}\right) \tag{2.3}$$

and β is the second Dundurs' parameter

$$\beta = \frac{1}{2}\frac{\mu_1(1-2v_2) - \mu_2(1-2v_1)}{\mu_1(1-v_2) + \mu_2(1-v_1)} \tag{2.4}$$

where the shear moduli are $\mu_1 = E_1/(2(1+v_1))$ and $\mu_2 = E_2/(2(1+v_2))$. The relation between the energy release rate and the magnitude $|K|$ of stress intensity factors is

$$G = \frac{1}{2}(1-\beta^2)\left[\frac{1-v_1^2}{E_1} + \frac{1-v_2^2}{E_2}\right]|K|^2, \quad |K| = \sqrt{K_I^2 + K_{II}^2} \tag{2.5}$$

With a reference length L chosen to characterize the remote field an L-dependent measure of mode mixity ψ is defined by

$$\tan\psi = \frac{\text{Im}[(K_I + iK_{II})L^{i\epsilon}]}{\text{Re}[(K_I + iK_{II})L^{i\epsilon}]} \tag{2.6}$$

which reduces to the more familiar measure, $\tan\psi = K_{II}/K_I$, when $\epsilon = 0$. By using (2.2) in (2.6) it is seen that $\tan\psi = \sigma_{12}/\sigma_{22}$ at $r = L$ on the interface. The displacement components associated with the singularity field, with amplitude $|K|$, are specified in Tvergaard and Hutchinson [1].

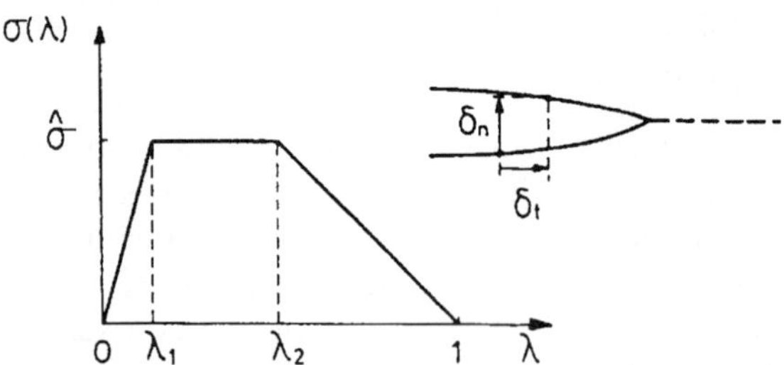

Fig. 2. Specification of traction–separation relation.

A special version of a traction–separation law proposed by Needleman [13] and generalized by Tvergaard [14] was used by Tvergaard and Hutchinson [1]. Here, δ_n and δ_t denote the normal and tangential components of the relative displacement of the crack faces across the interface in the zone where the fracture processes are occurring (Fig. 2). When δ_n^c and δ_t^c are critical values of these displacement components and a single non-dimensional separation measure is defined as $\lambda = [(\delta_n/\delta_n^c)^2 + (\delta_t/\delta_t^c)^2]^{1/2}$ the tractions drop to zero when $\lambda = 1$. With $\sigma(\lambda)$ displayed in Fig. 2, a potential from which the tractions are derived is defined as

$$\Phi(\delta_n, \delta_t) = \delta_n^c \int_0^\lambda \sigma(\lambda')d\lambda' \tag{2.7}$$

The normal and tangential components of the traction acting on the interface in the fracture process zone are given by

$$T_n = \frac{\partial \Phi}{\partial \delta_n} = \frac{\sigma(\lambda)}{\lambda} \frac{\delta_n}{\delta_n^c}, \quad T_t = \frac{\partial \Phi}{\partial \delta_t} = \frac{\sigma(\lambda)}{\lambda} \frac{\delta_t}{\delta_t^c} \frac{\delta_n^c}{\delta_t^c} \tag{2.8}$$

The peak normal traction under pure normal separation is $\hat{\sigma}$, and the peak shear traction is $(\delta_n^c/\delta_t^c)\hat{\sigma}$ in a pure tangential separation. The work of separation per unit area of interface is given by Eq. (2.7) with $\lambda = 1$, and for the separation function $\sigma(\lambda)$ in Fig. 2 the work is

$$\Gamma_0 = \tfrac{1}{2}\hat{\sigma}\delta_n^c(1 - \lambda_1 + \lambda_2) \tag{2.9}$$

It has been found by Tvergaard and Hutchinson [1,15] that the details of the shape of the separation law are relatively unimportant, and that the two most important parameters characterizing the fracture process in this model are Γ_0 and $\hat{\sigma}$.

Based on (2.5) and (2.9) a reference stress intensity factor is defined as

$$K_0 = \left[\frac{1 - v_1^2}{E_1} + \frac{1 - v_2^2}{E_2}\right]^{-1/2} \left(\frac{2\Gamma_0}{1 - \beta^2}\right)^{1/2} \tag{2.10}$$

Here, K_0 represents the value of $|K|$ needed to advance the interface crack in the absence of any plasticity. This value is independent of ψ since a potential is used to generate the relation of tractions to crack face displacements of the interface. A length quantity R_0, which scales with the size of the plastic zone in material No. 1 (when $|K| \cong K_0$), is defined as

$$R_0 = \frac{1}{3\pi}\left(\frac{K_0}{\sigma_{Y1}}\right)^2 = \frac{2}{3\pi}\left[\frac{1 - v_1^2}{E_1} + \frac{1 - v_2^2}{E_2}\right]^{-1} \frac{\Gamma_0}{(1 - \beta^2)\sigma_{Y1}^2} \tag{2.11}$$

While the mode mixity measure ψ refers to the distance L from the tip, it is natural to define a reference measure of mixity, ψ_0, based on the reference length R_0. The relation between ψ_0 and ψ is

$$\psi_0 = \psi + \epsilon \ln(R_0/L) \tag{2.12}$$

In the numerical analyses finite strains are accounted for, using a convected coordinate, Lagrangian formulation of the field equations, in which g_{ij} and G_{ij} are metric tensors in the reference configuration and the current configuration, respectively, with determinants g and G, and $\eta_{ij} = 1/2(G_{ij} - g_{ij})$ is the Lagrangian strain tensor. The contravariant components τ^{ij} of the Kirchhoff stress tensor on the current base vectors are related to the components of the Cauchy stress tensor σ^{ij} by $\tau^{ij} = \sqrt{G/g}\sigma^{ij}$. Then, in the finite-strain generalization of J_2-flow theory discussed by Hutchinson [16], an incremental stress–strain relationship is obtained of the form $\dot{\tau}^{ij} = L^{ijkl}\dot{\eta}_{kl}$ (e.g. see [1]).

The numerical solutions are obtained by a crack growth procedure as that used by Tvergaard and Hutchinson [1] and Tvergaard [2]. Thus, a finite-element approximation of the displacement fields is used in a linear incremental method, with a Cartesian coordinate system x^i as reference. The solution is based on the incremental principle of virtual work, with equilibrium correction terms applied.

An example of the mesh used for the computations is shown in Fig. 3, where it is seen that a uniform mesh region is used in the range where crack growth is studied. The length of one square element inside the uniform mesh is denoted Δ_0, and the initial crack tip is located at $x^1 = 0$. The computations are carried out with 120×6 quadrilaterals in the uniform mesh along the interface. The elements used are quadrilaterals each built-up of four triangular, linear displacement elements. The outer radius of the region analysed is chosen to be $A_0/\Delta_0 = 80\,000$, in order that the plastic zone size should not exceed $A_0/10$.

It is noted that the graded layer can in some cases extend outside the uniform mesh region shown in Fig. 3, depending on the values of D/Δ_0 and D_1/D. In all cases, the variation of the material parameters is represented by using the appropriate values in each integration point. Thus, the solution will depend on the mesh refinement, relative to D or R_0, but it is found that this mesh dependence is weak as long as the fracture process region in the cohesive zone is longer than $2\Delta_0$.

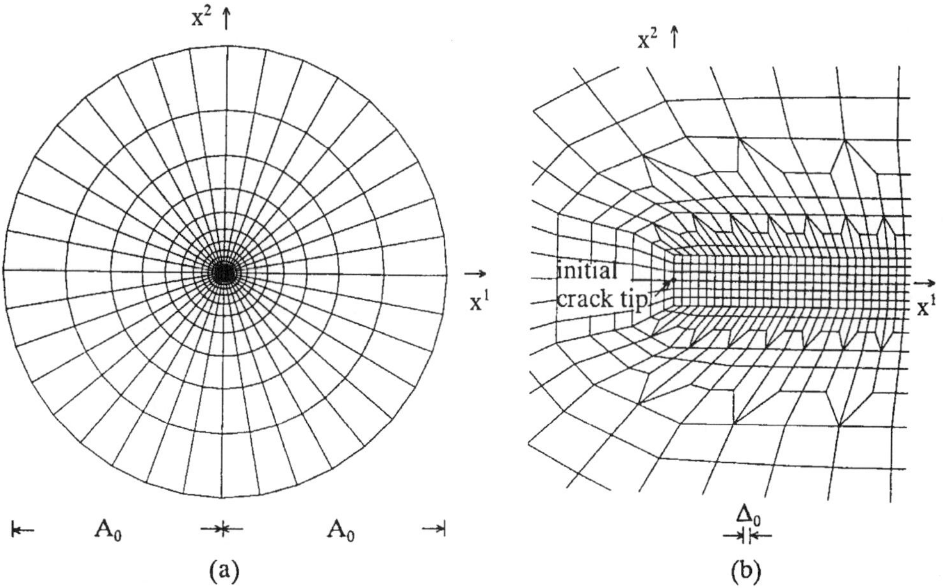

Fig. 3. Mesh used for some of the crack growth analyses.

While crack growth is prescribed to be along the symmetry plane of the mesh used for the analysis (Fig. 3), the material properties specified for each integration point are non-symmetrically distributed relative to the coordinate system, as illustrated in Fig. 1. The properties of material No. 1 are here specified as $\sigma_{Y1}/E_1 = 0.003$, $v_1 = 1/3$ and $N_1 = 0.1$, while different values are considered for the properties of material No. 2. In the traction–separation law the values $\delta_n^c/\delta_t^c = 1$, $\delta_n^c = 0.1\Delta_0$, $\lambda_1 = 0.15$ and $\lambda_2 = 0.50$ are used, while $\hat{\sigma}/\sigma_{Y1}$ is varied.

On the free crack surfaces, for $x^2 = 0$ and $x^1 < 0$, zero tractions, $T^1 = T^2 = 0$, are prescribed, while on the remaining part of the interface, $x^2 = 0$ and $x^1 > 0$, the displacements and tractions are related by the traction–separation law (2.8). During the initial part of the crack growth resistance curve an increment of $|K|$ is prescribed, but when $|K|$ approaches its asymptote, a Rayleigh–Ritz finite-element method [17] is needed to ensure a monotonic increase in displacement differences across the crack tip. It is noted that full finite strain effects are accounted for so that crack tip blunting can be represented, and this is important if the peak stress $\hat{\sigma}$ of the debonding model is near or above the maximum possible stress during blunting (see discussion in [1]).

At some stages of the deformation the value of the J-integral [18,19] is calculated on a number of contours around the crack tip to check agreement with the prescribed amplitude $|K|$ of the edge displacements, which represents the $|K|$ field in the elastic material far away from the tip. The finite strain expression for the J-integral is evaluated along contours outside the uniform mesh region, through rings of quadrilaterals as those seen in the outer mesh in Fig. 3a. It is noted that the standard J-expression applies here, as the crack is parallel to the graded material layer. The path independence of J is confirmed by these calculations, and good agreement is found with the value of the energy release rate G given by (2.5) in terms of $|K|$.

3. Results

In the cases to be analysed here both materials Nos. 1 and 2 are taken to be elastic–plastic. The values of the material parameters above the interface are in all cases taken to be $\sigma_{Y1}/E_1 = 0.003$, $v_1 = 1/3$ and $N_1 = 0.1$. The substrate has $v_2 = 1/3$ and $N_2 = 0.1$, while different values are considered for the ratios E_2/E_1 and σ_{Y2}/σ_{Y1}. In the first cases analysed, (Figs. 4 and 5), these values are taken to be $E_2/E_1 = 2.0$ and $\sigma_{Y2}/\sigma_{Y1} = 2.0$, and the peak stress in the cohesive zone model is specified by $\hat{\sigma}/\sigma_{Y1} = 4.0$. Furthermore, the

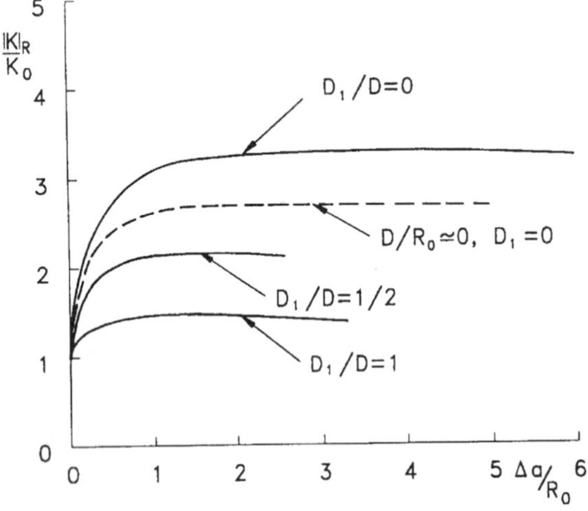

Fig. 4. Crack growth resistance curves for different locations D_1/D of the crack plane. Material parameters are $\hat{\sigma}/\sigma_{Y1} = 4.0$, $\sigma_{Y1}/E_1 = 0.003$, $E_2/E_1 = 2.0$, $\sigma_{Y2}/\sigma_{Y1} = 2.0$, $D/R_0 = 0.277$, with the near-tip mode mixity specified by $\psi_0 = 3.95°$.

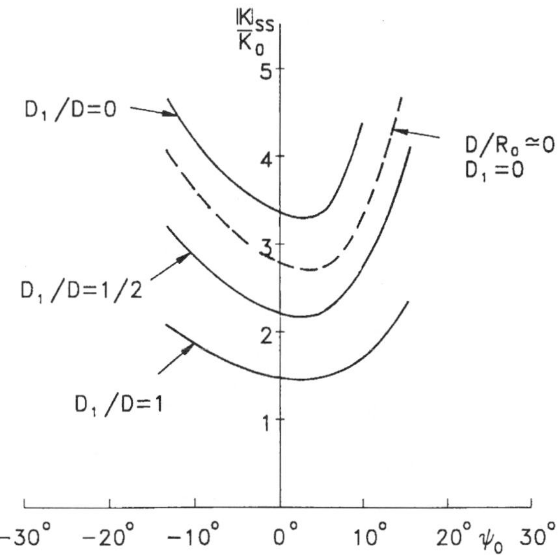

Fig. 5. Steady-state toughness as a function of the local mixity measure ψ_0, for different locations D_1/D of the crack plane. Material parameters are $\hat{\sigma}/\sigma_{Y1} = 4.0$, $\sigma_{Y1}/E_1 = 0.003$, $E_2/E_1 = 2.0$, $\sigma_{Y2}/\sigma_{Y1} = 2.0$ and $D/R_0 = 0.277$.

width of the functionally graded layer between the two materials is specified by $D/R_0 = 0.277$, where the mesh size in the uniform part of the mesh is chosen such that $D/\Delta_0 = 4.0$.

The three solid curves in Fig. 4 show crack growth resistance curves corresponding to three different locations of the crack plane, when the value of the local mode mixity near the crack tip is taken to be specified by $\psi_0 = 3.95°$. The highest fracture toughness is found for $D_1/D = 0$, where the crack runs along the upper edge of the graded layer, while the toughness is much smaller when the crack runs along the lower edge of the graded layer, for $D_1/D = 1$. The crack growth resistance curve corresponding to a sharp interface is included for comparison as a dashed curve in Fig. 4, computed by taking $D_1/D = 0$ and D negligibly small, as indicated by $D/R_0 \simeq 0$. It is seen that the fracture toughness for the sharp interface is well above that corresponding to crack growth in the centre of the graded layer $(D_1/D = 1/2)$. The results here are based on the assumption that the values of the peak stress $\hat{\sigma}$ and the work of separation Γ_0, characterizing the fracture process, are the same for each of the crack planes considered in the graded layer. This is done here to obtain a parametric understanding. But it is noted that in practice the values of the fracture process parameters will most likely vary somewhat with the grading of the other material parameters, and this would be important if the analysis was extended to account for crack growth deviation from the initial crack plane. If the fracture process parameters are independent of grading, as considered in Fig. 4, it could be expected that under near mode 1 conditions the crack would actually want to grow towards the substrate, where the crack growth resistance is smaller.

The influence of the mode mixity, ψ_0, is illustrated in Fig. 5 by showing the steady-state toughness $|K|_{SS}/K_0$. Here, the peak values of $|K|_R$ on resistance curves as those in Fig. 4 are used to define $|K|_{SS}$. It is seen that the points for $\psi_0 = 3.95°$, as considered in Fig. 4, are close to the minima of the curves in Fig. 5. The dashed curve corresponding to a sharp interface shows the result found earlier [1,2] that increasing mode II contribution near the crack tip (ψ_0 deviating from zero) gives a strong increase of the fracture toughness. The curves corresponding to the graded interface layer, for three different values of D_1/D, all show a very similar dependence on varying the near tip mode mixity.

Computations have also been carried out for a four times larger width of the graded interface layer, $D/R_0 = 1.109$, but otherwise for the same material parameters as those considered in Figs. 4 and 5. The

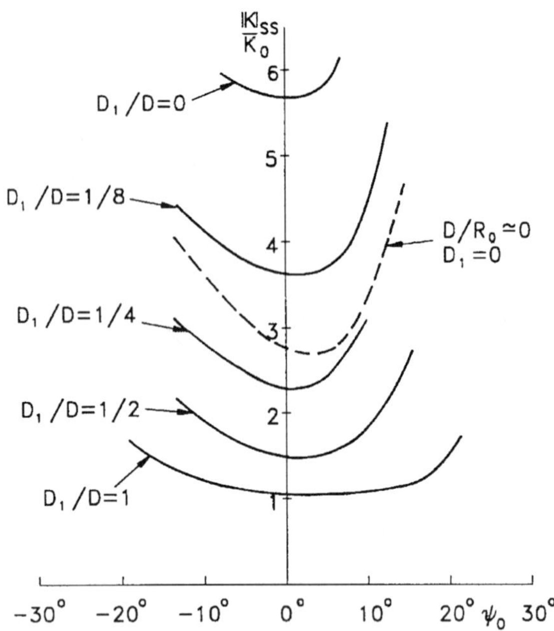

Fig. 6. Steady-state toughness as a function of the local mixity measure ψ_0, for different locations D_1/D of the crack plane. Material parameters are $\hat{\sigma}/\sigma_{Y1} = 4.0$, $\sigma_{Y1}/E_1 = 0.003$, $E_2/E_1 = 2.0$, $\sigma_{Y2}/\sigma_{Y1} = 2.0$ and $D/R_0 = 1.109$.

uniform mesh size relative to the reference plastic zone size R_0 is unchanged, so that here $D/\Delta_0 = 16$. Fig. 6 shows curves for the influence of the mode mixity on the steady-state toughness, analogous to the curves in Fig. 5, and it is noted that the dashed reference curves for a sharp interface are identical in these two figures. The main difference from the previous figure is that in Fig. 6 the curves for $D_1/D = 1$ and $D_1/D = 1/2$ show much lower toughness while the curve for $D_1/D = 0$ shows much higher toughness than found in Fig. 5. With the larger width of the graded layer relative to the reference plastic zone size a crack growing along the upper or lower edge of the graded layer will not see so big difference from a crack growing inside either material No. 1 or 2. Since $\sigma_{Y2}/\sigma_{Y1} = 2.0$, the reference plastic zone size for material No. 2 is only $0.25R_0$, with $\hat{\sigma}/\sigma_{Y2} = 2.0$ in the case of Fig. 6, and this is the reason why the curve for $D_1/D = 1$ shows only a small increase of the fracture toughness above the reference value K_0. The rather large differences between the toughness levels found for D_1/D equal to 1/4, 1/8 or 0 agree with the strong sensitivity to the ratio of the peak stress $\hat{\sigma}$ and the local yield stress at the crack plane found in earlier investigations [1,2,13].

Fig. 7 shows results analogous to those in the two previous figures, but for a different set of material parameters $E_2/E_1 = 1.0$, $\sigma_{Y2}/\sigma_{Y1} = 1.8$ and $\hat{\sigma}/\sigma_{Y1} = 4.25$. It is noted that this interface is of the type considered by Kim et al. [10], with a graded plasticity mismatch between two adjacent materials, but with no elastic mismatch. Relevant material examples are mentioned in [10], such as a ferritic steel joined to an austenitic steel by explosion cladding, but the focus in [10] is on a crack perpendicularly approaching the interface. In Fig. 7 the width of the functionally graded layer relative to the uniform mesh size is $D/\Delta_0 = 4$, with $D/R_0 = 0.350$. Also in this figure the reference curve corresponding to a sharp interface is shown dashed. It is seen that for these material parameters the difference between the toughness increases found for D_1/D equal to 1 or 0 is larger than that in Fig. 5 but smaller than that in Fig. 6. The variation of the toughness with mode mixity has shown a clear non-symmetry in Figs. 5 and 6, so that the minima occur for a small positive value of ψ_0. In Fig. 7, where the two materials have different yield stresses but equal elastic

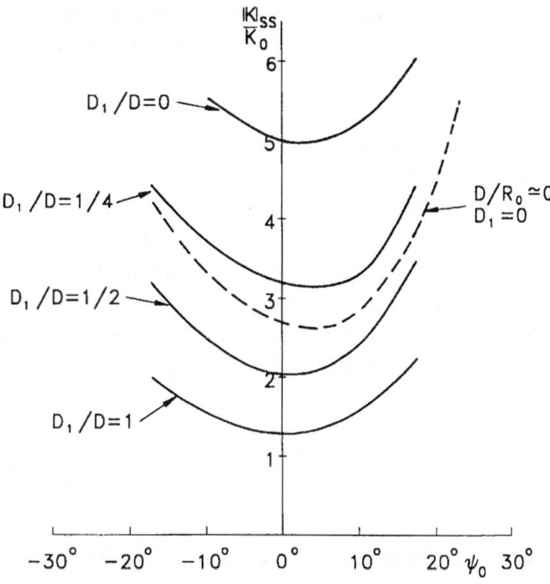

Fig. 7. Steady-state toughness as a function of the local mixity measure ψ_0, for different locations D_1/D of the crack plane. Material parameters are $\hat{\sigma}/\sigma_{Y1} = 4.25$, $\sigma_{Y1}/E_1 = 0.003$, $E_2/E_1 = 1.0$, $\sigma_{Y2}/\sigma_{Y1} = 1.8$ and $D/R_0 = 0.350$.

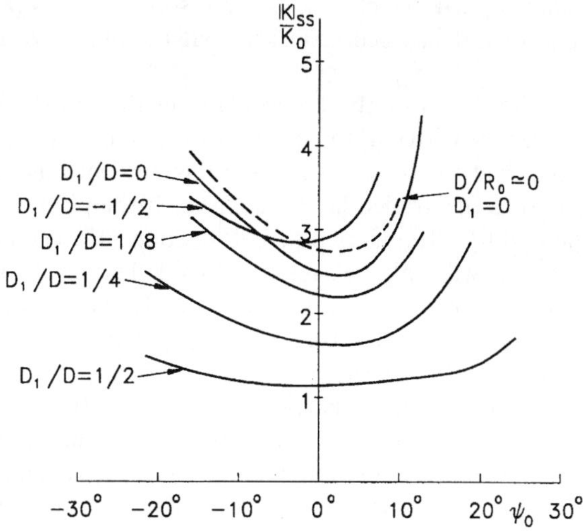

Fig. 8. Steady-state toughness as a function of the local mixity measure ψ_0, for different locations D_1/D of the crack plane. Material parameters are $\hat{\sigma}/\sigma_{Y1} = 3.5$, $\sigma_{Y1}/E_1 = 0.003$, $E_2/E_1 = 6.0$, $\sigma_{Y2}/\sigma_{Y1} = 6.0$ and $D/R_0 = 0.961$.

properties, this non-symmetry is also visible on the higher curves, where plasticity gives a large contribution to the fracture toughness, but the lower curves for D_1/D equal to 1 or 1/2 are nearly symmetric.

In Fig. 8 the mismatch is taken to be larger, $E_2/E_1 = 6.0$ and $\sigma_{Y2}/\sigma_{Y1} = 6.0$, with $\hat{\sigma}/\sigma_{Y1} = 3.5$. Here the elastic mismatch is large enough to represent a metal bonded to a ceramic, and the grading could result from a graded metal ceramic composite composition in an interlayer, as e.g. discussed in [8]. The assumed large mismatch of the initial yield stresses means that there is actually no plastic yielding in the substrate, so that plasticity is limited to material No. 1 and the upper part of the graded layer (see Fig. 1). Thus, if the crack plane is close to the substrate (D_1/D close to unity) there will hardly be any plasticity at all that could lead to resistance curve behaviour and therefore $|K|_{\mathrm{SS}}/K_0 \simeq 1$. Indeed, the crack closest to the substrate, $D_1/D = 1/2$, considered in Fig. 8, results only in a small increase of the fracture toughness above K_0. As the crack plane approaches the upper material the fracture toughness ratio increases, but even for $D_1/D = 0$ the ratio is not quite as high as that corresponding to a sharp interface, in contrast to the results shown in the three previous figures. When comparing with results in [2] for a sharp interface between an elastic–plastic material and an elastic substrate it is noted that a large elastic mismatch, as that in Fig. 8, results in a significantly larger increase of the fracture toughness ratio than that corresponding to the smaller elastic mismatches in Figs. 5–7. The additional curve included in Fig. 8, for $D_1/D = -1/2$, represents a crack growing fully inside material No. 1, parallel to the graded layer. This gives a small increase of the minimum relative to that for $D_1/D = 0$, but more noticeable is the fact that the minimum is somewhat shifted towards a small negative value of ψ_0.

4. Discussion

The approximation of prescribing the boundary conditions on the outer edge of the region analysed according to the singularity fields for a sharp interface between dissimilar elastic materials is based on the assumption that the outer radius A_0 of the region is much larger than the thickness D of the functionally graded layer. Then, in the outer part of the region the difference between the solution for the graded interface and that for a sharp interface will be hardly noticeable at all. In the present computations $A_0/D = 5000$ is the smallest value applied, which is considered sufficiently large. At the same time, to satisfy the small scale yielding assumptions, it has been checked that the plastic zone size never reaches a value near $A_0/10$.

The results here for different locations of the crack plane in the graded layer show that the relative toughness increase, $|K|_{\mathrm{SS}}/K_0$, is very sensitive to the location. A smaller plastic region when the crack grows near the material with the larger yield stress results in less toughness increase. But this conclusion depends on the plastic zone size compared to the graded layer thickness. If the plastic zone size is much larger than D, the plastic zone will only be slightly affected by different locations of the crack plane in the graded layer, and therefore the toughness increase will only be slightly affected. The results in Figs. 5 and 6 are obtained for identical material parameters, apart from a four times smaller layer thickness D in Fig. 5, and it is clearly seen that the smaller layer thickness for fixed reference plastic zone size gives less variation of the fracture toughness increase with the value of D_1/D.

Whether or not an assumed grading of material properties is realistic in a layer joining dissimilar materials depends on available processing methods, as discussed in [7]. One kind of grading is that obtained from a graded metal ceramic composite composition in an interlayer between a metal and a ceramic, as has been studied in [8]. The material parameters in Fig. 8 have been chosen such that they could represent this type of graded interface. Another kind can be obtained by joining different metals, as e.g. a ferritic steel joined to an austenitic steel by explosion cladding [10], which will give a plasticity mismatch but no mismatch of elastic properties, as in the case of Fig. 7.

As in previous crack growth analyses based on cohesive zone models [1,2,15] it has been assumed here that the growing crack remains in the initial crack plane. In practice this would only be true if the crack growth resistance is smaller along this plane. In a more general study where possible deviations of the crack path from the initial crack plane are to be determined, it would be necessary to know the dependence of the fracture process parameters on deviations from straight ahead crack growth, which is not straightforward in the case of plastic failure mechanisms. In addition, the crack growth direction would depend on the current mode mixity locally at the crack tip.

In the present investigation the material properties have been taken to vary linearly through the graded layer. However, other distributions are possible, and this could have a noticeable influence on the toughness of the joint, as has been found for an elastic crack problem in [9], by considering various elastic property distributions in a graded interface layer.

References

[1] Tvergaard V, Hutchinson JW. The influence of plasticity on mixed mode interface toughness. J Mech Phys Solids 1993;41:1119–35.

[2] Tvergaard V. Resistance curves for mixed mode interface crack growth between dissimilar elastic–plastic solids. J Mech Phys Solids 2001;49:2689–2703.

[3] Cao HC, Evans AG. An experimental study of the fracture resistance of bimaterial interfaces. Mech Mater 1989;7:295–304.

[4] Liechti KM, Chai YS. Asymmetric shielding in interfacial fracture under in-plane shear. J Appl Mech 1992;59:295.

[5] O'Dowd NP, Stout MG, Shih CF. Fracture toughness of alumina/niobium interfaces: experiments and analyses. Phil Mag A 1992;66:1037.

[6] Reiter T, Dvorak GJ, Tvergaard V. Micromechanical models for graded composite materials. J Mech Phys Solids 1997;45:1281–302.

[7] Mortensen A, Suresh S. Functionally graded metals and metal–ceramic composites: Part 1 Processing. Int Mater Rev 1995;40:239–65.

[8] Suresh S, Mortensen A. Functionally graded metals and metal–ceramic composites: Part 2 Thermomechanical behaviour. Int Mater Rev 1997;42:85–116.

[9] Erdogan F. Fracture mechanics of functionally graded materials. Compos Eng 1995;5:753–70.

[10] Kim AS, Suresh S, Shih CF. Plasticity effects on fracture normal to interfaces with homogeneous and graded compositions. Int J Solids Struct 1997;34:3415–32.

[11] England AH. A crack between dissimilar media. J Appl Mech 1965;32:400–2.

[12] Rice JR. Elastic fracture mechanics concepts for interfacial cracks. J Appl Mech 1988;55:98–103.

[13] Needleman A. A continuum model for void nucleation by inclusion debonding. J Appl Mech 1987;54:525–31.

[14] Tvergaard V. Effect of fibre debonding in a whisker-reinforced metal. Mater Sci Eng A 1990;125:203–13.

[15] Tvergaard V, Hutchinson JW. The relation between crack growth resistance and fracture process parameters in elastic–plastic solids. J Mech Phys Solids 1992;40:1377–97.

[16] Hutchinson JW. Finite strain analysis of elastic–plastic solids and structures. In: Hartung RF, editor. Numerical Solution of Nonlinear Structural Problems. New York: ASME; 1973. p. 17.

[17] Tvergaard V. Effect of thickness inhomogeneities in internally pressurized elastic–plastic spherical shells. J Mech Phys Solids 1976;24:291–304.

[18] Rice JR. A path independent integral and the approximate analysis of strain concentration by notches and cracks. J Appl Mech 1968;35:379–86.

[19] Eshelby JD. In: Kanninen MF et al., editors. Inelastic Behaviour of Solids. New York: McGraw-Hill; 1970. p. 77–115.

PERGAMON

Engineering Fracture Mechanics 69 (2002) 1647–1665

Engineering Fracture Mechanics

www.elsevier.com/locate/engfracmech

R-curve behavior in alumina–zirconia composites with repeating graded layers

Robert J. Moon [a,*], Mark Hoffman [a], Jon Hilden [b], Keith J. Bowman [b], Kevin P. Trumble [b], Jürgen Rödel [c]

[a] *School of Materials Science and Engineering, The University of New South Wales, Sydney, NSW 2052, Australia*
[b] *School of Materials Engineering, Purdue University, West Lafayette, IN 47907, USA*
[c] *Institute of Materials Science, University of Technology, Darmstadt, D-64287 Darmstadt, Germany*

Received 19 February 2001; received in revised form 14 January 2002; accepted 18 January 2002

Abstract

The single-edge-V-notched-beam testing geometry was used to measure the crack growth resistance (R-curve) behavior of multilayer graded alumina–zirconia composites for crack extensions parallel to the graded direction. Fracture mechanics weight function analysis was applied to explain the R-curve behavior of a compositional and grain-size graded microstructure. The results were then used to differentiate the influence of residual stress from other closure stresses, attributed to crack bridging, on the measured R-curve behavior.
© 2002 Elsevier Science Ltd. All rights reserved.

Keywords: Bridging stresses; Functionally graded material; R-curve; Residual thermal stresses; Weight function

1. Introduction

The concept of a compositional gradient within a material to attain specific functional properties has resulted in the development of functionally graded materials (FGM). This concept is not a new one, for example, case hardening of metals is a long-used process involving a microstructural and consequently hardness gradient. Recent research in this field has concentrated upon systems that involve stiffness and toughness gradients, which is usually a consequence of changes in other functional properties, e.g. refractoriness.

Ceramic coatings are usually considered brittle. However, a range of structural ceramics have enhanced toughness because interlocking grains bridge the crack in the region behind the crack tip leading to crack growth resistance or R-curve behavior [1–5]. Alternatively, by layering different structural ceramic materials the apparent fracture toughness can be increased through either enhancement of toughening mechanisms or residual stress effects [5–11]. Various processing techniques may also lead to a microstructural gradient

---

* Corresponding author.
E-mail address: moon@materials.unsw.edu.au (R.J. Moon).

within ceramic layers or coatings which is usually associated with a change in grain size and, hence presumably, a change in the crack growth resistance behavior [12,13].

A theoretical investigation by Jin and Batra [14] calculated the *R*-curve behavior of cracks extending parallel to the graded direction in a ceramic–metal FGM. The calculations showed that there was a "strong" *R*-curve behavior when a crack grows from the ceramic rich region toward the metal-rich region. The results of the calculations were confirmed by experiments conducted on alumina/aluminum (ceramic/metal) gradient structure [15] where *R*-curves were measured for cracks extending parallel to the graded direction, from the ceramic rich to metal rich regions in the gradient. Additionally, several compositional profiles were considered which resulted in different crack growth resistance behavior, demonstrating that the bridging by metallic ligaments of different graded volume fraction can affect the *R*-curve behavior. A special weight function analysis was used which accounted for the notable elasticity gradient in the sample [16]. However, for an elasticity difference across the graded region of $E_1/E_2 = 1.33$, the discrepancy in stress-intensity factors that were calculated using a weight function which considered an elasticity gradient and one which did not was <10% [15]. These works demonstrate that the effects of crack growth resistance toughening by crack bridging, attributable to the compositional change, appear to have a far greater influence upon the crack growth behavior than the resultant elasticity gradient.

A well recognized effect of producing composite materials with different thermal expansion behavior is that residual stresses form within the composite. This is especially significant when high temperatures are used during processing. The measured fracture toughness and crack growth resistance behavior, which depend on both microstructure and residual stress, become functions of position within composites having microstructural or macroscopic residual stress distributions [6–11,17].

When measuring the *R*-curve behavior of either multilayered or gradient composites, separating microstructure-related toughening mechanisms (i.e. crack bridging, kinking, transformation toughening, etc.) from residual stress based mechanisms is difficult. It has already been demonstrated that the weight function analysis may be used to differentiate the influence of a macroscopic residual stress distribution (stepwise change in residual stress) from that of other microstructural mechanisms on the measured *R*-curve behavior of a layered alumina–zirconia composite [10]. Experimental results showed that the macroscopic residual stress distribution significantly influenced the measured *R*-curve behavior.

The purpose of the present study is to distinguish the effects of residual stress distributions from the crack bridging closure stresses on the measured *R*-curve behavior of composites containing multiple graded layers. In this work the alumina–zirconia system is considered.

2. Experimental procedure and analysis

2.1. Sample preparation

Multilayered alumina–zirconia composites were produced by sequential centrifugal consolidation [12,13,18,19] of coagulated [12,20] aqueous alumina–zirconia slurries. Further details of the centrifugal procedure used can be found elsewhere [21]. The composite green bodies were dried and then fired at 1600 °C for 4 h in air with 250 °C/h heating and cooling ramp rates. Two sintered discs were produced measuring 60 mm diameter by 10 mm thick and containing <5% porosity.

Sintered discs were surface-ground flat using a 600 grit diamond wheel and multiple bend bars measuring 4 mm × 3 mm × 35 mm were then cut from the center of each disc. A V-notch was cut across the 3 mm × 35 mm face, perpendicular to the length of the bend bar as described in previous work [9,22]. The V-notch radii, ρ, were measured to be 5–10 μm using optical microscopy. The 4 mm × 35 mm side-surfaces were additionally polished to 15 μm diamond abrasives to facilitate observation of crack growth.

Table 1
List of samples tested

Sample	Layer thickness (µm)	Testing orientation	Sample cross-section (mm)		V-notch depth (µm)
			B	W	a_0
Monolithic	–	–	3.01	4.02	1050
1	~700	1	2.97	4.02	920
2	~700	1	2.96	4.03	935
3	~700	2	2.97	4.03	1045
4	~700	2	3.03	4.02	1340

Both monolithic (homogeneous properties) and graded samples were produced (Table 1). The monolithic sample, having a composition of 80 vol.%-alumina–20 vol.%-zirconia (80Al), was used to measure the R-curve behavior without the influence of bulk thermal residual stresses or grain size variations. The graded samples (1,2,3 and 4) were produced by centrifuging slurries having a solids composition of 76 vol.% alumina [1] + 19 vol.% zirconia [2] + 5 vol.% platelike alumina. [3] During centrifugation, the larger platelike alumina particles (average particle diameter, $D_{50} = 9$ µm) preferentially settled, producing a microstructural gradient as seen in Fig. 1(a). The ASTM E562 standard point counting method (900 points per location) was used to determined the composition and porosity, while the ASTM E112-96e1 standard (duplex microstructure line intercept method) was used to estimate the average grain size. The results of these calculations are shown in Fig. 1(b) and (c), respectively. The Young's modulus, $E(x)$, Poisson's ratio, $v(x)$, and coefficient of thermal expansion (CTE), $\alpha(x)$, as a function of position within the layer were estimated by taking the geometric average of the upper (equal strains) and lower (equal stress) bounds of the rule-of-mixtures models using the material properties shown in Table 2. Fig. 1(d) shows the results of this calculation for Young modulus and CTE across a single layer.

The graded samples tested in this investigation consisted of ~6 layers, a macrograph of a bend bar side surface is shown in Fig. 2(a). The R-curve behavior was tested in two different orientations with respect to the gradient within the layers. In orientation 1 (samples 1 and 2), the cracks were extended in the particle settling direction, i.e. from small to large grain size, and in orientation 2 (samples 3 and 4) the cracks were from large to small grain size.

2.2. Mechanical testing

Direct observations of crack initiation and extension were made on the bend bar side surfaces using a specialized four-point bend fixture placed on the stage of an optical microscope [9,21,23]. Loading was achieved using a piezoelectric translator and measured with a miniature load cell. A fluorescent dye penetrant (Met-L-Clek FP 90, Helling KG, GMBH, Hamburg, Germany) and an ultraviolet light source were used to measure crack lengths. The fluorescent dye was in contact with the V-notch tip during crack initiation and was within the crack during every subsequent crack extension. For slow crack extensions, the fluorescent dye immediately penetrated into the newly extended crack allowing crack extensions to be observed. There was no measurable influence of the fluorescent dye on crack extension [24].

Specimens were tested under displacement control where subcritical crack initiation and further crack extensions (~10 µm increments) were achieved by loading at a slow rate. The incremental loading technique used is described elsewhere [9,21]. The applied load to instigate crack propagation, P_c, and total flaw length,

---

[1] A16SG, 99.8%, $D_{50} = 0.4$ µm, ALCOA, Bauxite, AR.
[2] CEZ-12, 98%, $D_{50} = 0.4$ µm, American Vermiculite Corp., Marietta, GA.
[3] PWA 9, $D_{50} = 9$ µm Fujimi America Inc., Wilsonville, OR.

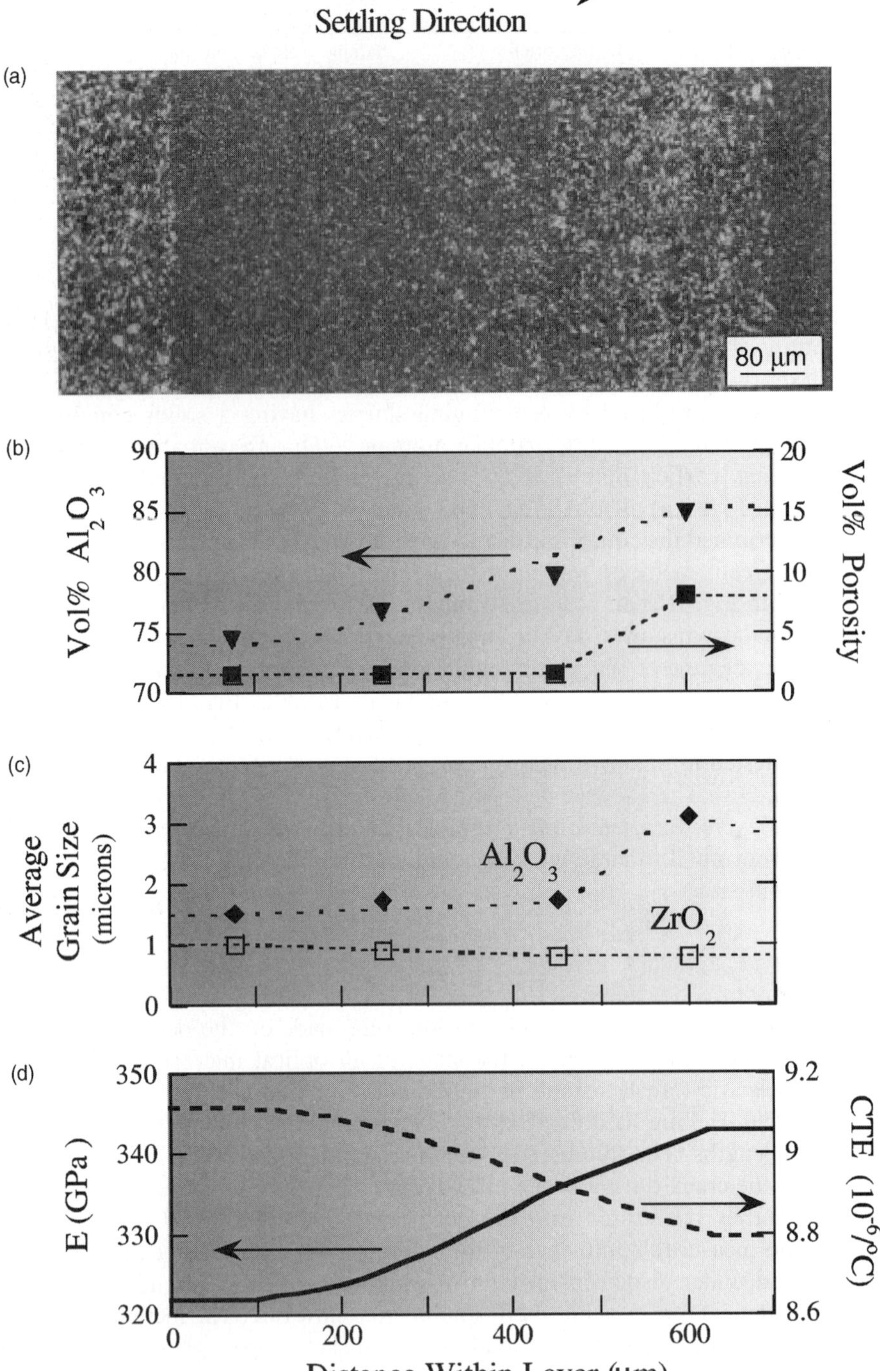

Fig. 1. Orientation 1: The graded properties within each layer: (a) optical micrograph, (b) measured alumina vol.% (▼) and porosity (■), (c) the average grain size, and (d) estimated Young's modulus (—) and CTE (- - -). Note in (b) and (c) that the symbols mark measured data while the plotted curves are the estimated profiles used.

Table 2
Material properties of composite constituents

Material	E (GPa)	v	α $(10^{-6}/°C)$
Al_2O_3	380[a]	0.25[b]	8.39[a]
$t\text{-}ZrO_2$	205[a]	0.32[a]	11.5[a]

[a] Ref. [39].
[b] Ref. [40].

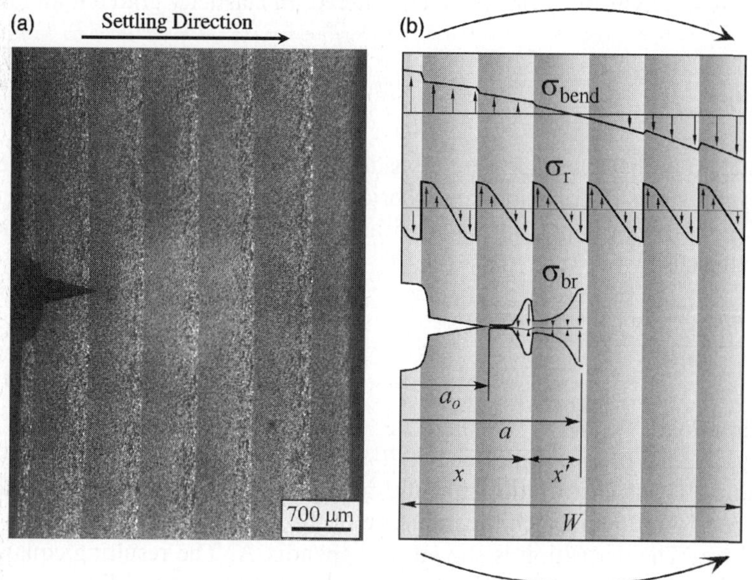

Fig. 2. Orientation 1: (a) Optical micrograph of the post-cracked single-edge-V-notched-beam sample showing the layer stacking with respect to the V-notch. (b) A schematic of the three independent stress distributions acting within a specimen: bending $(\sigma_{bend}(x))$, residual thermal $(\sigma_r(x))$, and bridging stress $(\sigma_{br}(x,a))$.

a, were measured and for each crack extension the apparent fracture toughness, K_R, was calculated as outlined below.

2.3. Weight function analysis

Bueckner [25] showed that the stress-intensity factor for an edge crack of depth a can be calculated by intergrating over the crack length the product of a weight function, $h(x,a)$, and any stress distribution $\sigma(x)$ acting normal to the fracture plane:

$$K = \int_0^a h(x,a)\sigma(x)\,dx \tag{1}$$

where x is the distance along the crack measured from the surface. The weight function used must be derived for a specific crack-component configuration.

In the work of Fett and Munz [26–28] a weight function for a single-edge-notched-beam (SENB) sample and notch geometry is given by:

$$h(x,a) = \sqrt{\frac{2}{\pi a}} \frac{1}{\sqrt{1 - \frac{x}{a}\left(1 - \frac{a}{W}\right)^{1.5}}} \left[\left(1 - \frac{a}{W}\right)^{1.5} + \sum A_{\nu\mu}\left(1 - \frac{x}{a}\right)^{\nu+1}\left(\frac{a}{W}\right)^{\mu}\right] \tag{2}$$

where a is the total flaw length measured from the bend bar tensile surface and W is the sample width as shown in Fig. 2(b). The values of the coefficients $A_{\nu\mu}$ and the exponents ν and μ are given in Refs. [26,27]. Note that for small changes in modulus across the graded region that the decrepancy in stress-intensity factors calculated using a weight function which considered an elasticity gradient and one which did not is <10% [15,16].

2.4. Stress distribution

Three independent stress distributions were considered in this study: an applied bending stress distribution, residual thermal stress distribution, and a bridging stress distribution. The applied bending stress distribution for the monolithic sample, $\sigma^{h}_{\text{bend}}(x,P)$, was estimated using a standard bending-stress formula for homogeneous materials:

$$\sigma^{h}_{\text{bend}}(x,P) = \frac{1.5P(S_{o} - S_{i})}{BW^2}\left(1 - \frac{2x}{W}\right) \tag{3a}$$

where P is the applied load, S_{o} and S_{i} are the outer and inner support spans of the four-point bending fixture (in this study: 20 and 10 mm respectively), B is the sample thickness, and W is the sample width.

The applied bending stress distribution for the graded samples, $\sigma^{g}_{\text{bend}}(x,P)$, was estimated using a modified bending-stress formula in which the influence of the elasticity variation across the bend bar cross-section is accounted for. The derivation is shown in Appendix A. The resulting equation can be used for any elasticity gradient and sample size for samples tested in four-point bending:

$$\sigma^{g}_{\text{bend}}(x) = \frac{P}{4B}(S_{o} - S_{i})E'(x)\frac{E_2 - E_1 x}{E_2^2 - E_1 E_3} \tag{3b}$$

where $E'(x)$ is the plane strain variation of elasticity across the sample cross-section and E_1, E_2 and E_3 are given in Appendix A.

The residual thermal stress distribution within the graded composites, $\sigma_{r}(x)$, was estimated from the thermal expansion mismatch strains resulting from the variations of $E(x)$, $\nu(x)$, $\alpha(x)$ across the sample. The derivation is shown in Appendix A.

$$\sigma_{r}(x) = \Delta T E'(x)\left[\alpha(x) + \frac{-A_1 E_2 x + A_2 E_1 x + A_1 E_3 - A_2 E_2}{E_2^2 - E_1 E_3}\right] \tag{4}$$

where A_1, A_2, E_1, E_2 and E_3 are given in Appendix A. A $\Delta T = 1225\ °C$ was used for the temperature range over which the residual thermal stress develops. When cooling from the sintering temperature (1600 °C) to ~1250 °C, the thermal stresses developed within the sample were believed to be alleviated via creep [29,30].

The bridging stress, σ_{br}, for the samples tested in the current study was believed to be caused by frictional pullout of interlocking grains, which resulted from the predominantly intergranular fracture mode [9,10].

X-ray diffraction of the fracture surface did not detect phase transformation of the 15–25 vol.% tetragonal zirconia [24], thus, transformation toughening was assumed not to contribute to the crack closure stresses.

The bridging stress distribution that acts along a given crack length will be dependent on the crack opening displacement (COD) along the crack length. For homogeneous materials the bridging stress distribution, $\sigma_{br}(u)$, can be considered to have an empirical power-law form similar to that which has been used previously [1–4,31–33]:

$$\sigma_{br}^{h}(u) = \sigma_{max}\left(1 - \frac{u}{u^*}\right)^n \tag{5}$$

where $2u$ is the COD, σ_{max} is the maximum stress supported by the bridging zone, n is a softening coefficient, and $2u^*$ is the critical COD* in which closure stresses resulting from interlocking grains cease contributing to the bridging stress.

The weight function calculation used in this study required that the bridging stress function in Eq. (5) be modified so that the bridging stress was a function of distance behind the crack tip rather then a function of COD. The COD was related to the distance behind the crack tip using the Irwin K-field plane-strain displacement relation, $u(x')$, developed by Barenblatt [34] for homogeneous materials:

$$\text{COD} = 2u(x') = 2\left[\left(\frac{8x'}{\pi}\right)^{1/2}\frac{K_o}{E/(1-v^2)}\right] \tag{6}$$

where x' is the distance behind the crack tip and K_o is the intrinsic stress-intensity factor that represents a lower bound for COD [4]. The COD estimate was simplified by using $K_o = 3.6$ MPa m$^{1/2}$ and $E = 335$ GPa, as obtained from the 80Al composition monolithic sample. Wake effects resulting from crack bridging will in fact reduce the COD, however, due to the complexity of estimating these effects for graded materials, they were not considered.

Eq. (5) was modified to be a function of distance behind the crack tip by substituting in Eq. (6) for both u and u^*. Where, for the u substitution, x' was replaced with $a - x$, and for the u^* substitution, x^* was replaced with L. The resulting equation relates the bridging stress as a function of distance behind the crack tip, $\sigma_{br}^{h}(x, a)$:

$$\sigma_{br}^{h}(x, a) = \sigma_{max}\left(1 - \left(\frac{a-x}{L}\right)^{1/2}\right)^n \tag{7}$$

where $a - x$ is the distance from the crack tip and L is the steady-state bridging zone length, i.e. the critical distance behind the crack tip in which the COD is large enough that closure stresses resulting from interlocking grains are zero, x^*. Values of $\sigma_{max} = 19$ MPa, $n = 0.6$, and $L = 900$ μm were determined from experimentally measured data for a 80Al composition monolithic sample, which has been described previously [10].

For the graded samples the changing composition and grain size will influence the bridging stress distribution acting along a crack, and Eqs. (5) and (7) were modified to account for such changes. The bridging stress distribution was believed to be dominated by the changing grain size, whereas the compositional change of ~10 vol.% Al$_2$O$_3$ across the graded region (Fig. 1(b)) was assumed not to have an influence. The later assumption was based on the observation that the initial R-curve (up to 300 μm) of the 90 vol.%-alumina–10 vol.%-zirconia (90Al) layered sample reported in Ref. [21] was accurately estimated with values of σ_{max}, n, and L calculated from the 80Al monolithic sample.

The influence of grain size on σ_{max} and n is unclear, thus, were considered independent of grain size. Steinbrech et al. [3] showed that R-curves predicted with fixed σ_{max} and n values could accurately estimate the experimentally measured R-curve for alumina samples having 4, 9 or 16 μm grain sizes, suggesting that

$\sigma_{\max}$ and n are not functions of grain size. However, Sohn et al. [32] have reported that n is dependent upon the grain size distribution within the material and that $\sigma_{\max}$ was also influenced by the grain size.

The influence of the changing grain size, $GS(x')$, on u^* or L, was believe to dominate the bridging stress distribution acting along the cracks in the graded samples. It is recognized that a large grain size requires greater CODs before separation of bridging grains occurs [3,33]. This critical COD* can be estimated if one considers that a single grain can only contribute to the bridging toughening mechanism if the COD at this given grain is less than one half the grain size, thus, $2u^*(GS) = GS/2 = COD^*$. For the case where the grain size changes with position along the crack the critical COD* that is necessary to end the bridging stresses for each x' location along the crack will be different. The bridging stress distribution as a function of grain size, $\sigma_{br}^{g}(GS(x'))$, is given by:

$$\sigma_{br}^{g}(GS(x')) = \sigma_{\max}\left(1 - \frac{u(x')}{\dfrac{GS(x')}{4}}\right)^{n} \tag{8}$$

where x' is the distance behind the crack tip. Eq. (8) was modified to make it a function of distance behind the crack tip rather than a function of COD. Initially, Eq. (6) was set equal to $GS/2$, solved for x', where x' being the critical distance behind the crack tip in which complete grain separation occurs for a given grain size was renamed to, x^*:

$$x^* = \frac{\pi}{8}\left(\frac{E/(1-v^2)}{K_o}\frac{GS}{4}\right)^{2} \tag{9}$$

If the material had a homogeneous grain size this critical crack length would be the "steady-state" bridging zone length L. However, for composites with graded grain size regions there is no steady-state bridging zone length and Eq. (9) becomes a function of distance behind the crack tip, $x^*(x')$. For each location along the crack length the local grain size, $GS(x')$, influences the size of the bridging stress that will be applied at that location. Substituting $x^*(x')$ into Eq. (7) for L and replacing x' with $a - x$ the resulting equation calculates the bridging stress as a function of distance behind the crack tip:

$$\sigma_{br}^{g}(GS(a-x)) = \sigma_{br}^{g}(x,a) = \sigma_{\max}\left(1 - \left(\frac{a-x}{x^*(a-x)}\right)^{1/2}\right)^{n} \tag{10}$$

Fig. 2(b) shows the distribution of these three stress contributions, σ_{bend}^{g}, σ_{r} and σ_{br}^{g}, within a graded sample loaded according to orientation 1.

2.5. Measured stress-intensity factors

The weight function analysis was used to calculate the apparent stress-intensity factor, K_R, from experimental data: the critical applied load to further extend a crack, P_c, and the average total flaw length, $a = (a_A + a_B)/2$, measured on both sides (A and B) of the sample. Note that a_A and a_B are the sum of the V-notch depth, a_0, and the crack length as measured on sides "A" and "B" of the bend bar [9].

For the monolithic sample a standard bending-stress formula, Eq. (3a), was substituted into Eq. (1) and for each P_c vs. a data obtained from the experiment, the apparent stress-intensity factor, K_R, was calculated. The procedure for the gradient samples was much the same as for the monolithic sample but the modified bending-stress formula, Eq. (3b), was substituted into Eq. (1) to account for the influences of the changing modulus across the sample.

2.6. Calculated stress-intensity factors

The stress-intensity factors associated with each stress distribution were independently defined using the weight function analysis. The applied stress-intensity factor for the monolithic sample, $K_a^m(x, P)$, was calculated with Eq. (3a) substituted in Eq. (1), whereas for the graded samples, $K_a^g(x, P)$, was calculated with Eq. (3b) substituted in Eq. (1). The stress-intensity factor resulting from the residual stress distribution acting along the total flaw length, $K_r(x)$, was calculated with Eq. (4) substituted in Eq. (1). Note that the range of integration was from 0 to a.

The stress-intensity factor resulting from the bridging stress distribution acting along the crack length for the monolithic sample, $K_{br}^m(x, a_0)$, was calculated with Eq. (7) substituted in Eq. (1), and for the graded samples the stress-intensity factor, $K_{br}^g(x, a_0)$, was calculated with Eq. (10) substituted in Eq. (1). Bridging stresses are considered to only act along the extended crack length, i.e. the region defined by $a_0 < x < a$, and thus the range of integration was from a_0 to a.

2.7. Superposition of stress-intensity factors

It is of particular interest to ascertain the effects of microstructural stresses due to crack bridging and residual stress upon the ultimate applied stress-intensity factor for crack propagation. To this end, the principle of superposition was used to sum K_a, K_r, and K_{br} resulting in the crack tip stress-intensity factor, K_{tip}, which is shown below:

$$K_{tip} = K_a + K_r + K_{br} \tag{11}$$

For estimating the stress-intensity factor for crack propagation as a function of total flaw length, a, Eq. (11) was solved for K_a, which was relabeled to K_R. Note that K_{tip} was replaced with the intrinsic crack tip toughness, K_o ($K_{tip} = K_o$ for crack extension). The resulting equation is shown below:

$$K_R(a, a_0) = K_o(a) - K_r(a) - K_{br}(a, a_0) \tag{12}$$

For the monolithic sample, $K_o(a)$ was estimated by extrapolating the measured R-curve to the y-axis (located at a_0), resulting in $K_o^{80Al} = 3.6$ MPa m$^{1/2}$. The stress intensity obtained using this technique is believed to give a reasonable estimate for K_o because the effects of crack tip shielding would be minimized. Since the monolithic sample has a continuous composition, K_o remained constant for all values of a.

In an attempt to account for the modulus variations in the graded samples on the estimated crack tip toughness, $K_o(a)$, a procedure introduced by Lakshminarayanan et al. [6] was used. First, $K_a^g(a, P)$ was set equal to K_o for the corresponding composition within the specimen. Then the critical applied bending load necessary to cause crack extension, P_c, as a function of total flaw length, a, was calculated from this equality. The P_c vs. a results, incorporating the influence of the modulus variations, were then used in a standard formula for stress intensity for a SENB specimen [22,35] to calculate the $K_o(a)$ profile.

The crack tip toughness in the graded samples was assumed to vary as a function of composition across the layer where, for the alumina richer region (85Al), $K_o^{85Al} = 3.4$ MPa m$^{1/2}$, and for the alumina poor region (75Al), $K_o^{75Al} = 3.8$ MPa m$^{1/2}$. The estimated values were based on extrapolations of the 80Al and 90Al measured R-curves to the y-axis, in which the intrinsic crack tip toughness values of $K_o^{80Al} = 3.6$ MPa m$^{1/2}$ and $K_o^{90Al} = 3.2$ MPa m$^{1/2}$ were obtained, respectively [10,21].

The $K_r(a)$ and $K_{br}(a, a_0)$ profiles, being independent of the applied load, were simply calculated for several values of a.

3. Results

3.1. Bridging function

The bridging function for the graded samples was intended to account for the variation in crack growth resistance behavior resulting from the grain size gradient within each layer. The bridging stress distribution that acts along a given crack length will be dependent on the difference between the COD along the crack length and the critical COD*. Fig. 3(a) shows the COD profile for a homogeneous material as a function of distance behind the crack tip (Eq. (6)), and superimposed on this plot is the critical COD* for a homogeneous material with an average grain size of 1.9 μm (representative of the Al_2O_3 grain size for the 80Al monolithic sample). Note that the crack tip is located at the far right of the figure. For distances behind the crack tip where the critical COD* is greater than the COD of the crack, the grain size is large enough to bridge the crack and provide closure stresses. The example shown in Fig. 3(a), for the monolithic, sample suggests that bridging stresses act along the crack until ~900 μm behind the crack tip, and for further distances behind the crack tip, the COD is sufficiently large that crack bridging cannot occur.

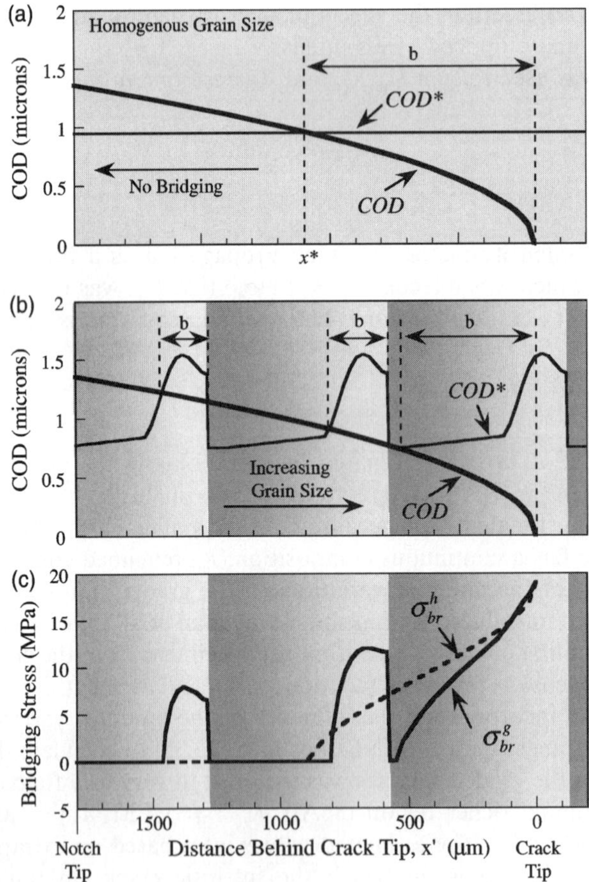

Fig. 3. The location behind the crack tip in which the microstructure can bridge the crack and apply closure stresses (regions labeled "b") for (a) the monolithic sample and (b) the graded sample in orientation 1. (c) The estimated bridging stress distribution acting along the 1800 μm crack length for the monolithic sample (- - -) and the graded sample (—).

The critical COD*, for graded grain sized materials will be directly related to the variation of grain size (Fig. 3(b)). For the graded samples investigated in this study the larger grain size in the platelike particle region of each layer results in a larger critical COD* and thus these regions can continue to contribute to the bridging stresses for much longer distances behind the crack tip (regions "b" in Fig. 3(b)).

The bridging stress distribution, $\sigma_{\mathrm{br}}(x, a)$, resulting from the monolithic and graded sample (sample 1) is shown in Fig. 3(c). For this figure the total flaw length was 2720 µm (920 µm notch depth + 1800 µm crack length), and the graph shows the bridging stress distribution acting along the crack as a function of distance behind the crack tip. For the monolithic sample the bridging stress distribution falls off continuously as a function of distance behind the crack tip and reduces to zero at $x^* \approx 900$ µm ($L = 900$ µm in Eq. (7)). For the graded sample the resulting stress distribution profile, as a function of distance behind the crack tip was directly related to −ve variation of the critical COD*. Within the platelike particle regions the bridging stresses are still being applied at large distances behind the crack tip where the amount of bridging stress from the platelike particle regions decreases as the distance from the crack tip increases.

3.2. Calculated stress intensity profiles

Fig. 4(a) and (b) show the calculated stress intensity profiles $K_\mathrm{o}(a)$, $K_\mathrm{r}(a)$ and $K_{\mathrm{br}}(a, a_0)$ for the graded samples with a crack extending from the V-notch tip, length a_0, in orientation 1 and 2, respectively. Note that the interface between layers was treated as a 5 µm wide region where K_o changed linearly from 3.4 to 3.8 MPa m$^{1/2}$. This estimate was a simplification, but was considered a reasonable estimate in regards to the focus of this investigation. From Fig. 4, $K_{\mathrm{br}}(a, a_0)$ is seen to reduce the stress at the crack tip for all crack extensions, whereas for $K_\mathrm{r}(a)$, the stress at the crack tip can be reduced or increased depending on the residual stress distribution. Additionally, the extent of influence of each component; $K_\mathrm{o}(a), K_\mathrm{r}(a)$ and $K_{\mathrm{br}}(a, a_0)$ on the resulting $K_R(a, a_0)$ profile can also be seen.

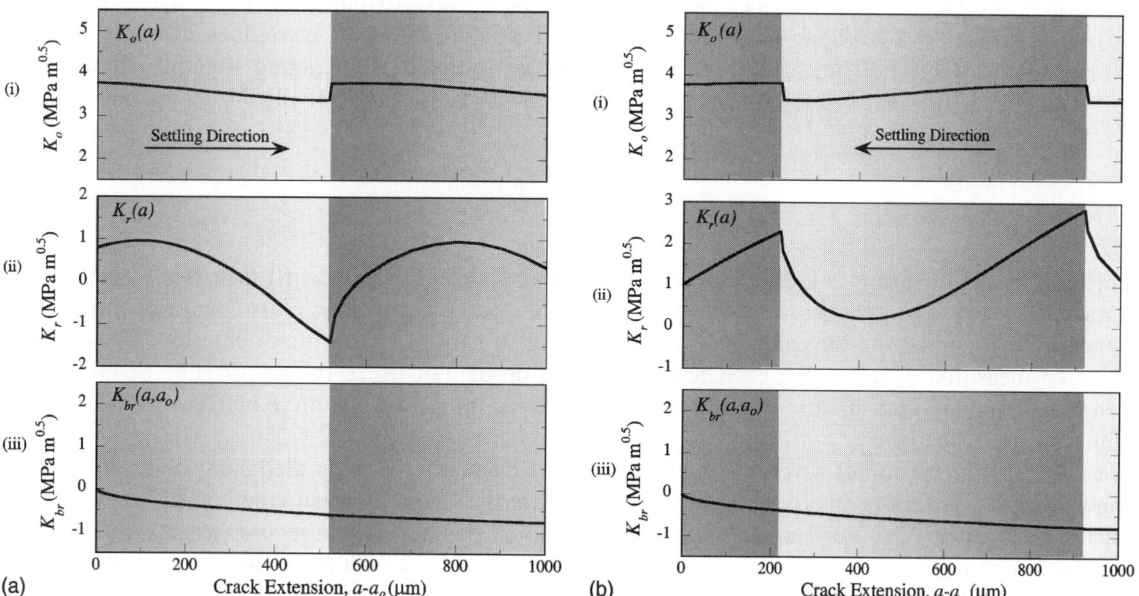

Fig. 4. A schematic of the estimated stress-intensity factors as a function of position within the graded specimen in (a) orientation 1 and (b) orientation 2: (i) K_o, (ii) K_r, (iii) K_{br}.

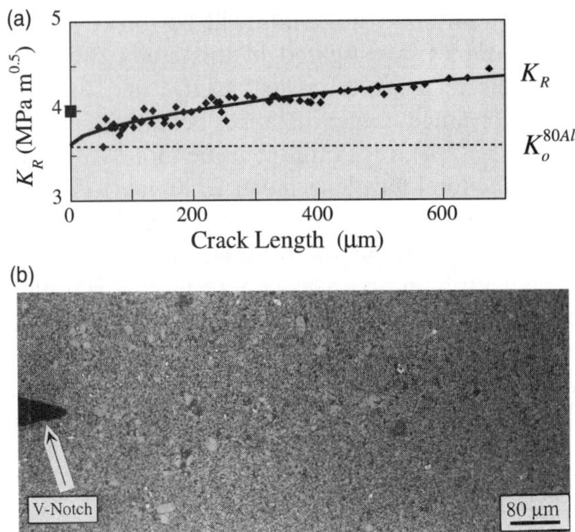

Fig. 5. Monolithic sample (a) The experimentally measured R-curve (♦) and the calculated $K_R(a, a_0)$ and K_o^{80Al} profiles. (b) An optical micrograph of the post-tested sample showing the microstructure that the crack propagated through. The ■ symbol on the K_R-axis represents the K_i for crack initiation from the V-notch tip.

3.3. Monolithic sample R-curve

The microstructure and R-curves for the monolithic sample are shown in Fig. 5. The plot consists of three curves: the measured R-curve, the calculated $K_R(a)$ profile and the estimated intrinsic crack tip toughness, $K_o^{80Al} = 3.6$ MPa m$^{1/2}$. The measured stress intensity for crack initiation from the V-notch, K_i was 4.0 MPa m$^{1/2}$, in which the discrepancy between K_i and K_o^{80Al} was attributed to the relatively blunt ($\rho = 5$ μm) starter notch tip. The R-curve had a shallow rise where a 0.8 MPa m$^{1/2}$ increase in K_R occurred after a $\sim$700 μm crack extension. The calculated $K_R(a, a_0)$ profile, using Eq. (7) bridging stress distribution with $\sigma_{max} = 19$ MPa, $n = 0.6$, and $L = 900$ μm, estimated the R-curve behavior of the 80Al material.

3.4. Graded sample R-curve

The microstructure and R-curves for the graded samples tested in orientation 1 (samples 1 and 2) and in orientation 2 (samples 3 and 4) are shown in Figs. 6 and 7, respectively. The plots consist of three curves: two measured R-curves and the calculated $K_R(a, a_0)$ profile for the given testing orientation. The similarity of the two measured R-curves for each testing orientation demonstrates the consistency of the R-curve measurement technique, while the proposed model describes the general R-curve behavior of both sample configurations.

The direct superposition of several R-curves measured from layered or graded samples on to the same graph can be inappropriate due to the influences of starting the R-curve measurement at different locations within the microstructure. For samples 1 and 2 the V-notch position (R-curve starting position) within a given layer were nearly identical, whereas the sample 3 V-notch was $\sim$20 μm closer to the bottom of the next layer as compared to sample 4. This problem was partially alleviated by lining up the layer interface for both samples through subtracting 20 μm from the measured crack length for each data point of the sample 4 R-curve (for graphing purposes only).

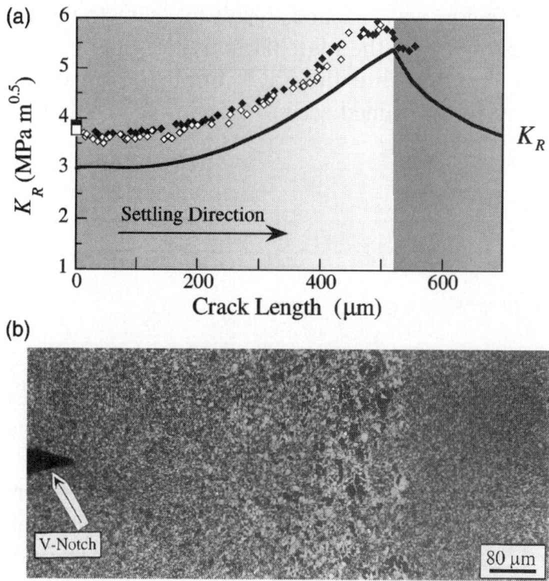

Fig. 6. Orientation 1: (a) The experimentally measured sample 1 (◆) and 2 (◇) *R*-curves and the calculated $K_R(a, a_0)$ profile. (b) A representative microstructure through which the crack extended. The (■ and □) symbols on the K_R-axis represents the K_i for crack initiation from the V-notch tip.

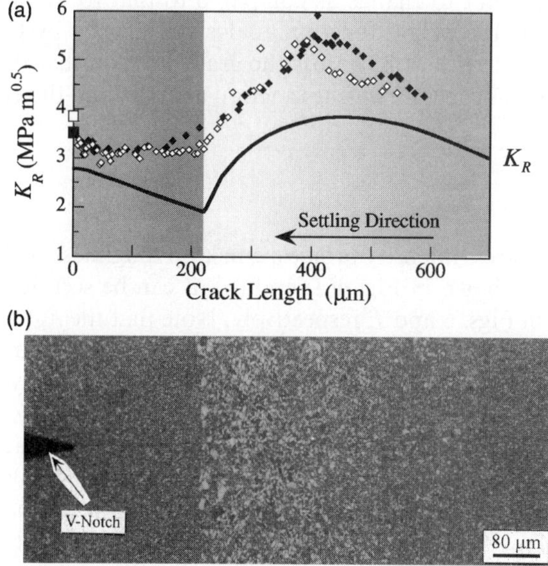

Fig. 7. Orientation 2: (a) The experimentally measured sample 3 (◆) and 4 (◇) *R*-curves and the calculated $K_R(a, a_0)$ profile. (b) A representative microstructure through which the crack extended. The (■ and □) symbols on the K_R-axis represent the K_i for crack initiation from the V-notch tip.

The gradient microstructure resulted in an asymmetry in the measured *R*-curves and in the calculated $K_R(a, a_0)$ profile for orientations 1 and 2. The orientation 1 *R*-curves have a gradual increase in slope until the end of the platelike particle region was reached, resulting from the gradual increase in residual compressive stress and bridging stresses within the platelike particle region of the layer. The orientation 2 R-curves initially have a near flat R-curve until the crack impinged on the bottom of the next layer, the resulting steeper rise in K_R was due to the immediate increase in the residual compressive stress and bridging stresses.

4. Discussion

4.1. Stress-intensity calculation errors

Using a weight function, $h(x, a)$, that was developed for elastically homogeneous samples for calculating the stress intensities for elastically graded samples, will result in errors. However, due to the small elasticity change ($E_1/E_2 = 1.06$) in the graded samples investigated in this study, the extent of theses errors was believed to be <10%, and was considered acceptable.

The studies of Erdogan [36] and Jin and Batra [14,37], demonstrated that the extent of the elasticity change across the graded region significantly influenced the size of the deviation between stress intensity calculations that accounted for the elasticity variations and those that did not. For changes in elasticity of $E_1/E_2 = 1.36$ across the graded region Jin and Batra [14] have shown that the deviation between stress intensity calculations that account for the elasticity variations and the ones that do not to be ~10%.

The weight function that was used in this investigation was developed by Fett and Munz [26–28], and they have recently investigated methods for incorporating the influences of elasticity variations using a specialized weight function analysis [16]. They report that the procedures using the weight function analysis for graded materials were found not to change as compared to that for homogeneous materials [16]. The study by Chung et al. [15], using the weight function analysis developed by Fett and Munz, demonstrated that for a elasticity variation of $E_1/E_2 = 1.33$ across the graded region that the difference in the calculated stress intensities between the homogeneous weight function analysis and the elastic graded weight function analysis was <10%.

4.2. Residual stress influence

By comparing the calculated stress intensity profiles; crack tip toughness ($K_o(a)$), residual thermal ($K_r(a)$) and crack bridging ($K_{br}(a, a_0)$), shown in Fig. 4(a) and (b), it can be seen how each add up to obtain the final $K_R(a, a_0)$ profiles shown in Figs. 6 and 7, respectively. Note that the $K_R(a, a_0)$ profile across the graded region was dominated by the $K_r(a)$ profile having a variation of 2.7 MPa m$^{1/2}$, whereas the $K_o(a)$ stress intensity variation was only 0.4 MPa m$^{1/2}$, and the $K_{br}(a, a_0)$ stress intensity variation was 0.8 MPa m$^{1/2}$ after 700 μm crack extension.

It should be noted that for orientation 1 the $K_R(a, a_0)$ profile underestimated the measured R-curve by ~0.75 MPa m$^{1/2}$ over short crack lengths. This is unexpected since toughening by bridging is small at short crack lengths. For distances <10 μm from the notch tip both curves should have similar K_R values (Fig. 6). This difference was also apparent in samples 3 and 4, which were tested in orientation 2. Incorrect estimates of the elastic moduli and CTE across each layer may account for these deviations and are addressed shortly.

4.3. Bridging stress influence

The resulting bridging stress distribution shown in Fig. 3 demonstrated that the bridging stress relation derived for gradient materials, Eq. (10), could be used to estimate the variable stress distribution associated with a changing grain size and with the distance behind the crack tip. This suggests if $\sigma_{max}(x)$, $n(x)$ and

$x^*(a-x)$ are well estimated, for a particular gradient, that a reasonable estimate of the bridging stress distribution acting along the crack can be calculated and the resulting K_{br} profile will describe the bridging toughening component.

In the current study, the bridging stress distribution between the monolithic sample and the graded samples were significantly different, however, the resulting K_{br} profiles were essentially the same. The bridging stress distribution for the graded samples was not significant enough to account for the discrepancies between the measured R-curve and the calculated $K_R(a, a_0)$ profile. It is possible that $\sigma_{max}(x)$ and $n(x)$ may also vary as a function of position within the gradient, thus, altering the bridging stress distribution. Additionally, it should be noted that the estimated average grain size for the large grain sized regions of the graded layers ($\sim$3.1 μm) is a simplification, as some of the platelike alumina grains were nearly 10 μm long, suggesting that the COD* and $x^*(a-x)$ would actually be larger than the estimates used.

4.4. Influence of material property estimates on K_R

To evaluate the influence of the material properties estimates on the calculated $K_R(a, a_0)$ behavior, a comparison was made between using $E(x)$, $v(x)$, and $\alpha(x)$ estimated by either the upper (equal strain), lower (equal stress), or geometric average of the upper and lower bounds of the rules-of-mixture. These three estimating schemes resulted in different material properties values across the graded region, thus influencing the K_a, K_r, and K_{br} profiles.

Each material property estimating scheme resulted in a different $K_R(a, a_0)$ profile and there was also an influence of the testing orientation. In Figs. 6(a) and 7(a) the $K_R(a, a_0)$ profiles were calculated using material properties that were estimated by the geometric average. For orientation 1 the upper and lower bound estimates resulted in a 0–0.25 MPa m$^{1/2}$ increase in the $K_R(a, a_0)$ profile shown in Fig. 6(a) Whereas in orientation 2 the upper bound estimate resulted in a 0–0.25 MPa m$^{1/2}$ increase in the $K_R(a, a_0)$ profile and the lower bound estimate resulted in a 1–1.25 MPa m$^{1/2}$ increase in the $K_R(a, a_0)$ profile shown in Fig. 7(a). This asymmetry caused by the different sample orientations was primary due to the changes in the residual thermal stress distribution within the tile.

5. Conclusions

1. The addition of a layered microstructure having a grain size and composition gradient within each layer was found to significantly alter the R-curve behavior of alumina–zirconia composites. Additionally, the measured R-curve was influenced by the direction of crack extension with respect to the layered microstructure.
2. The weight function analysis demonstrated that the macroscopic residual stress distribution acting within a specimen can have a significant influence on the measured R-curve behavior.
3. For the samples investigated in this study, the spatial change in grain size along a crack influences the resultant bridging stress distribution, however, the effect upon the final R-curve was not significant.

Acknowledgements

The authors thank Martin Stech and Emil Aulbach for experimental assistance. This work was supported by the United States Army Research Office: MURI grant number DAAH04-96-1-0331 and the Australian Research Council.

Appendix A

A.1. Bending stress distribution

This analysis estimates the bending stress distribution within graded samples tested in four-point bending (Fig. 8). It is assumed that the applied forces in the y direction will be zero, and that there is an applied bending moment as shown below:

$$F = \int_{x_o}^{x_f} \varepsilon(x)E'(x)B\,\mathrm{d}x = 0 \tag{A.1}$$

$$M = \int_{x_o}^{x_f} \varepsilon(x)E'(x)Bx\,\mathrm{d}x = \frac{P}{4B}(S_o - S_i) \tag{A.2}$$

where $E'(x)$ is the elastic modulus corrected for plane strain, B is the sample thickness, P is the applied load, and S_o and S_i are the outer and inner loading span lengths for four-point bending.

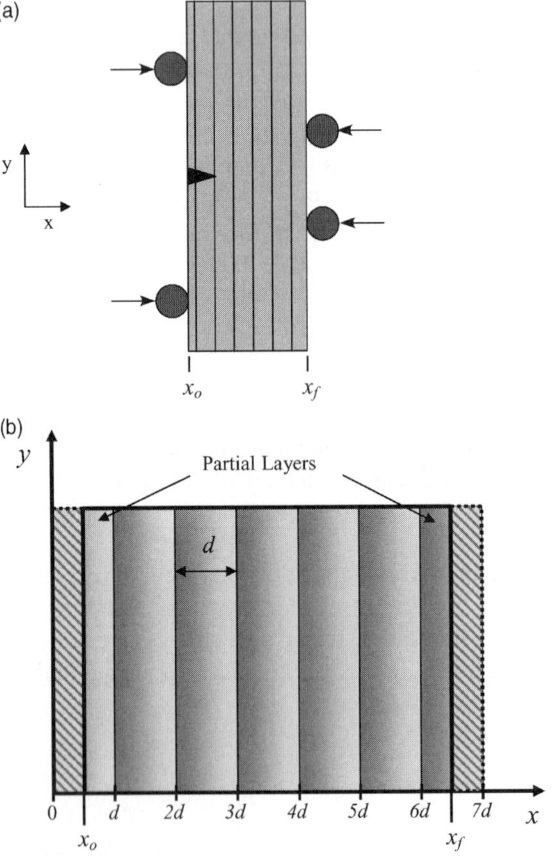

Fig. 8. Schematic bend bar with graded layers. (a) Layer configuration with respect to the four-point bending testing geometry, (b) the layer configuration within the bend bar in which there are five complete layers (thickness d) and two partial layers.

Due to variations in $E(x)$ across the sample the local strains were assumed to vary as:

$$\varepsilon(x) = a_{\mathrm{b}}x + b_{\mathrm{b}} \tag{A.3}$$

where a_{b} and b_{b} are constants. By substituting Eq. (A.3) into Eqs. (A.1) and (A.2), the constants a_{b} and b_{b} are obtained:

$$a_{\mathrm{b}} = -\frac{P}{4B}(S_{\mathrm{o}} - S_{\mathrm{i}})\frac{E_1}{E_2^2 - E_1 E_3} \tag{A.4}$$

$$b_{\mathrm{b}} = \frac{P}{4B}(S_{\mathrm{o}} - S_{\mathrm{i}})\frac{E_2}{E_2^2 - E_1 E_3} \tag{A.5}$$

where

$$E_1 = \int_{x_{\mathrm{o}}}^{x_{\mathrm{f}}} E'(x)\,\mathrm{d}x \tag{A.6}$$

$$E_2 = \int_{x_{\mathrm{o}}}^{x_{\mathrm{f}}} xE'(x)\,\mathrm{d}x \tag{A.7}$$

$$E_3 = \int_{x_{\mathrm{o}}}^{x_{\mathrm{f}}} x^2 E'(x)\,\mathrm{d}x \tag{A.8}$$

Substitute Eqs. (A.4) and (A.5) into Eq. (A.3) and using the relation $\sigma(x) = \varepsilon(x)E'(x)$, a bending stress relation is obtained:

$$\sigma_{\mathrm{bend}}^{\mathrm{g}}(x) = \frac{P}{4B}(S_{\mathrm{o}} - S_{\mathrm{i}})E'(x)\left[\frac{E_2 - E_1 x}{E_2^2 - E_1 E_3}\right] \tag{A.9}$$

A.2. Residual thermal stress distribution

The analysis completed here is similar to the one used by Ravichandran [38]. It is assumed for the graded samples in this investigation that the variations in elastic modulus, Poisson's ratio, and thermal expansion coefficient are periodic (the same in each layer). When the tile is subjected to a temperature change ΔT, the variations in CTE and elastic modulus within each layer will produce variations in local strain. The strain at any point is given by:

$$\varepsilon(x) = \varepsilon^{\mathrm{th}}(x) + \varepsilon^{\mathrm{el}}(x) \tag{A.10}$$

where $\varepsilon^{\mathrm{th}}(x)$ is the thermal contraction, $(\alpha(x)\Delta T)$, $\varepsilon^{\mathrm{el}}(x)$ is the elastic response for equilibrium, $(ax + b)$, which accounts for warping of the tile resulting from non-symmetric variations in $\alpha(x)$ across the sample, and ΔT is positive on cooling. The values, a and b, are constants and must be determined such that the net load and bending moment exerted by the tile are zero. The net load, F, and bending moment, M, are given by:

$$F = \int_{x_{\mathrm{o}}}^{x_{\mathrm{f}}} \varepsilon(x)E'(x)B\,\mathrm{d}x = 0 \tag{A.11}$$

$$M = \int_{x_{\mathrm{o}}}^{x_{\mathrm{f}}} \varepsilon(x)E'(x)Bx\,\mathrm{d}x = 0 \tag{A.12}$$

Solving for a and b one obtains:

$$a = -\Delta T \left(\frac{A_1 E_2 - A_2 E_1}{E_2^2 - E_1 E_3} \right) \tag{A.13}$$

$$b = \Delta T \left(\frac{A_1 E_3 - A_2 E_2}{E_2^2 - E_1 E_3} \right) \tag{A.14}$$

where E_1, E_2 and E_3 are defined by Eqs. (A.6)–(A.8), respectively, and A_1 and A_2 are defined as follows:

$$A_1 = \int_{x_o}^{x_f} \alpha(x) E'(x) \, dx \tag{A.15}$$

$$A_2 = \int_{x_o}^{x_f} x \alpha(x) E'(x) \, dx \tag{A.16}$$

The residual thermal stress relation is obtained by substituting Eqs. (A.13) and (A.14) into Eq. (A.10) and using the relation $\sigma(x) = \varepsilon(x) E'(x)$, to obtain:

$$\sigma_r(x) = \Delta T E'(x) \left[\alpha(x) + \frac{-A_1 E_2 x + A_2 E_1 x + A_1 E_3 - A_2 E_2}{E_2^2 - E_1 E_3} \right] \tag{A.17}$$

References

[1] Mai Y-W, Lawn BR. Crack-interface grain bridging as a fracture resistance mechanism in ceramics: II. Theoretical fracture mechanics model. J Am Ceram Soc 1987;70(4):289–94.

[2] Bennison S, Lawn B. Role of Interfacial grain-bridging sliding friction in the crack-resistance and strength properties of nontransforming ceramics. Acta Metall 1989;37(10):2659–71.

[3] Steinbrech R, Reichl A, Schaarwächter W. *R*-curve behavior of long cracks in alumina. J Am Ceram Soc 1990;73(7):2009–15.

[4] Rödel J, Kelly JF, Lawn BR. In situ measurements of bridged crack interfaces in the scanning electron microscope. J Am Ceram Soc 1990;73(11):3313–8.

[5] Harmer MP, Chan HM, Miller GA. Unique opportunities for microstructural engineering with duplex and laminar ceramic composites. J Am Ceram Soc 1992;75(7):1715–28.

[6] Lakshminarayanan R, Shetty D, Cutler R. Toughening of layered ceramic composites with residual surface compression. J Am Ceram Soc 1996;79(1):79–87.

[7] Marshall D, Ratto J, Lange F. Enhanced fracture toughness in layered microcomposites of Ce–ZrO$_2$ and Al$_2$O$_3$. J Am Ceram Soc 1991;74(12):2979–87.

[8] Rao MP, Sánches-Herencia AJ, Beltz GE, McMeeking RM, Lange FF. Laminar ceramics that exhibit a threshold strength. Science 1999;286:102–5.

[9] Moon R, Bowman K, Trumble K, Rödel J. Fracture resistance curve behavior of multilayered alumina–zirconia composites produced by centrifugation. Acta Mater 2001;49:995–1003.

[10] Moon R, Hoffman M, Hilden J, Bowman K, Trumble K, Rödel J. A weight function analysis on the *R*-curve behavior of multilayered alumina–zirconia composites. J Am Ceram Soc 2002;85(6):1505–11.

[11] Blattner AJ, Lakshminarayanan R, Shetty DK. Toughening of layered ceramic composites with residual surface compression: Effects of layer thickness. Eng Fract Mech 2001;68:1–7.

[12] Chang JC, Velamakanni BV, Lange FF, Pearson DS. Centrifugal consolidation of Al$_2$O$_3$ and Al$_2$O$_3$/ZrO$_2$ composite slurries vs. interparticle potentials: Particle packing and mass segregation. J Am Ceram Soc 1991;74(9):2201–4.

[13] Huisman W, Graule T, Gauckler LJ. Alumina of high reliability by centrifugal casting. J Euro Ceram Soc 1995;15:811–21.

[14] Jin Z-H, Batra RC. *R*-curve and strength behavior of a functionally graded material. Mater Sci Eng 1998;A242:70–6.

[15] Chung TJ, Neubrand A, Rödel J, Fett T. Fracture toughness and *R*-curve behavior of Al$_2$O$_3$/Al FGMs. Ceramic Transactions 2001;114:789–96.

[16] Fett T, Munz D, Yang YY. Direct adjustment procedure for weight function analysis of graded materials. Fat Fract Eng Mater Struct 2000;23:191–8.

[17] Moon R, Hoffman M, Hilden J, Blanton W, Bowman K, Trumble K, Rödel J. A weight function analysis of R-curve behavior in gradient alumina–zirconia composites. Ceramic Transactions 2001;114:781–8.

[18] Chang J, Lange F, Pearson D. Pressure sensitivity for particle packing of aqueous Al_2O_3 slurries vs. interparticle potential. J Am Ceram Soc 1994;77(5):1357–60.

[19] Sbaizero O, Lucchini E. Influence of residual stresses on the mechanical properties of a layered ceramic composite. J Euro Ceram Soc 1996;16:813–8.

[20] Velamakanni BV, Chang JC, Lange FF, Pearson DS. New method for efficient colloidal particle packing via modulation of repulsive lubrication hydration forces. Langmuir 1990;6(7):1323–5.

[21] Moon R, Bowman K, Trumble K, Rödel J. A comparison of R-curves from SEVNB and SCF fracture toughness test methods on multilayered alumina–zirconia composites. J Am Ceram Soc 2000;83(2):445–7.

[22] Kübler J. Fracture toughness using the SEVNB method: Preliminary results. Ceram Eng Sci Proc 1997;18(4):155–62.

[23] Stech M, Rödel J. Method for measuring short-crack R-curves without calibration parameters: Case studies on alumina and alumina/aluminum composites. J Am Ceram Soc 1996;79(2):291–7.

[24] Moon RJ. Static fracture behavior of multilayer alumina–zirconia composites. PhD Thesis, Purdue University, 2000.

[25] Bueckner HF. A novel principle for the computation of stress intensity factors. Z Angew Math Mech 1970;50:529–46.

[26] Fett T, Munz D. Influence of crack–surface interactions on stress intensity factor in ceramics. J Mater Sci Lett 1990;9:1403–6.

[27] Fett T, Munz D. Determination of fracture toughness at high temperature after subcritical crack extension. J Am Ceram Soc 1992;75(11):3133–6.

[28] Fett T. Determination of residual stresses in components using the fracture mechanics weight function. Eng Fract Mech 1996;55(4):571–6.

[29] French JD, Zhao J, Harmer MP, Chan HM, Miller GA. Creep of duplex microstructures. J Am Ceram Soc 1994;77(11):2857–65.

[30] Cai PZ, Green DJ, Messing GL. Constrained densification of alumina/zirconia hybrid laminates, II: Viscoelastic stress computation. J Am Ceram Soc 1997;80(8):1940–8.

[31] Gilbert CJ, Ritchie RO. On the quantification of bridging tractions during subcritical crack growth under monotonic and cyclic fatigue loading in a grain-bridging silicon carbide ceramic. Acta Metall 1998;46(2):609–16.

[32] Sohn K-S, Lee S, Baik S. Analytical modeling for bridging stress function involving grain size distribution in a polycrystalline alumina. J Am Ceram Soc 1995;78(5):1401–5.

[33] Hay JC, White KW. Grain-bridging mechanisms in monolithic alumina and spinel. J Am Ceram Soc 1993;76(7):1849–54.

[34] Barenblatt GI. The mechanical theory of equilibrium cracks in brittle fracture. Adv Appl Mech 1962;20(3):55–129.

[35] Srawley JE, Gross B. Side-cracked plates subject to combined direct and bending forces. Cracks and fracture. ASTM STP 601 1976:559–79.

[36] Erdogan F. Fracture mechanics of functionally graded materials. Compos Eng 1995;5(7):753–70.

[37] Jin Z-H, Batra RC. Some basic fracture mechanics concepts in functionally graded materials. J Mech Phys Solids 1996;44(8):1221–35.

[38] Ravichandran KS. Thermal residual stress in a functionally graded material system. Mater Sci Eng 1995;A201:269–76.

[39] Hillman C, Suo Z, Lange FF. Cracking of laminates subjected to biaxial tensile stresses. J Am Ceram Soc 1996;79(8):2127–33.

[40] Munro RG. Evaluated material properties for a sintered α-alumina. J Am Ceram Soc 1997;80(8):1919–28.

PERGAMON

Engineering Fracture Mechanics 69 (2002) 1667–1678

Engineering Fracture Mechanics

www.elsevier.com/locate/engfracmech

Effects of residual stress and geometry on crack kink angles in graded composites

J. Chapa-Cabrera, I.E. Reimanis *

Department of Metallurgical and Materials Engineering, Colorado Center for Advanced Ceramics, Colorado School of Mines, Golden, CO 80401, USA

Received 9 February 2001; received in revised form 5 July 2001; accepted 10 July 2001

Abstract

Effects of residual stress, geometry and applied load on the stress intensity factors of cracks in discretely layered, graded composites were investigated using finite element techniques. Elastic gradient profiles introduce mixed mode conditions for cracks oriented perpendicular to the gradient. Residual stresses further affect the in-plane stress intensity factors such that the amount of mode-mixity depends on the applied load. The effect of residual stress and applied load on the mode-mixity depends on the specimen geometry. Implications of the residual stress and geometric-load effects on the predicted crack kink angle are discussed.
© 2002 Elsevier Science Ltd. All rights reserved.

Keywords: Graded materials; Mixed mode; Fracture; Residual stress

1. Introduction

Many optimized structural applications demand a spatial variance in properties within a given component. Such a variance may be achieved by utilizing a compositionally graded material. Predicting the fracture behavior of graded materials has been the subject of various theoretical and experimental studies [1–9], some of which have been concerned with predicting crack paths [2–5]. As these studies have shown, a crack located asymmetrically within a graded material, experiences a mixed mode state of stress, even when loaded under nominally mode I conditions. For brittle materials, in which cracks only propagate in a direction of vanishing shear stress, the crack path is a direct reflection of the mode-mixity. A crack situated asymmetrically with respect to the gradient is deflected towards the more compliant material, the crack path being determined by the elastic mismatch between constituent materials as well as geometrical and gradient architectural parameters [2–4]. Some general trends predicting crack deflection for cracks oriented in a direction perpendicular to the gradient have been revealed, but a complete, systematic study is lacking.

A further complication is introduced by the presence of residual stress. Graded composites typically experience significant residual stress fields when they are processed at elevated temperatures. Factors such

---

* Corresponding author. Tel.: +1-303-273-3549; fax: +1-303-273-3057.
E-mail address: reimanis@mines.edu (I.E. Reimanis).

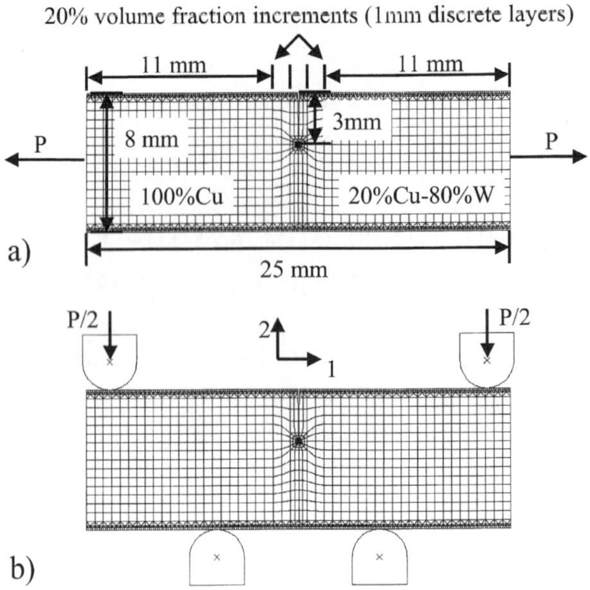

Fig. 1. Mesh and loading for the (a) SENT specimen, and (b) 4PB specimen.

as thermal expansion coefficient mismatch of the different constituents and non-uniform sintering give rise to residual stresses, with magnitude and distribution determined by the geometry and the gradient profile. There are no studies combining effects of elastic mismatch and residual stress for cracks oriented perpendicular to the gradient.

Graded composites are typically either discretely layered or continuously graded. The present study considers cracks embedded in a layered elastic medium whose elastic modulus and thermal expansion coefficient varies in the different layers. The crack is oriented perpendicular to the gradient as illustrated in Fig. 1, and is located within a particular layer. Thus, interface fracture is not considered. As part of a larger study that includes plasticity effects, the materials copper and tungsten were chosen for ease of experimental studies. However, the present paper focuses only on elastic effects, though the results apply generally to any elastic system. Crack tip stress fields are analyzed and the stress intensity factors are obtained for three different specimen geometries. Results are presented in terms of predicted crack deflection angles based on a maximum principal stress criterion.

2. Problem formulation

Three different specimen geometries were chosen to study the geometrical effects; all three were modeled with the same finite element analysis (FEA) mesh and only the loading and boundary conditions were altered. The first two specimens comprise a single edge notch geometry loaded in tension (SENT) and in four point bending (4PB). The third specimen is a center cracked plate loaded in tension (CCT) perpendicular to the notch. All the notches are oriented perpendicular to the gradient. Sequential FEA is used, allowing the application of mechanical loading to a body that is already deformed by residual stresses. Fig. 1 shows the mesh and the variations to model the different specimens. To model the SENT specimen, tension is applied on the exterior surfaces perpendicular to the crack as shown in Fig. 1a. Results for the CCT specimen are obtained by adding symmetry boundary conditions on the top surfaces of the SENT specimen. The 4PB specimen is supported with two pins in the bottom and equal loads are applied on each of the two upper pins as shown in Fig. 1b. Plain strain conditions were assumed for all of the models, and thus, two-dimensional analyses was used.

The mesh is 25 mm long and 8 mm high. The discrete compositional gradient is formed by 11 mm of pure copper, followed by 1 mm layers of 80%Cu–20%W, 60%Cu–40%W and 40%Cu–60%W, and by an 11 mm section with 20%Cu–80%W; all percentages are volumetric. A crack 3 mm long is cut perpendicular to the gradient in the center of the 60%Cu–40%W middle layer, as shown in Fig. 1. Crack tip vicinity stresses resulting from two different conditions are superimposed. First, the thermal residual stresses resulting from thermal expansion coefficient mismatch and a change in temperature are obtained. Second, the stresses resulting from the applied load are evaluated. The residual stresses were obtained by applying an initial temperature of 300 °C to all the nodes and subsequently cooling down uniformly over the whole model to 25 °C. 300 °C was chosen as part of another study in which the plastic properties of the copper are also considered [10]. Loading was such that the applied stresses were 10, 20, 40, 80 and 160 MPa. For the 4PB geometry the applied stresses reported correspond to the maximum stress on the tensile surface of a homogeneous material without a crack. Stress intensity factors were obtained from FEA for the residual stresses and for each applied load with and without residual stresses.

During the thermal step that generates residual stresses, the different geometries are simply supported on two points, and they are free to deform. The CCT model has the extra constraint of symmetric boundary conditions on the top surfaces. Tension was applied perpendicular to the crack on the vertical free surfaces of the SENT and CCT models. The nodes on the loaded surfaces were constrained to obtain fixed displacement conditions. The bottom pins are completely fixed in position for the 4PB model. The upper pins are constrained to zero displacement on direction 1 and concentrated loads are applied on direction 2. The node located on the center of the bottom surface is fixed in direction 1. No friction was assumed for the bar–pin contact surfaces.

Rules of mixtures (ROMs) used in previous studies [11–13] were applied in this work to estimate the constitutive properties of the different compositions. Table 1 provides the properties used. Details of the applied rules of mixtures to estimate the mechanical properties are presented in [11], where the impact of the constitutive assumptions was also studied. Following Konda and Erdogan [1], who showed that the effect of the Poisson's ratio gradient is negligible, a value of 0.3 was used for all compositions. The thermal expansion coefficient was estimated using the following relation from [14]:

$$\alpha = \frac{\sum_i \alpha_i V_i K_i}{\sum_i V_i K_i} \tag{1}$$

where α_i is the thermal expansion coefficient of material i, V_i is the volume fraction and K_i is the bulk modulus. This model assumes that no shear is transmitted along the boundaries of the different phases and has been shown to provide useful representation of experimental data for several metal matrix systems [14].

Table 1
Material properties

	E (GPa)	v	CTE ($10^{-6}/$°C @20 °C)
Cu	124	0.3	16.6
80%Cu–20%W	144	0.3	11.3
60%Cu–40%W	173	0.3	8.5
40%Cu–60%W	214	0.3	6.7
20%Cu–80%W	280	0.3	5.5

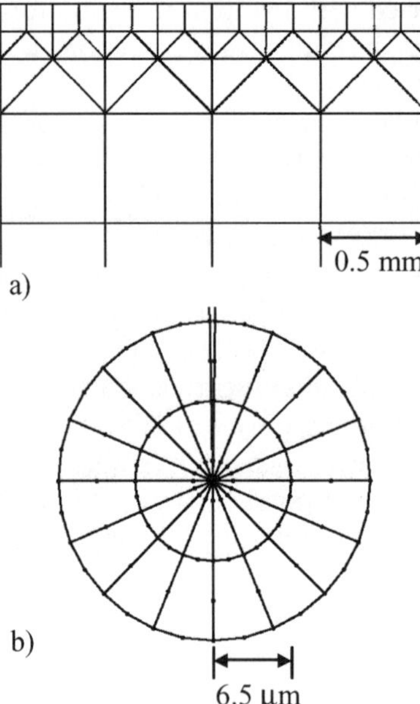

Fig. 2. Detail of the mesh refinement around (a) a pin-loaded surface, and (b) the crack tip.

The commercial finite element analysis code ABAQUS [15] was used to conduct the study. Plane-strain quad second order elements were used throughout the model except close to the top and bottom surfaces, where plane-strain second order triangular elements were used to transition to a finer mesh with 0.125 mm elements as shown in Fig. 2a. The mesh was refined to model the contact between the loading pins. Coarse elements of 0.5 mm × 0.5 mm were used refining the mesh to 6.25 μm size elements around the crack tip. To confirm that the mesh density away from the crack tip was dense enough, the element density was doubled, and several analyses were completed. The difference between the coarse and fine mesh was at most 0.2%, and thus, the coarse mesh was used. The mesh refinement close to the crack tip is shown in Fig. 2b. Quad second order elements with three nodes collapsed on the crack tip were used. The midside node immediately away from the tip was positioned to a 1/4 of the element length to make the element square root singular [16].

The maximum tangential stress criterion [17] was used to predict the crack kink angle θ_m. The stress components close to the crack tip are shown in Fig. 3. To obtain these stress components as a direct output from the FEA code, cylindrical coordinates with origin at the crack tip were used. The stresses at 12.5 μm from the crack tip were plotted for angles in the range of −90° to 90°. The distance of 12.5 μm corresponds to the nodes common to the second and third sets of elements away from the crack tip.

To obtain the mode I and mode II stress intensity factors, K_I and K_{II}, values from ABAQUS the J integral was first obtained for a total of seven contours around the crack tip. Then the J obtained from the first contour, which was that closest to the crack tip, was discarded and an average J was calculated with the remaining six contour values. The displacement u_1 and u_2 of the first eight nodes in both, right and left crack faces were also extracted from the FEA. An average value of these displacements, except for the first node away from the tip, was used in the following calculations. These average values are virtually identical with those obtained by extrapolating a linear fit to $r = 0$. Using the obtained J, the elastic modulus E of the 60%Cu–40%W cracked layer, u_1 and u_2, the mode I and II stress intensity factors were calculated with the following relations [18]:

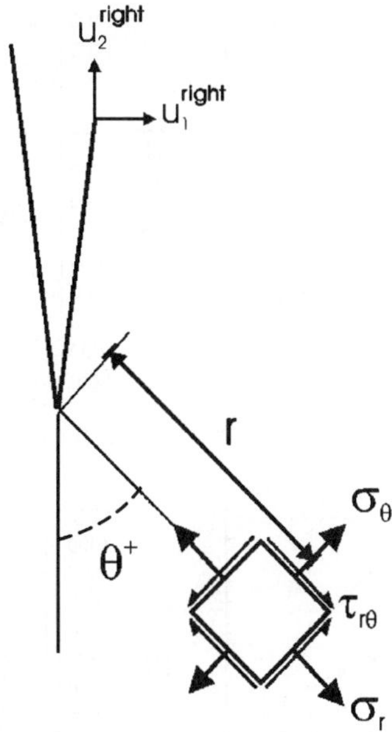

Fig. 3. Stress components around the crack tip in cylindrical coordinates.

$$K_{\mathrm{I}} = \sqrt{\frac{JE}{(1+R^2)(1-v^2)}}$$

$$K_{\mathrm{II}} = RK_{\mathrm{I}} \tag{2}$$

$$R = \lim_{r \to 0} \frac{u_2^{\text{right}} - u_2^{\text{left}}}{u_1^{\text{right}} - u_1^{\text{left}}}$$

3. Finite element analysis results

3.1. Single edge notch specimen loaded in tension

The shear and normal stresses, $\tau_{r\theta}$ and σ_θ, at 12.5 μm from the crack tip are plotted as a function of angle in Fig. 4. The angles at which σ_θ is maximum and minimum (Fig. 4a and c) may be accurately calculated by examining where $\tau_{r\theta}$ goes to zero (Fig. 4b and d). Thus, plots of these two stress components as a function of angle around the crack tip provides a prediction of the crack kink angle, assuming a maximum principal stress criterion. In the absence of residual stress the predicted crack kink angle θ_m is $-6.95°$ for all the applied loads as observed from the angle for which $\tau_{r\theta} = 0$ in Fig. 4b. The value of θ_m that is calculated from the crack face displacement and the J integral is $-7.3°$, revealing a difference of only 0.35° from the results obtained from the shear stress distribution (Fig. 4b). In the presence of residual stress, the stress fields are modified as shown in Fig. 4c and d. It is important to note that all the $\tau_{r\theta}$ curves in Fig. 4d still intercept on $-6.95°$ but θ_m has shifted towards the stiffer side of the gradient and has become load dependent. The variation of θ_m as a function of load is plotted in Fig. 5. The implications of this load dependence are discussed later.

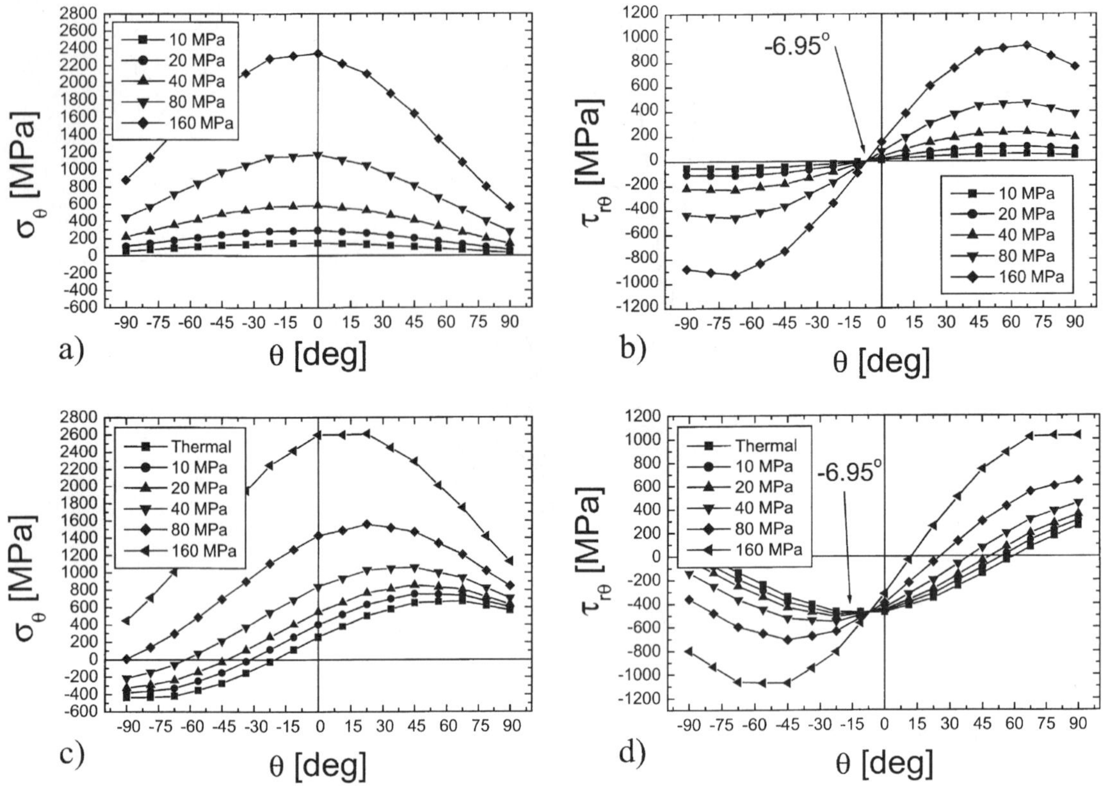

Fig. 4. SENT specimen stress vs. angle at 12.5 μm from the crack tip: (a) tangential and (b) shear stress components without residual stress; (c) tangential and (d) shear stress with residual stress.

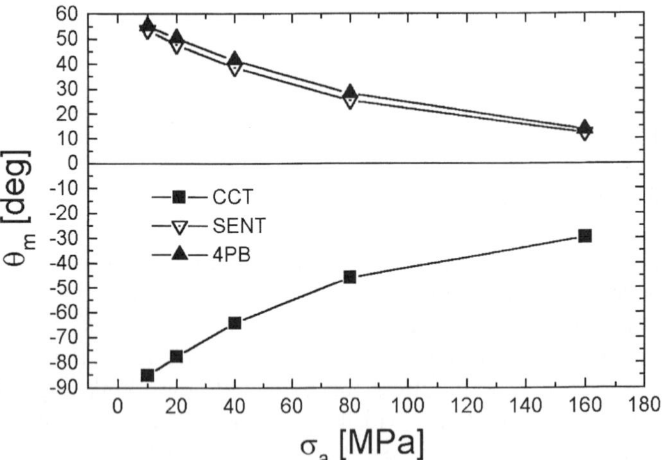

Fig. 5. Variation of the predicted crack kink angle with the applied load in presence of residual stresses for the different geometries.

Applying superposition, the in-plane stress intensity factors for both modes can be separated in a residual stress term and in an applied load term to quantify the amount of mode-mixity as illustrated in the following expression for the phase angle of loading:

Table 2
Stress intensity factors for the SENT

$K_{\mathrm{I}}^{\mathrm{Res}}$ (MPa m$^{1/2}$)	$K_{\mathrm{II}}^{\mathrm{Res}}$ (MPa m$^{1/2}$)	Applied load (MPa)	$K_{\mathrm{I}}^{\mathrm{App}}$ (MPa m$^{1/2}$)	$K_{\mathrm{II}}^{\mathrm{App}}$ (MPa m$^{1/2}$)	$K_{\mathrm{I}}^{\mathrm{Res}} + K_{\mathrm{I}}^{\mathrm{App}}$ (MPa m$^{1/2}$)	$K_{\mathrm{II}}^{\mathrm{Res}} + K_{\mathrm{II}}^{\mathrm{App}}$ (MPa m$^{1/2}$)	φ (deg)
2.42	−4.27	10	1.32	0.09	3.74	−4.19	−48.22
		20	2.65	0.17	5.07	−4.10	−39.00
		40	5.30	0.34	7.71	−3.93	−26.99
		80	10.60	0.68	13.01	−3.59	−15.41
		160	21.20	1.37	23.61	−2.90	−7.01

$$\varphi = \tan^{-1} \frac{K_{\mathrm{II}}}{K_{\mathrm{I}}} = \tan^{-1} \frac{K_{\mathrm{II}}^{\mathrm{Res}} + K_{\mathrm{II}}^{\mathrm{App}}}{K_{\mathrm{I}}^{\mathrm{Res}} + K_{\mathrm{I}}^{\mathrm{App}}} \tag{3}$$

where the superscripts Res and App denote residual and applied components of the stress intensity factor. Values for the computed K_{I} and K_{II} for the different applied stresses with and without residual stresses are shown in Table 2. The K^{Res} and K^{App} values were obtained from modeling the cooling down and the mechanical loading separately. Total stress intensity values obtained from superposition are reported in Table 2.

3.2. Four point bending specimen

Shear and normal stresses as a function of angle around the crack tip for the 4PB specimen under different applied loads are shown in Fig. 6. The shear stress $\tau_{r\theta}$ in the absence of residual stresses is plotted as a function of the angle for the different applied loads in Fig. 6b. It is noted that all the curves intercept and go to zero $\tau_{r\theta}$ at an angle of −9°.

In the presence of residual stresses the crack tip stress fields are modified in a manner similar to the SENT specimen, as shown in Fig. 6c and d. The predicted crack kink angle θ_{m}, shifts towards the stiff side of the gradient and decreases as the applied load is increased, as is apparent in Figs. 5 and 6d. Values for the computed K_{I} and K_{II} for the different applied stresses with and without residual stresses are shown in Table 3.

3.3. Center cracked specimen loaded in tension

Similar to the SENT and 4PB geometries, the predicted crack kink angle without residual stresses is negative as seen in Fig. 7b with a value of −8.8°.

In contrast to the results for the other specimens studied here, in the presence of residual stresses, the crack faces come in contact. The contact region is very small (∼1.6 μm) compared to any characteristic geometric dimension of the specimen. Because such contact may influence θ_{m}, and because the degree of influence may depend on friction, the sensitivity of the results to friction on the crack faces was examined using the Coulomb model and performing the analysis with friction coefficient values of 0.2 and 0.8. The default Coulomb model implemented in ABAQUS was applied. This model assumes that no relative motion occurs between the contact surfaces until the frictional stress reaches a critical value. It was observed that friction did not alter the predicted deflection angles, a result likely due to the limited area that is in contact. Other gradient profiles, geometries or different processing conditions may result in larger crack closure and consequently in higher sensitivity of the stress intensity factors to friction.

As a result of the crack faces closing, negative K_{I} values are obtained as seen in Table 4. Furthermore, the shear that the crack tip experiences as a consequence of the residual stress field is opposite in sign to that

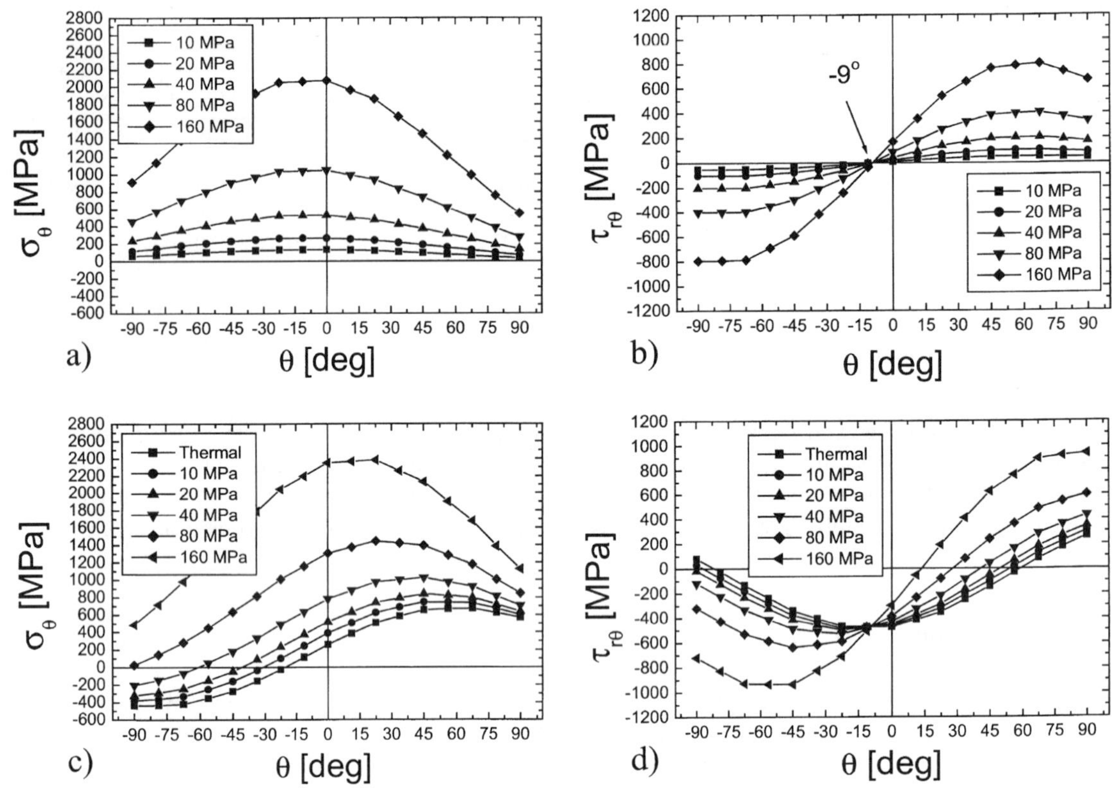

Fig. 6. 4PB specimen stress vs. angle at 12.5 μm from the crack tip: (a) tangential and (b) shear stress components without residual stress; (c) tangential and (d) shear stress with residual stress.

Table 3
Stress intensity factors for the 4PB specimen

$K_{\mathrm{I}}^{\mathrm{Res}}$ (MPa m$^{1/2}$)	$K_{\mathrm{II}}^{\mathrm{Res}}$ (MPa m$^{1/2}$)	Applied load (MPa)	$K_{\mathrm{I}}^{\mathrm{App}}$ (MPa m$^{1/2}$)	$K_{\mathrm{II}}^{\mathrm{App}}$ (MPa m$^{1/2}$)	$K_{\mathrm{I}}^{\mathrm{Res}} + K_{\mathrm{I}}^{\mathrm{App}}$ (MPa m$^{1/2}$)	$K_{\mathrm{II}}^{\mathrm{Res}} + K_{\mathrm{II}}^{\mathrm{App}}$ (MPa m$^{1/2}$)	φ (deg)
2.42	−4.29	10	1.22	0.10	3.63	−4.19	−49.05
		20	2.43	0.20	4.85	−4.09	−40.13
		40	4.85	0.39	7.26	−3.90	−28.23
		80	9.56	0.75	11.98	−3.53	−16.434
		160	18.92	1.52	21.34	−2.77	−7.38

on the SENT and 4PB specimens. The effect of the shear stress reversal is also apparent in the predicted crack kink angle plotted in Fig. 5.

4. Discussion

The predicted crack path, described here by the crack kink angle, θ_{m}, only represents the first increment in crack growth, and may in fact not represent a macroscopic crack path. Since the total stress field is non-uniform, it would be expected that the mode-mixity would change as the crack extends, resulting in a non-

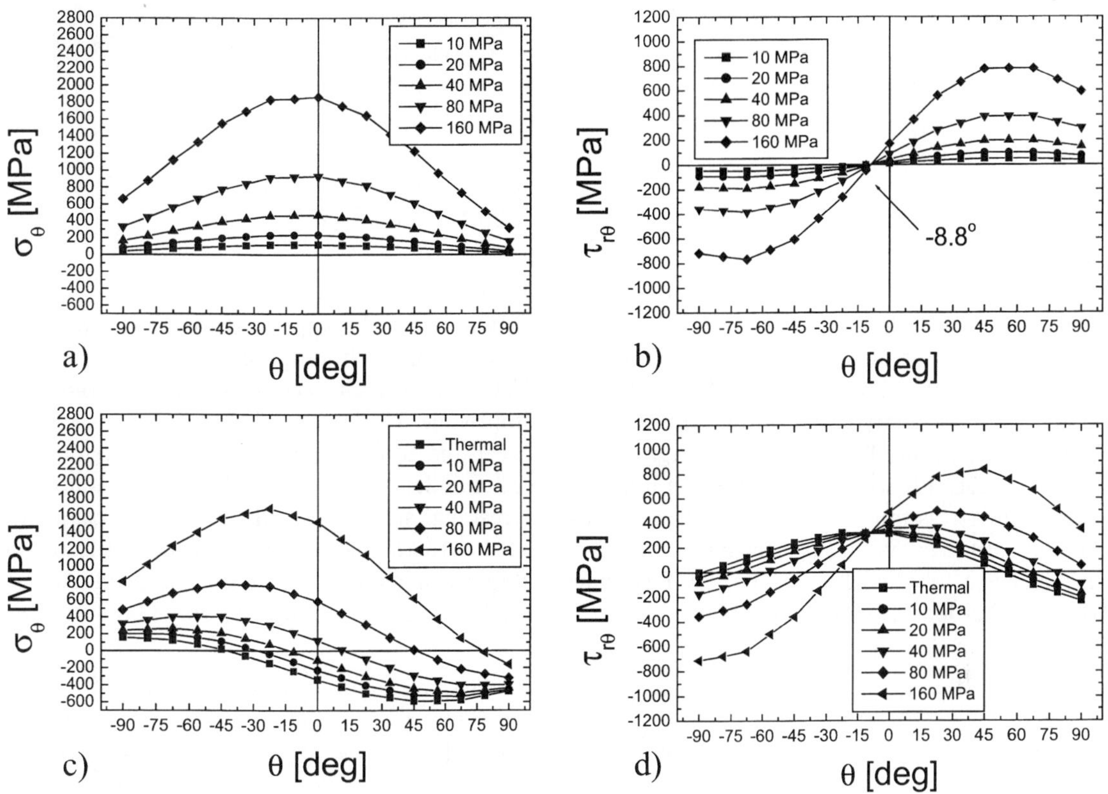

Fig. 7. CCT specimen stress vs. angle at 12.5 μm from the crack tip: (a) tangential and (b) shear stress components without residual stress; (c) tangential and (d) shear stress with residual stress.

Table 4
Stress intensity factors for the CCT specimen

$K_{\mathrm{I}}^{\mathrm{Res}}$ (MPa m$^{1/2}$)	$K_{\mathrm{II}}^{\mathrm{Res}}$ (MPa m$^{1/2}$)	Applied load (MPa)	$K_{\mathrm{I}}^{\mathrm{App}}$ (MPa m$^{1/2}$)	$K_{\mathrm{II}}^{\mathrm{App}}$ (MPa m$^{1/2}$)	$K_{\mathrm{I}}^{\mathrm{Res}} + K_{\mathrm{I}}^{\mathrm{App}}$ (MPa m$^{1/2}$)	$K_{\mathrm{II}}^{\mathrm{Res}} + K_{\mathrm{II}}^{\mathrm{App}}$ (MPa m$^{1/2}$)	φ (deg)
		10	1.05	0.09	−2.13	2.86	N/A
		20	2.10	0.19	−1.07	2.96	N/A
−3.18	2.77	40	4.20	0.37	1.03	3.14	71.90
		80	8.41	0.74	5.23	3.52	33.90
		160	16.81	1.49	13.64	4.26	17.35

linear crack path. Experimental validation would require high spatial resolution near the original crack. Despite the difficulty in confirming these predictions, the results illustrate the complexity associated with predicting crack paths in graded materials that contain residual stresses. In the SENT and 4PB specimen geometries, the presence of residual stresses resulted in crack kinking towards the stiffer side of the gradient, opposite to mechanics predictions that do not account for residual stresses [2–4]. The magnitude of this effect depends on the applied load at fracture, and hence the fracture toughness. Thus, the crack path depends not only on the elastic properties and the residual stress state, but the inherent fracture toughness of the materials, which is not necessarily constant through the gradient.

The maximum stress criterion assumes that crack extension takes place if the maximum stress σ_θ has the same value as the fracture stress in an equivalent mode I case. Therefore, it is reasonable to utilize an equivalent mode I stress intensity factor K_I^e [17,19]:

$$K_I^e = K_I \cos^3 \frac{\theta_m}{2} - 3K_{II} \cos^2 \frac{\theta_m}{2} \sin \frac{\theta_m}{2} \tag{4}$$

Fracture would occur when:

$$K_I^e = K_{IC} \tag{5}$$

where K_{IC} is the toughness of the cracked layer.

The equivalent mode I stress intensity factor K_I^e is plotted versus the applied stress in Fig. 8 with and without residual stresses for the different specimens. K_I for a linear, homogeneous elastic material (for which $\theta_m = 0$) is also shown in each figure; for the CCT and 4PB samples, this homogeneous K_I was calculated using an analytic fracture mechanics expression, while for the SENT sample, it was calculated using the numerical model. The excellent agreement between the graded specimens without residual stress and the homogeneous specimens provides a high degree of confidence in the graded model. It is noted that the equivalent stress intensity factor is increased by the residual stresses for all the applied loads on the SENT and 4PB (Fig. 8a and b) specimens, revealing that the fracture condition is reached at lower loads in the presence of residual stresses. The CCT specimen exhibits similar behavior at lower applied loads (<40 MPa). At higher loads, the presence of residual stress actually lowers K_I^e, though by only about 10% at the

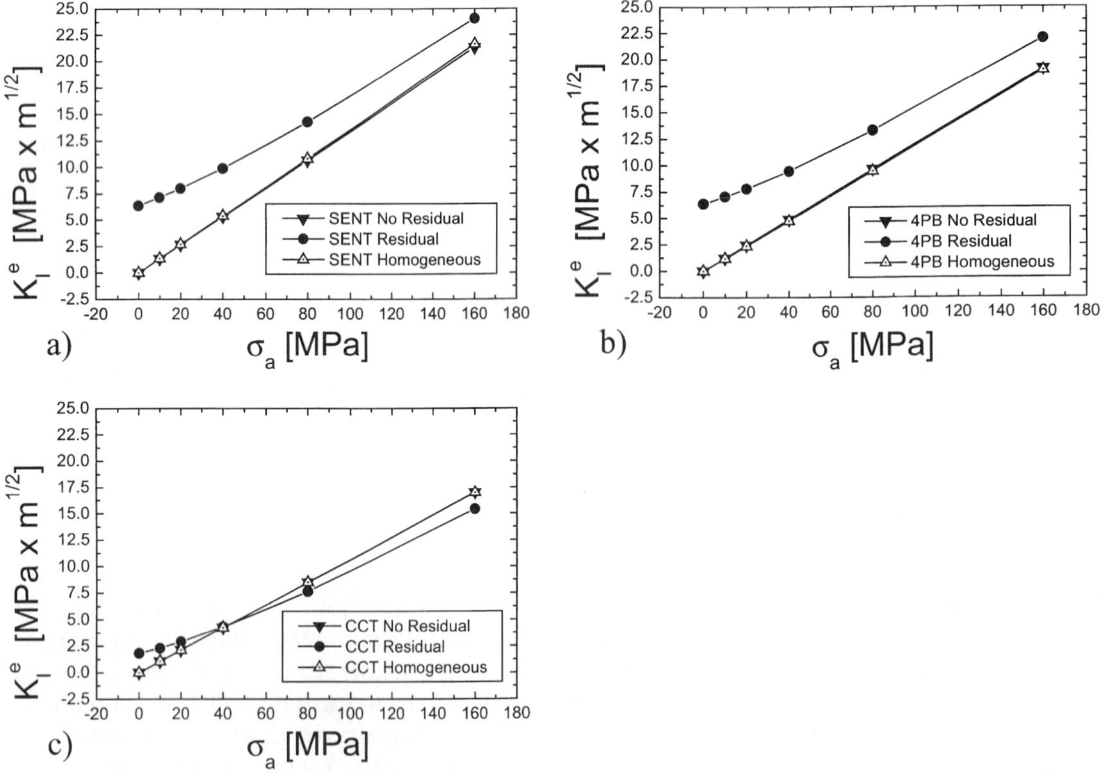

Fig. 8. Equivalent mode I stress intensity factor vs. applied load with and without residual stresses for the graded (a) SENT, (b) 4PB and (c) CCT geometries. The solution for a homogeneous bar is also plotted.

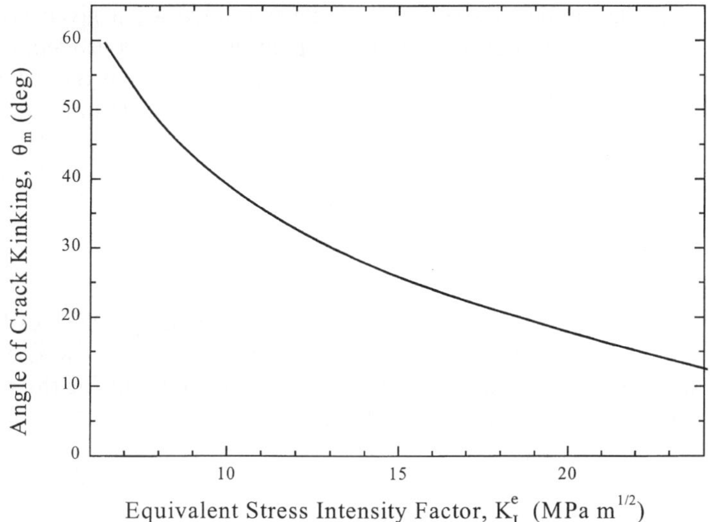

Fig. 9. The angle of crack kinking, θ_m, is plotted as a function of the equivalent stress intensity factor. The leftmost limit on the curve (e.g. $K_I^e \sim 7$ MPa m$^{1/2}$) represents the situation when the applied load is zero.

highest loads examined. Thus, the residual stress could be beneficial or detrimental, depending on the fracture toughness.

Fig. 9 shows how the crack kink angle, θ_m, varies as a function of K_I^e for the SENT geometry. The range of K_I^e plotted here represent typical toughness values for engineering ceramics and brittle matrix composites, which vary from about 2 to 25 MPa m$^{1/2}$. In this range of K_I^e, it is clear that the crack kink angle is very sensitive to relatively small changes in fracture toughness. At higher K_I^e, not shown here, the sensitivity decreases and the crack kink angle eventually becomes negative; it converges to the value calculated without residual stress, $\theta_m = -6.95°$ as the K_I^e approaches infinity.

Assuming that fracture occurs when $K_I^e = K_{IC}$, Fig. 9 predicts that when a crack is contained within a material that has a fracture toughness less than about 7 MPa m$^{1/2}$, the crack will propagate without any applied load (i.e. due to residual stresses alone), and at a kink angle near 60°. If the interlayer containing the crack is tougher, a higher applied load is required for crack propagation, and the crack kink angle is smaller. Though Fig. 9 is unique for the particular specimen geometry and material combinations examined here, it provides insight into the potential importance of residual stress in controlling crack kinking in graded materials.

5. Summary and conclusions

Geometry and residual stresses in linear elastic graded materials are shown to influence strongly the predicted crack kink angle according to the maximum stress criterion. Using superposition, the modes I and II of the stress intensity factors may be separated into residual stress and applied stress components, clarifying the effect of residual stresses on the predicted crack kink angle. It is seen that a crack in a simple layered, graded system may kink towards the stiffer side of the gradient when residual stresses are present. The angle of deflection depends on the applied load, and is also affected by the geometrical conditions. In contrast, in the absence of residual stress, for all three geometries examined here, the crack would always propagate towards the more compliant side of the gradient. As expected from a linear elastic analysis, in the absence of residual stress, the predicted crack kink angle is unique for a given geometry and does not depend on the applied load. Special attention should be placed on this dependency of θ_m on the applied load for fracture testing of graded materials. These results clearly show that the deflection angle may be a function of the material fracture toughness, thereby adding a microstructural aspect to crack path predictions.

Acknowledgements

The authors would like to acknowledge the US Department of Energy, Office of Basic Energy Sciences for funding this research under contract DE-FG03-96ER45575. JC further acknowledges the Mexican National Council for Science and Technology (CONACYT) and Fulbright for the scholarship support to conduct graduate studies at the Colorado School of Mines.

References

[1] Konda N, Erdogan F. The mixed mode crack problem in a nonhomogeneous elastic medium. Eng Fract Mech 1994;44:533–45.

[2] Gu P, Asaro RJ. Crack deflection in functionally graded materials. Int J Solids Struct 1997;34:3085–98.

[3] Gu P, Asaro RJ. Cracks in functionally graded materials. Int J Solids Struct 1997;34:1–17.

[4] Rousseau C-E, Tippur HV. Compositionally graded materials with cracks normal to the elastic gradient. Acta Mater 2000;48:4021–33.

[5] Becker TL, Cannon RM, Ritchie RO. Finite crack kinking and T-stresses in functionally graded materials, Int J Solids Struct 38 (2001) 5545–63.

[6] Kim AS, Suresh S, Shih CF. Plasticity effects on fracture normal to interfaces with homogenous and graded compositions. Int J Solids Struct 1997;34:3415–32.

[7] Marur PR, Tippur HV. Numerical analysis of crack-tip fields in functionally graded materials with a crack normal to the elastic gradient. Int J Solids Struct 2000;37:5353–70.

[8] Li H, Lambros J, Cheeseman AB, Santare MH. Experimental investigation of the quasi-static fracture of functionally graded materials. Int J Solids Struct 2000;37:3715–32.

[9] Anlas G, Santare MH, Lambros J. Numerical calculation of stress intensity factors in functionally graded materials. Int J Fract 2000;104:131–43.

[10] Chapa J. Effects of residual stress, geometry and plasticity on crack kink angles in graded composites. PhD Thesis (in progress), Colorado School of Mines, 2001.

[11] Williamson RL, Rabin BH, Drake JT. Finite element analysis of thermal residual stresses at graded ceramic–metal interfaces. Part I. Model description and geometrical effects. J Appl Phys 1993;74:1310–20.

[12] Drake JT, Williamson RL, Rabin BH. Finite element analysis of thermal residual stresses at graded ceramic–metal interfaces. Part II. Interface optimization for residual stress reduction. J Appl Phys 1993;74:1321–6.

[13] Chapa J. Modeling of thermal stresses of a graded Cu/W joint. MSc Thesis, Colorado School of Mines. 1998.

[14] Kingery WD, Bowen HK, Uhlmann DR. In: Introduction to ceramics. 2nd ed. New York: John Wiley & Sons; 1976. p. 603–11.

[15] ABAQUS/Online reference manuals Version 5.8. Hibbit: Karlsson & Sorensen; 2000.

[16] Anderson TL. In: Fracture mechanics, fundamentals and applications. Boca Raton: CRC Press; 1991. p. 685–92.

[17] Erdogan F, Sih GC. On the crack extension in plates under plane loading and transverse shear. J Basic Eng 1963;85:519–27.

[18] ABAQUS/Fracture mechanics USA: HKS; 1996. p. L6.55.

[19] Broek D. In: Elementary engineering fracture mechanics. 4th ed. Dordrecht, The Netherlands: Kluwer; 1991. p. 374–80.

PERGAMON

Engineering Fracture Mechanics 69 (2002) 1679–1693

**Engineering
Fracture
Mechanics**

www.elsevier.com/locate/engfracmech

Influence of elastic variations on crack initiation in functionally graded glass-filled epoxy

Carl-Ernst Rousseau [1], Hareesh V. Tippur [*]

Department of Mechanical Engineering, Auburn University, 202 Ross Hall, Auburn, AL 36849, USA

Received 21 February 2001; received in revised form 7 June 2001; accepted 11 June 2001

Abstract

Crack tip deformations and fracture parameters in functionally graded glass-filled epoxy beams are experimentally evaluated under static and dynamic loading conditions. Beams with unidirectional, monotonic elastic gradients and cracks along the gradient are examined. SEN samples with increasing or decreasing Young's modulus ahead of the crack tip are studied in symmetric four-point bending and one-point impact loading configurations. Optical method of coherent gradient sensing (CGS) is used to measure crack tip deformations prior to crack initiation. For impact loading experiments, CGS is used in conjunction with high-speed photography for recording instantaneous deformation fields. Stress intensity factors (SIF) or SIF-histories in functionally graded materials (FGM) based on locally homogeneous material descriptions in the immediate crack tip vicinity are evaluated and compared with companion finite element simulations. The influence of elastic gradients in FGM samples with cracks on the compliant and stiff sides of the beam are quantified relative to their homogeneous counterparts and with each other. Under static loading conditions, the crack tip located on the compliant side of the beam is elastically shielded when compared to the situation when the crack is on the stiffer side of the same FGM beam. Under dynamic conditions, however, elastic gradients affect crack initiation differently. Crack initiation in an FGM with a crack on the stiff side of the beam and impact occurring on the compliant edge is delayed when compared to the opposite configuration. Independent finite element simulations of FGMs with idealized elastic gradients with identical crack tip elastic properties suggest that lower crack tip loading rate in the former is responsible for the differences.
© 2002 Elsevier Science Ltd. All rights reserved.

Keywords: Functionally graded materials; Elastic gradients; Crack initiation; Stress intensity factor; Static loading; Dynamic loading; Optical interferometry; FEA

1. Introduction

Nonhomogeneous material systems with gradual variation in properties are collectively referred to as functionally graded materials or FGMs. Gradual variation of material properties in FGMs, unlike abrupt

---

[*] Corresponding author. Tel.: +1-334-844-3327; fax: +1-334-844-3307.

E-mail address: htippur@eng.auburn.edu (H.V. Tippur).

[1] Present address: California Institute of Technology, Pasadena, CA.

changes encountered in discretely layered systems, is known to improve failure performance [1] while preserving the intended thermal, tribological, and/or structural benefits of combining dissimilar materials. Accordingly, FGMs are considered ideal for applications involving high strain rate and thermal shock loading. Assessing the influence of compositional and hence material property gradients on the failure behavior is central for understanding FGMs. In this article, the influence of elastic gradients on the static and dynamic crack initiation in FGMs is examined. *Of specific interest to this work are FGMs with cracks parallel to the elastic gradient.*

Some of the early works on the fracture mechanics of FGMs include those by Atkinson and List [2], and Delale and Erdogan [3]. An inverse $\sqrt{r}$ stress singularity at the crack in FGMs was suggested in these articles and was later confirmed by Eischen [4] using asymptotic analysis. Jin and Noda [5] concluded likewise, independently of the crack orientation relative to the property gradient. In a paper summarizing recent advances in fracture mechanics of FGMs, Erdogan [6] has presented theoretical results for cracks oriented along the elastic gradient, and subjected to various loading conditions. Among the dynamic investigations on FGMs, transient (stress intensity factor) SIFs for a mode-III crack lying in an elastic media with spatially varying elastic properties normal to crack surfaces has been studied analytically by Babaei and Lukasiewicz [7]. They have found SIF to vary with crack length to layer thickness ratio. Dynamic crack propagation in functionally graded particulate is numerically studied by Nakagaki et al. [8] for shock loading to determine the effect of grading on crack severity as the crack propagates in the FGM. Parameswaran and Shukla [9] have shown experimentally that increasing toughness in the direction of crack growth reduces crack jump distance in discretely layered FGMs. Chiu and Erdogan [10] have evaluated the effect of material nonhomogeneity on one-dimensional wave propagation in FGMs having gradation in the direction of the incident pulse. Considerable wave distortions are reported as a rectangular pressure pulse propagates in the material.

Among the few experimental mechanics investigations on FGMs reported to date, Marur and Tippur [11] have developed an elastic impact method for characterizing the elastic properties of glass-filled epoxy FGMs. FGM processing, characterization and optical evaluation of SIFs are reported by Butcher et al. [12]. They have studied an FGM with a crack normal to the direction of the compositional gradient and observed enhancement of crack initiation toughness in graded systems when compared to bimaterials. A particulate FGM made of cenospheres dispersed in polyester matrix has been developed by Parameswaran and Shukla [13]. A UV-irradiated polymeric FGM has been prepared by Li et al. [14] to study quasi-static crack growth along the elastic and fracture toughness gradients. Boundary value measurements are used in finite element simulations to demonstrate toughening behavior of the FGM relative to homogeneous materials when the crack propagated from the stiffer side to compliant side. Finally, Marur and Tippur [15] have performed strain gage measurements and finite element simulations to evaluate dynamic performance of graded interface having cracks normal to the elastic gradient with respect to those of bimaterial. They have measured lower rate of crack tip loading in FGMs compared to the bimaterial counterparts.

In the present study, a comparative study of the influence of unidirectional elastic variations on crack initiation in FGMs subjected to static loading and low velocity impact is provided. FGMs with cracks oriented along the direction of the elastic gradient are considered. Although local fracture toughness dominantly affects crack initiation and growth in FGMs, there are instances [13,16] where material processing issues (resulting in porosity and other microdefects) could result in a relatively constant fracture toughness with monotonically varying elastic properties. Accordingly, attention is focussed on isolating the influence of elastic gradient on crack initiation in this paper. Following this introduction, a brief background of the elastic crack tip fields for FGMs is given. Section 3 describes material characteristics, optical technique and the method of extraction of the SIFs. Section 4 provides details of the complementary numerical simulations performed in this study. Finally, Section 5 summarizes the results for both static and dynamic experiments.

2. Elastic crack tip fields in FGMs: continuum models

For a crack oriented along the direction of the elastic gradient in a nonhomogeneous, isotropic planar body, using asymptotic analysis, Eischen [4] has shown that the crack tip stresses, say sum of the in-plane normal stresses $(\sigma_x + \sigma_y)$, in an FGM for mode-I conditions can be expressed as,

$$(\sigma_x + \sigma_y) = (C_0 r^{-1/2} f_0^{\mathrm{I}}(\theta) + C_1 r^0 f_1^{\mathrm{I}}(\theta)) + \mathrm{O}(r^{1/2}), \tag{1}$$

where (r, θ) denote crack tip polar coordinates, f_0^{I}, f_1^{I} are angular functions, and C_0 and C_1 are the coefficients of the expansion with $C_0 = K_{\mathrm{I}}/\sqrt{2\pi}$, K_{I} being the mode-I SIF. The angular functions have been shown to be identical to the ones for a crack in a homogeneous body. Accordingly, stress field description is identical to the homogeneous counterpart for the first two terms of the expansion. Moreover, Rousseau [17] has determined experimentally that addition of an expression of the form $C_2 r^{1/2} f_2^{\mathrm{I}}(\theta)$ can be used adequately as a third term in the asymptotic expansion, where C_2 is an unknown coefficient, and f_2^{I} is defined as the same angular function used for homogeneous materials. Expressions for in-plane displacements in FGMs were also developed by Eischen [4]. It was found that asymptotic displacement terms proportional to $r^{1/2}$, r^0 and r^1 are identical to those of homogeneous materials, thus are independent of gradient. The experimental method used in the current study measures surface slopes, and hence expressions for out-of-plane displacements, w, are necessary. Assuming plane stress conditions and constant thickness-wise strain, out-of-plane displacement can be obtained as,

$$w \approx \frac{-vB}{2}\left(A_0 r^{-1/2} f_0^{\mathrm{I}}(\theta) + A_1 r^0 f_1^{\mathrm{I}}(\theta) + A_2 r^{1/2} f_2^{\mathrm{I}}(\theta)\right) + \mathrm{O}(r), \tag{2}$$

with $A_j = C_j/E_0$ where the subscript E_0 denotes the value of crack tip Young's modulus. In the above, the Poisson's ratio v is assumed to be constant in the FGM.

Near tip expressions for stress fields in nonhomogeneous planar bodies that include the influence of elastic property gradients are reported by Erdogan [6]. For an exponential variation of elastic modulus of the form,

$$E(x) = E_0 \mathrm{e}^{\alpha x} = E_0 \mathrm{e}^{r\alpha\cos\theta}, \tag{3}$$

where E_0 is the Young's modulus at the origin, and α is a scalar dependent on the terminal values of the modulus and the length of the graded region, the near tip mode-I stresses are shown to be of the form,

$$\sigma_{ij}(r, \theta) \cong \mathrm{e}^{r\alpha\cos\theta}\left(\frac{K_{\mathrm{I}}}{\sqrt{2\pi r}} f_{ij}^{\mathrm{I}}(\theta)\right), \quad (i,j = x, y). \tag{4}$$

As $r \to 0$, Eq. (4) reduces to the K-dominant terms of Eq. (2).

In the absence of explicit expressions for stress fields for dynamically loaded stationary cracks, it could be inferred that Eq. (2) could be adequately extended for dynamic cases where inertial effects enter the coefficients without modifying the overall form of the expressions (i.e., $A_i \equiv A_i(t)$).

3. Experiments

3.1. The material and its elastic characteristics

Compositionally graded samples used in the current research comprised of an epoxy matrix (Young's modulus ~ 3 GPa, Poisson's ratio ~ 0.35) in which varying quantities of solid A-glass spheres (mean diameter ~ 42 μm) were dispersed. Thus the composite is *microscopically heterogeneous* but will be treated as

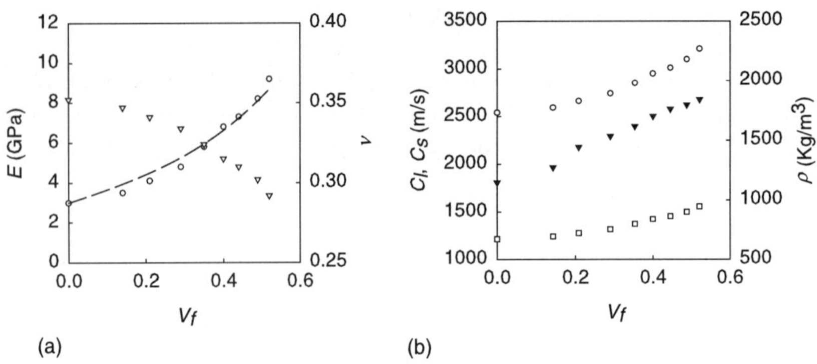

(a)   (b)

Fig. 1. Material properties of glass-filled epoxy particulate composites: (a) static Young's modulus and Poisson's ratio with filler volume fraction, (b) wave speeds and density with filler volume fraction.

an *isotropic, nonhomogeneous material at macroscopic scales* for measurement and analysis using continuum models. (That is, the material characteristics and optical measurements, to be discussed, are averaged over length scales that are 1–2 order of magnitude larger than the mean particle size.)

Prior to discussing the material properties of the FGMs used in the experiments, properties of homogeneous particulate glass-epoxy composites with different volume fraction of filler are described. Fig. 1(a) shows measured Young's modulus and Poisson's ratio variation of macroscopically *homogeneous mixtures* having different but constant volume fractions of the filler in the matrix. Young's modulus from about 3 to 9 GPa (static conditions) and a Poisson's ratio variation from 0.35 to 0.29 when the volume fraction of the filler is increased from 0 to 0.5. Also included in the figure are micromechanics predictions of the elastic modulus for two-phase mixtures [18] for different volume fractions V_f. Evidently, the predictions follow the experimental measurements rather closely over the entire range of filler volume fractions. (It should also be noted that these particle filled compositions show a linear load–displacement behavior [12] and can be treated to be nominally linear elastic.) In Fig. 1(b), the dynamic material properties namely longitudinal and shear wave speeds and density of the composites are shown. Evidently, each of these properties increases monotonically with the filler volume fraction in the composite.

Gravity casting was used for producing FGMs with monotonic Young's modulus variation between approximately 3 and 9 GPa under static conditions (measured using cantilever beam tests) or, approximately 4–12 GPa, under dynamic conditions (measured using ultrasonic pulse-echo measurements). These values of Young's moduli correspond to that of pure epoxy and a composite with a volume fraction of glass spheres in an epoxy matrix of ∼0.5, respectively. Details of the methods used in determining the elastic characteristics can be found in Butcher et al. [12]. Fig. 2 shows the modulus variation in a typical FGM sheet. The graded region in these sheets were machined into beam samples (dimensions 120 mm × 20 mm × 6 mm for static experiments and 150 mm × 37 mm × 6 mm for dynamic experiments). Fig. 2 shows typical elastic variation in a casting of which a portion was machined to obtain beam samples having the necessary elastic properties. The sample surfaces were then deposited with a thin layer of aluminum to obtain specular surface necessary for optical measurements. Edge notches (root radius 75 μm and length of 6 mm) were cut into the samples using a high-speed diamond impregnated circular saw.

Homogeneous samples (without compositional gradients) having elastic moduli equal to the ones at the crack tips of the FGMs were also prepared for comparative study. The geometry of these homogeneous samples was identical to that of the FGM samples.

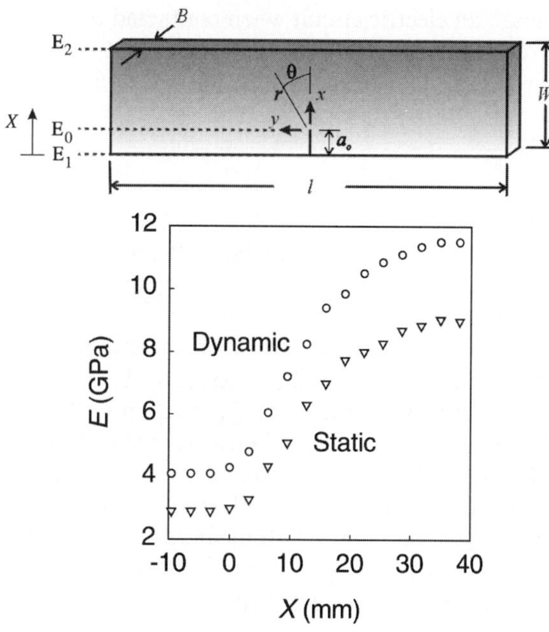

Fig. 2. Typical Young's modulus variations in cast FGM sheets.

3.2. Optical technique: coherent gradient sensing

The optical method of *reflection* coherent gradient sensing (CGS) was used for measuring crack tip deformations. The working principle of CGS has been reported by Tippur et al. [19]. An expanded and collimated beam of laser light (typically 50 mm in diameter) was used to interrogate the *specular* surface of the specimen. The reflected light beam, or the object wave front, contained information about the surface deformations. Optical shearing of the object wave front was used to decipher surface deformations in the form of interference patterns representing contours of constant surface slopes. In the present work, surface slopes along the crack orientation (*x*-axis) were determined as,

$$\frac{\partial w}{\partial x} = \frac{Np}{2\Delta}, \quad N = 0, \pm 1, \pm 2, \ldots,$$

(5)

where N denotes fringe orders, p is the pitch of the gratings (25 μm), and Δ is the grating separation distance. For plane stress conditions, the out-of-plane displacement w can be related to the in-plane stress components using expression for average out-of-plane strain, $\varepsilon_z \cong 2w/B$, or,

$$w \cong -\frac{\nu B}{2}\left(\frac{\sigma_x + \sigma_y}{E}\right),$$

(6)

where B is the undeformed thickness of the sample.

Interference fringes were recorded in real-time using a conventional camera during static experiments. The experimental set-up for dynamic study, however, was more elaborate. It included an impactor for stress wave loading, a pulse-laser as a light source, a CGS interferometer and a continuous access high-speed camera. During the experiment, a pneumatically operated impactor with a steel cylindrical head was launched towards the specimen. During its descent, the hammer first triggered open an electronic shutter of the high speed rotating mirror camera pre-spun to the desired speed, allowing light to reach its internal cavity. The specimen was subjected to one-point symmetric impact. An adhesive-backed copper tape placed on the top edge of the beam closed an electric circuit when contacted by the hammer. This in turn triggered the laser to begin pulsing at a repetition rate of 5 µs, with a pulse width of 50 ns, for a time duration of less than or equal to a single sweep of the light beam on the stationary film track of the high-speed camera.

3.3. Optical measurements

3.3.1. Statically loaded stationary cracks

The FGM and homogeneous beam samples were quasi-statically loaded in four-point bending using an Instron-4465 machine in displacement control mode (cross-head speed ~ 0.25 mm/min). The samples were subjected to symmetric bending moment acting over a 60 mm length in the mid-span of each beam. The resulting crack tip fringe patterns in FGMs with a crack on the stiff ($E_2/E_1 < 1$) and the compliant sides ($E_2/E_1 > 1$) of the beam are shown in Fig. 3(a) and (c), respectively. In these cases, the crack tip Young's moduli were 6.8 and 4.8 GPa, respectively. Further, the far-field applied stress σ_∞ ($= 6M/BW^2$ where M is the moment, W and B are the beam height and thickness, respectively) at which the deformations were recorded in the two cases are the same. Homogeneous samples having crack tip Young's moduli same as (to within $\pm 3\%$) that of the FGMs were also optically investigated. The crack tip deformation patterns for the homogeneous beams corresponding to the FGMs with the crack on the stiff and compliant side are shown in Fig. 3(b) and (d), respectively. Note that, the fringe patterns from the homogeneous samples are for the same load level as their FGM counterparts. The sensitivity of the optical measurements is $\sim 0.015°$/fringe. In each case, the fringes are nominally symmetric about the crack plane indicating mode-I behavior. However, qualitative differences exist between the fringe patterns in terms of relative fringe sizes between

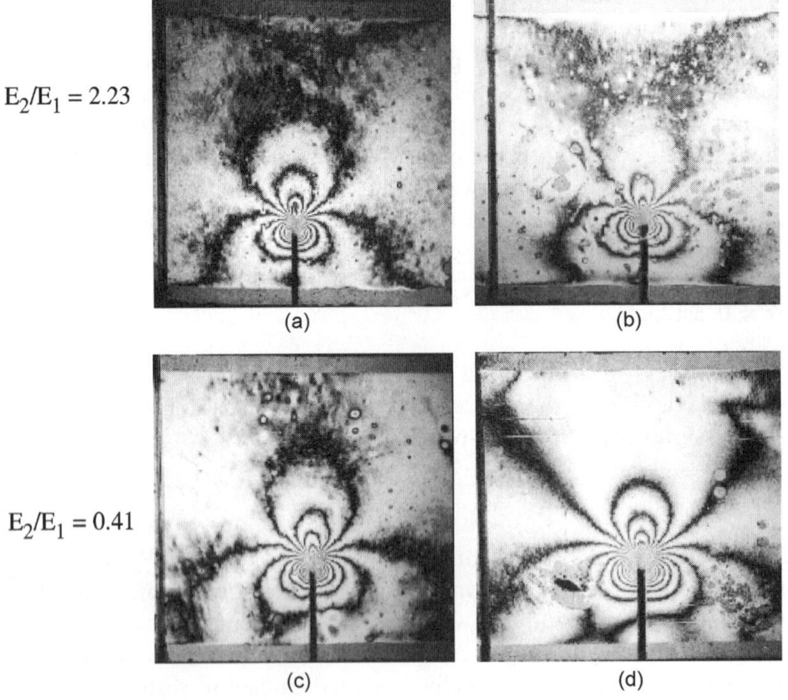

Fig. 3. Crack tip interference representing $\delta w/\delta x$ contours for a far-field bending stress $6M/BW^2 = 10.2$ MPa: (a) FGM with $E_2/E_1 = 0.41$, (b) homogeneous counterpart of (a), (c) FGM with $E_2/E_1 = 2.23$, (d) homogeneous counterpart of (c) (center-to-center distance from crack to drawn line is approximately 10 mm).

front and rear crack tip lobes in FGMs and, between the ones in FGMs and the equivalent homogeneous beams. These differences were quantified by extracting crack tip SIFs from the fringe patterns.

3.3.2. Dynamically loaded stationary cracks

The FGM and homogeneous beam samples were impact loaded in one-point symmetric loading configuration with a velocity of 5.3 m/s. The resulting interference patterns representing contours of dw/dx were recorded at a rate of 200 000 frames/s. For brevity, two select interferograms are shown for each experiment, at times 25 and 75 μs after impact, the latter being close to crack initiation in each case. CGS fringes for the homogeneous material are shown in the first column of Fig. 4. Fringes for the FGMs with crack located on compliant and stiff sides of the gradient are shown in the second and third columns of Fig. 4, respectively. The homogeneous beam had a uniform volume fraction of 0.42, and an elastic modulus of 9.6 GPa. The FGM with crack on the compliant side ($E_2/E_1 \sim 2.9$) has elastic modulus variation between 4 and 11 GPa with a value of $E_0 \sim 5.6$ GPa at the crack tip. Terminal values of Young's moduli are nearly the same for the FGM with crack on the stiff side ($E_2/E_1 \sim 0.45$), with crack tip $E_0 \sim 10$ GPa.

The structure of the fringe patterns in the crack tip vicinity for the two FGMs qualitatively resemble those in the homogeneous material. At the crack tip, one set of frontal and two sets of equally prominent lateral fringe lobes are present. Further examination of crack tip fringes in the FGM with crack on the compliant side ($E_2/E_1 > 1$, Fig. 4) show some deviation relative to the homogeneous case. Crack tip fringe lobes ahead of the crack show a lateral spread normal to impact direction. In this case, the compressive stress waves progressively slow down as they encounter more compliant material when they propagate away from the stiff edge. The opposite occurs when the tensile waves are reflected from the cracked edge.

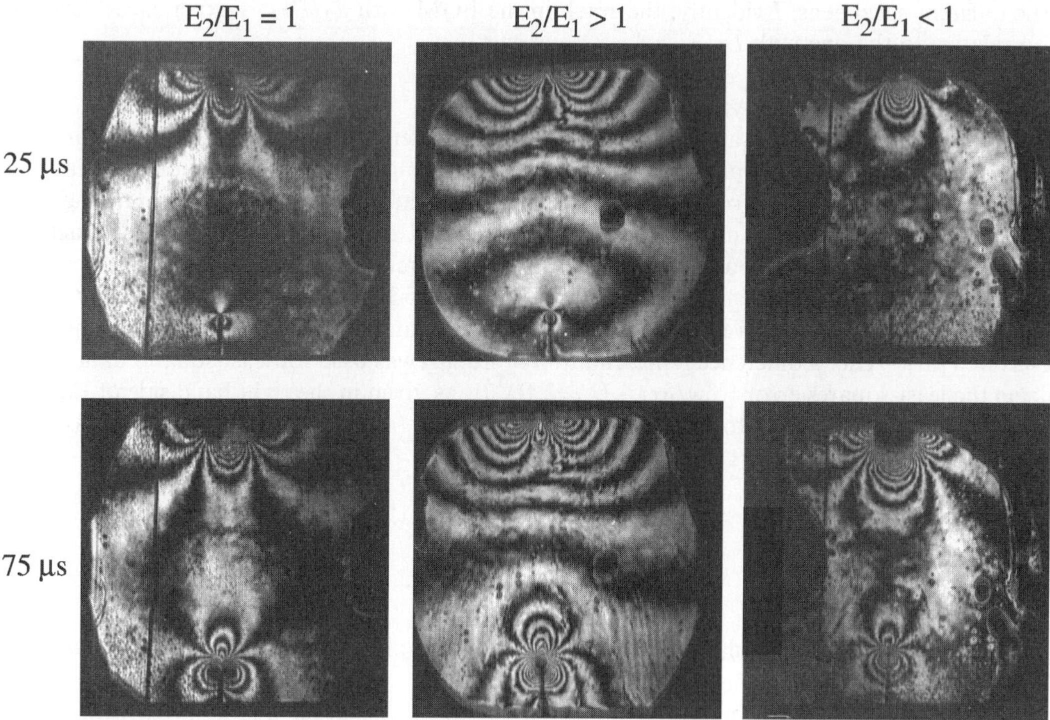

Fig. 4. Representative crack tip interference representing surface slope $\delta w/\delta x$ contours for homogeneous and FGM samples (fringe sensitivity 0.015°/fringes).

Crack tip fringes for the FGM with crack on the stiff side ($E_2/E_1 < 1$, Fig. 4) have a similar appearance to those for the homogeneous material. Here, the compressive stress waves generated at the compliant edge progressively encounter stiffer material and hence speed up during propagation away from the compliant edge. Nearly equi-sized front and rear fringe lobes around the crack tip are present. Additionally, a relatively uniform far-field is evident in this specimen.

For the homogeneous sample ($E_2/E_1 = 1$) with a longitudinal velocity of $\sim$2980 m/s, approximately 12.5 μs are required for the initial compressive waves to reach the bottom edge of the beam. The reflected tensile waves load the crack tip $\sim$ 15 μs after impact. Following that, crack tip fringes progressively enlarge and become more numerous up to crack initiation at $\sim$85 μs. Average longitudinal wave speed ($\overline{C}_l = 1/W \int_{-a}^{W-a} C_l(x)\,\mathrm{d}x$) for the FGM specimen with crack on the compliant side ($E_2/E_1 > 1$) is $\sim$2910 m/ s. In this case, initial propagation times of compressive and reflected tensile waves are similar to those of the homogeneous material. It is important to remember here that compressive waves progressively slow down as the crack tip is approached while the opposite is true for reflected tensile waves. The crack initiates $\sim$80 μs after impact. The FGM with the crack on the stiff side ($E_2/E_1 < 1$) has an average longitudinal velocity of $\sim$2900 m/s over the height of the beam. Again, this velocity corresponds to times of $\sim$12.5 μs for a compressive wave to reach the bottom edge, and $\sim$2.5 μs for a tensile wave to reflect to the crack tip. In this case, the incoming compressive waves from the contact point travel progressively faster towards the crack tip while the reverse occurs for the reflected tensile waves. The size of the crack tip fringes increases gradually until initiation at about 90 μs. An observation pertinent to FGM and homogeneous specimens can be made: compressive and tensile waves arrive at the original crack tips at approximately the same times for all the specimens since the average longitudinal wave speeds are nearly same. Therefore, initial crack tip loading occurs at nearly the same time. Local properties at individual crack tips, as well as nonuniform wave reflections from the edges will contribute to varying conditions in crack initiation between the different specimens. Evidently, the crack in the FGM with $E_2/E_1 < 1$ experiences delayed crack initiation relative to the one with $E_2/E_1 > 1$.

3.3.3. Extraction of SIFs from interferograms

Overdeterministic least-squares analyses of optical data [19] were used for extracting instantaneous SIFs. Briefly, the method consisted of digitizing the interferograms around the crack tip to obtain fringe order N and location (r, θ) data. In view of 3-D effects in the immediate crack tip vicinity, optical data within $r/B < 0.4$ were not considered in the analysis. This was based on existing knowledge of crack tip field triaxiality in finite size homogeneous samples [20]. Further, for mode-I, it has been shown using CGS that [19], data in the sector $90° < |\theta| < 135°$ have the least amount of 3-D effects and closely follow plane stress behavior. (Similar conclusions have been drawn in case of bimaterial interfacial cracks in finite thickness samples as well [21].) Also, a difference representation of Eq. (5) was used for least-squares data analysis, by minimizing the least-squares error $((\delta w/\delta x) - (Np/2\Delta))^2$ (w as given in the right hand side of Eq. (2)) at all digitized data points with respect to the unknown coefficients $A_0(\propto K_{\mathrm{Id}})$, A_1 and A_2 the SIF was evaluated.

4. Numerical simulations

4.1. Elasto-static computations

Companion finite element simulations of cracked FGM samples were carried out. *Plane stress* elasto-static simulations of cracked beams subjected to pure bending were performed for both FGMs and homogeneous materials. Owing to the symmetry, only one half of the beam was modeled with appropriate force and displacement boundary conditions. In the case of FGM beams, measured Young's modulus gradient from the samples was prescribed for FGMs using a novel technique developed by Rousseau and

Tippur [16]. The method essentially consists of a two step process in which first a thermo-mechanical problem is solved with a temperature distribution that corresponds to the required Young's modulus gradient to be imposed and coefficient of linear expansion set equal to zero over the entire domain. In the second step, Young's modulus is described as a function of the temperature gradient and the mechanical problem is solved. The Poisson's ratio was assumed constant and equal to a value corresponding to that of the crack tip, throughout the model. Separate simulations were carried out for situations with crack located on the stiff and the compliant sides of FGM beams. The corresponding crack tip elastic properties were used in separate homogeneous beam simulations. A typical model consisted of 8600 eight-noded iso-parametric elements with 26 000 nodes and two degrees of freedom per node, as shown in Fig. 5. The crack tip region had a square mesh with a typical size of 50–100 μm.

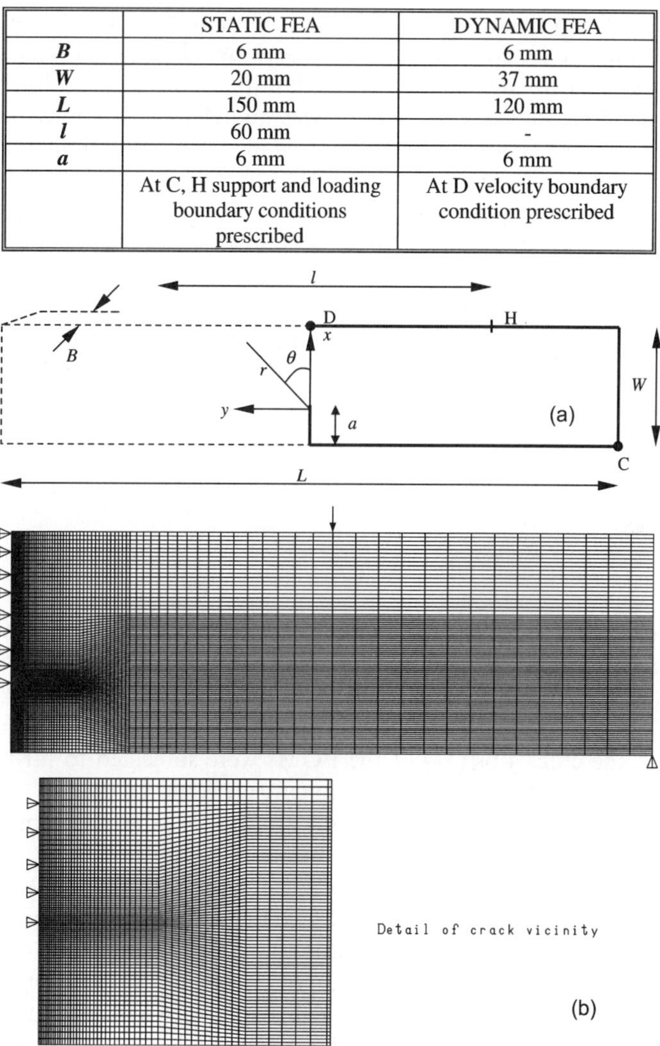

	STATIC FEA	DYNAMIC FEA
B	6 mm	6 mm
W	20 mm	37 mm
L	150 mm	120 mm
l	60 mm	-
a	6 mm	6 mm
	At C, H support and loading boundary conditions prescribed	At D velocity boundary condition prescribed

Fig. 5. Schematic of the geometry and loading configurations used in FEA models (a). Typical FGM finite element mesh (b).

In the immediate vicinity of the crack tip, assuming K-dominance, the expressions for crack opening displacements were truncated to the leading term and used for determining SIFs in FGMs in the limit $r \to 0$. Thus, for a locally homogeneous material behavior, crack opening displacement for an FGM is,

$$\delta_y(t)|_{\theta=\pm\pi, r\to 0} = \frac{8\overline{K}_1(t)}{E_0}\sqrt{\frac{r}{2\pi}}. \tag{7}$$

Here, E_0 is the crack tip Young's modulus. The variation of $\overline{K}_1(r, \pm\pi)$ obtained from the finite element solution was first plotted and its extrapolated value at the crack tip was identified as the SIF:

$$K_{\mathrm{I}} = \lim_{r\to 0} \overline{K}_{\mathrm{I}}. \tag{8}$$

4.2. Elasto-dynamic computations

Companion *plane stress* elasto-dynamic finite element simulation of all the three categories of experiments, homogeneous and the two FGMs, were performed. A two-dimensional quadrilateral mesh of eight-noded isoparametric elements with two degrees of freedom per node was used in the analysis. Again because of the symmetry, only one half of each sample was modeled, with approximately 10000 nodes and 3300 elements. The crack tip vicinity was highly refined as in static simulations for accurate evaluation of crack tip parameters. (The model was first benchmarked against elasto-dynamic finite element results for a mixed-mode stationary crack problem in homogeneous material. The details are avoided here for brevity.) Simulations were conducted as a one-point symmetric impact of an elastic planar medium by a rigid impactor, by imposing the downward velocity of 5.3 m/s (equal to the impact velocity of the experiments) to the topmost node in the plane of the crack. The Newmark time-integration scheme was used with a minimum time step of 0.04 µs. As in the static simulations, instantaneous mode-I SIFs were determined by regression analysis of crack tip opening displacements at discrete time intervals of every 5 µs after impact until crack initiation.

5. Results and discussion

5.1. Static loading

FGM specimens were subjected to far-field bending stress $(6M/BW^2)$ of 10.2 MPa. The FGM sample with a crack located on its compliant side had a crack tip modulus $E_0 = 4.8$ GPa, and a modulus ratio (E_2/E_1) between its two edges of 2.2. The other FGM sample, with an edge crack on the stiff side had a crack tip modulus $E_0 = 6.8$ GPa, and a modulus ratio of 0.4. The homogeneous samples with moduli corresponding to those at the crack tips (E_0) of the FGMs were subjected to far-field stresses identical to those of their FGM counterparts, so that a direct comparison could be made.

A comparison of the mode-I SIFs obtained from the analysis of interferograms is presented in Table 1. Under the premise of K-dominance, the experimental values of K_{I} for the FGMs are about 20% lower than those predicted by the finite element calculations. When least-squares analysis including nonsingular contributions to the singular field was carried out with two terms of the expansion field (Eq. (5)), the values of SIFs improved and the error between the numerical and experimental results reduced to 10% or less.

The influence of elastic gradient on statically loaded FGM beams can be inferred from SIF results. The beam with a crack on the glass-rich (stiff) side $(E_2/E_1 < 1)$ experience higher value of K_{I} when compared to its homogeneous counterpart subjected to identical far-field loading. On the other hand, the FGM beam with a crack on the epoxy-rich (compliant) side $(E_2/E_1 > 1)$ of the beam experiences a lower K_{I} compared to

Table 1
Comparison of computed and measured SIFs for a static far-field bending stress $(6M/BW^2) = 10.2$ MPa

Method of calculation	FGM $E_2/E_1 = 2.23$, $E_0 = 4.8$ GPa	Homogeneous $E_2/E_1 = 1$, $E_0 = 4.8$ GPa	FGM $E_2/E_1 = 0.41$, $E_0 = 6.8$ GPa	Homogeneous $E_2/E_1 = 1$, $E_0 = 6.8$ GPa
FEA (plane stress)	1.27	1.51	1.77	1.45
Optically measured (K-dominant description)	1.06	1.26	1.44	1.20
Optically measured (asymptotic expansion)	1.13	1.44	1.80	1.50

its homogeneous counterpart. Thus in the specific case studied, the results suggest that elastic gradients shield a crack on the compliant side and lower the SIF by a factor of about 1.5 when compared to the one with the crack on the stiff side.

5.2. Dynamic loading

Instantaneous SIFs, $K_{Id}(t)$, were evaluated up to crack initiation. Symbols in Fig. 6(a) and (b) represent optically determined dynamic stress factor history for the two FGMs with $E_2/E_1 < 1$ and $E_2/E_1 > 1$, respectively. In these plots, time after impact is normalized with respect to the dilatational wave speed at the crack tip location and the specimen height: $T = tC_l W$. Finite element results are also plotted as a solid line, up to crack initiation, t_I (determined experimentally). Good agreement between the two is evident. Note the absence of experimental data points for certain frames because of either loss of optical information and/or, during early stages of fringe formation, the fringes reside within the region of dominant triaxiality where they can not be reliably analyzed.

For comparison purposes, normalized SIFs are plotted against normalized time for the two FGMs and homogeneous sample in Fig. 6(c). In this plot, SIFs are normalized by the corresponding value at initiation. In each case, monotonic increase in $K_{Id}(t)$ is observed. Crack initiation occurred at $T \sim 6.8$ after initiation for the homogeneous material. Evidence of crack motion was not observed until $T \sim 6.2$ and 7.2 for the FGMs with $E_2/E_1 > 1$ and $E_2/E_1 < 1$, respectively. Note that the behavior of the homogeneous material lies between those of the two FGMs. Interestingly, crack initiation in FGMs with $E_2/E_1 < 1$ are delayed relative to the opposite FGM configuration and the homogeneous sample. Also, evident from the plots are

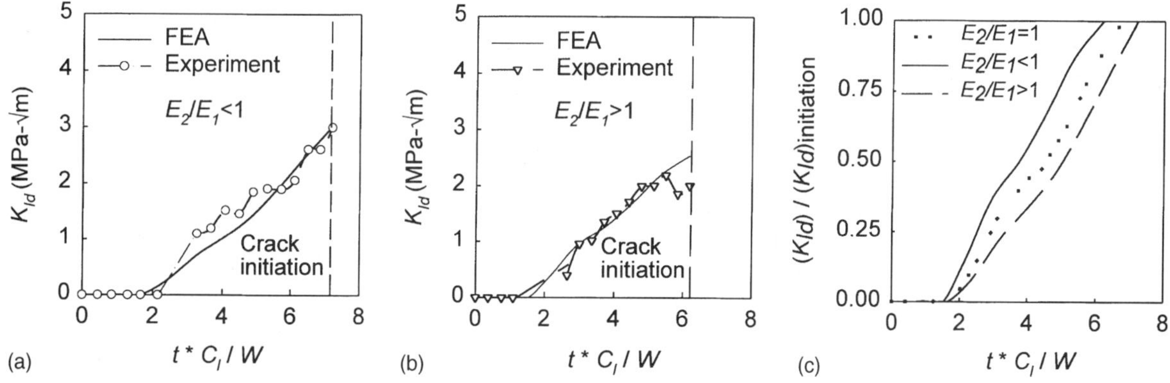

Fig. 6. Measured mode-I dynamic SIF history compared with FEA for FGM with $E_2/E_1 < 1$ (a), $E_2/E_1 > 1$ (b). Normalized SIF for FGMs and homogeneous materials (c).

the differences in the crack tip loading rate ($\Delta K_{\mathrm{Id}}/\Delta t$) with lower rate seen in case of the FGM with the crack on the stiff side of the beam.

In the results discussed above, in addition to the local fracture toughness difference at the crack tips, it should be noted that crack initiation event is affected by combination of differences in the average values of wave speeds, crack tip elastic modulus and crack tip elastic gradients ($\mathrm{d}(E/\rho)/\mathrm{d}X$) between samples. Accordingly finite element simulations were carried out to isolate the effect of elastic gradients on crack initiation. Homogeneous beams and FGM beams with assumed linear elastic variations were simulated. Further, material variations were identified in terms of E/ρ variations and Poisson's ratio was kept constant ($v = 0.34$). Material properties implemented were close to those realized in the actual glass/epoxy FGMs used in the experiments. The Young's moduli were varied between 4 and 12 GPa, and densities between 950 and 1850 kg/m^3. Individual variations of E and ρ were implemented using a 'look-up' chart relating E, ρ, (and E/ρ) to filler volume fraction in actual glass-filled epoxy. The prescribed elastic gradients consisted of linearly increasing and decreasing values of E/ρ for FGMs, and constant E/ρ representing homogeneous materials, all having identical properties at he crack tip (Fig. 7(a)). By selecting an $a/W = 0.5$, the crack tip Young's moduli, average value of the elastic wave speeds and $|\mathrm{d}(E/\rho)/\mathrm{d}X|$ were identical in all three situations. Thus, the influence of increasing and decreasing elastic gradients could be isolated.

Plot of K_{Id} history after impact is shown in Fig. 7(b) for the three cases. The variation in SIF for the FGM with $E_2/E_1 > 1$ displays the highest rate of increase whereas the FGM with $E_2/E_1 < 1$ has the lowest rate of increase. The homogeneous material response lies in between the two FGMs. Rates of increase in $\mathrm{d}K_{\mathrm{Id}}/\mathrm{d}t$ (Fig. 7(c)) are also lowest for FGM with $E_2/E_1 < 1$, followed by the homogeneous material. The FGM with $E_2/E_1 > 1$ has the steepest increase up to $T \sim 5$. Beyond that point, an inflection in the curve drops the increase in stress intensification below that of the homogeneous material but still higher than that for FGM with $E_2/E_1 < 1$. Material properties at the crack tip being the same in all the three cases, for an *assumed* critical value of SIF of, say, 3 MPa $\sqrt{\mathrm{m}}$, crack initiation occurs (Fig. 7(b)) at $T \sim 5.8$, 6.6, and 7.9 for FGM with increasing gradient, homogeneous material, and FGM with decreasing gradient, respectively. These trends are identical to the ones observed experimentally. Thus, purely based on elastic considerations, crack initiation in the latter case is significantly delayed, by a factor of ~ 1.4, when compared to the former. Behavior of the homogeneous material, as with all other parameters, lies between the two FGMs.

To further understand these differences, additional dynamic impact simulations on *uncracked* beams (FGM and homogeneous) were performed for monitoring stress histories. Contours of $(\sigma_x + \sigma_y)$ for all

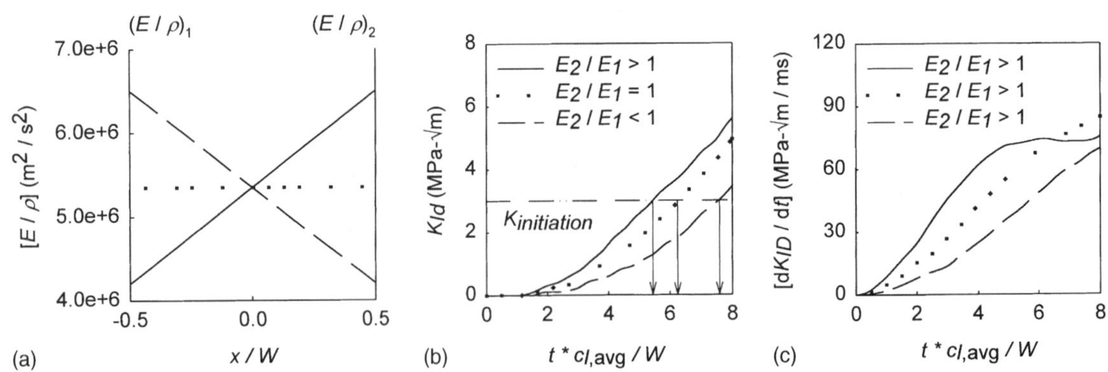

Fig. 7. SIF history in homogeneous beams and FGMs with linearly varying gradients (crack tip located at $X/W = 0.5$ or $x/W = 0$). (a) Elastic variations used in the FEA, (b) SIF histories, and (c) crack tip loading rate histories (W = height of the sample).

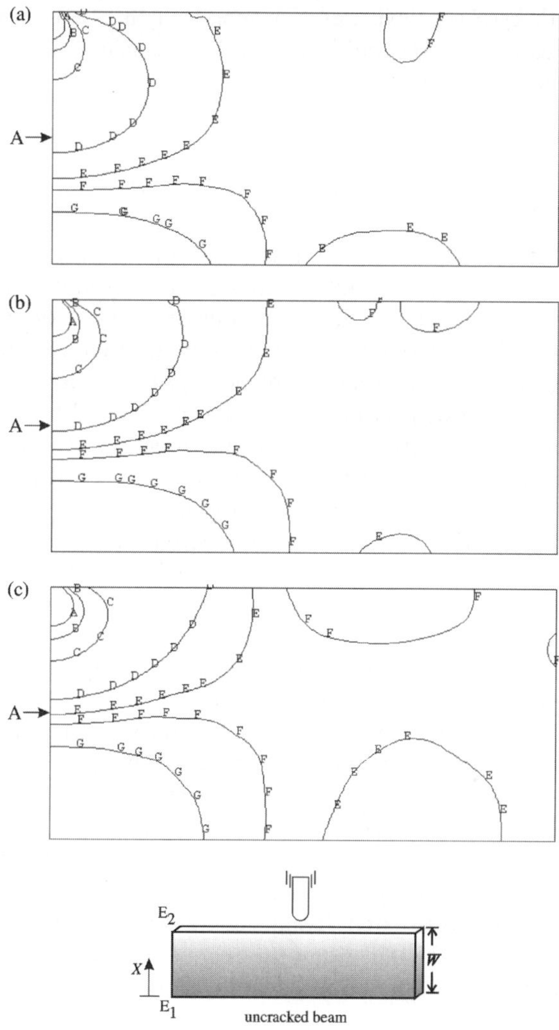

Fig. 8. $(\sigma_x + \sigma_y)$ contours at nondimensional time $T = 5.7$ $(A = -5, B = -3.5, C = -2, D = -0.5, E = -0.1, F = 0.1, G = 0.5) * 10$ MPa. (a) $E_2/E_1 < 1$, (b) $E_2/E_1 = 1$, $E_2/E_1 > 1$. Point A corresponds to $(X/W = 0.5, y = 0)$. Note that only the right-half of the beam is modeled.

three materials are shown in Fig. 8 after several wave reflections $(T = 5.7)$ between the top and bottom edges. Relative to the homogeneous beam (Fig. 8(b)), compressive waves emanating from the impact point for the FGM with $E_2/E_1 > 1$ (Fig. 8(c)) are constricted vertically, whereas those for the FGM with $E_2/E_1 < 1$ (Fig. 8(a)) are elongated in the same direction. In the former case, as compressive waves (for example, contour level D) travel from the top to the bottom edge, they are delayed by material property variations in that direction. Conversely, for the FGM with $E_2/E_1 < 1$, material property variation promotes faster wave propagation downwards, thereby elongation of $(\sigma_x + \sigma_y)$ contour (again, level D) along the beam height is evident. The reflected tensile waves accelerate upward for FGMs with $E_2/E_1 > 1$, decelerate for FGMs with $E_2/E_1 < 1$, and propagate at a constant speed in the homogeneous case. As seen in Fig. 8, at time $T = 5.7$, a point A (located at $X/W = 0.5$) along the line of symmetry of the beam experiences tensile stresses for the FGM with $E_2/E_1 > 1$ while the corresponding points in the homogeneous beam and the FGM beam with $E_2/E_1 < 1$ experience progressively higher compressive stresses, respectively. Accordingly, the latter two cases are elastically shielded from tensile stresses for a longer duration and would experience delayed crack initiation.

6. Concluding remarks

Influence of elastic variations on crack initiation in compositionally graded beams was studied experimentally and numerically. Beam samples with unidirectional elastic gradients subjected to static four-point bending and dynamic one-point impact loading were considered for experimentation. Compositionally graded epoxy with solid-microsphere glass filler was the material of choice for producing elastic gradients by continuously varying filler volume fractions over the height of the beam samples. Macroscopic Young's modulus variations of about 1:3 and density of about 1:2 were realized using gravity casting. The optical method CGS was used for measuring surface slopes in the direction of elastic gradients. The high-speed photography was utilized for dynamic crack initiation studies for capturing instantaneous deformation field histories. A locally homogeneous material behavior in the immediate vicinity of the FGM crack tip is assumed for extracting SIFs in both static and dynamic situations using continuum models for crack stresses. Companion plane stress finite element simulations for both static and dynamic loading situations were performed and SIFs were determined using regression analysis of crack tip opening displacements. The experimentally determined values of SIFs are in good agreement with the ones from computations.

Under static loading conditions, the elastic gradients offer crack tip shielding. In the specific examples studied, the crack tip located on the compliant side of a beam subjected to pure bending moment experienced SIF 1.5 times lower than that with a crack on the stiff side in a beam with an identical elastic variation. In the dynamic experiments, crack initiation in situation when the crack is located on the stiffer side of the beam were delayed when compared to the opposite situation when the crack is located on the compliant. However, this was a combined effect of the crack tip variables such as elastic properties, elastic gradient and the local fracture toughness. Accordingly, computations on dynamically loaded stationary cracks were performed for idealized situations where identical crack tip elastic properties were enforced and the effect of elastic gradients alone was isolated. Results confirmed the experimental observation that crack initiation in an FGM with crack on the stiff side of a beam subjected to symmetric one-point impact is delayed considerably when compared to the opposite configuration due to lower rate of crack tip loading.

Acknowledgements

The support of this research by ARO Solid Mechanics Program (DAAG55-97-1-0110) and NSF Materials Program (CMS-9622055) is gratefully acknowledged.

References

[1] Nino M, Hirai T, Watanabe R. The functionally gradient materials. J Jpn Soc Compos Mater 1987;13:257.
[2] Atkinson C, List RD. Steady state crack propagation into media with spatially varying elastic properties. Int J Eng Sci 1978;16:717–30.
[3] Delale F, Erdogan F. The crack problem for a nonhomogeneous plane. ASME J Appl Mech 1983;50:609–14.
[4] Eischen JW. Fracture of nonhomogeneous materials. Int J Fract 1987;34:3–22.
[5] Jin Z-H, Noda N. Crack-tip singular fields in nonhomogeneous materials. ASME J Appl Mech 1994;61:738–40.
[6] Erdogan F. Fracture mechanics of functionally graded materials. Compos Eng 1995;7:753–70.

[7] Babaei R, Lukasiewicz SA. Dynamic response of a crack in a functionally graded material between two dissimilar half-planes under anti-plane shear impact load. Eng Fract Mech 1998;60:479–87.

[8] Nakagaki M, Wu YD, Hagihara S. Dynamically propagating crack in graded particle dispersed composites. Fract Strength Sol 1998;145:333–42.

[9] Parameswaran V, Shukla A. Dynamic fracture of a functionally gradient material having discrete property variation. J Mater Sci 1998;33:3303–11.

[10] Chiu TC, Erdogan F. One-dimensional wave propagation in a functionally graded elastic medium. J Sound Vib 1999;222:453–87.

[11] Marur PR, Tippur HV. Evaluation of mechanical properties of functionally graded materials. J Testing and Evaluation 1998;26:539–45.

[12] Butcher RJ, Rousseau C-E, Tippur HV. A functionally graded particulate composite: preparation measurements and failure analysis. Acta Mater 1999;47:259–68.

[13] Parameswaran V, Shukla A. Processing and characterization of a model functionally gradient Material. J Mater Sci 2000;35:21–9.

[14] Li H, Lambros J, Cheeseman B, Santare MH. Experimental investigation of quasi-static fracture of functionally graded material. Int J Solids Struct 1999;37:3715–32.

[15] Marur PR, Tippur HV. Dynamic response of bimaterial and graded interface cracks under impact loading. Int J Fract 2000;103:103–9.

[16] Rousseau C-E, Tippur HV. Compositionally graded materials with cracks normal to the elastic gradient: examination of fracture parameters relative to bimaterials. Acta Mater 2000;48:4021–33.

[17] Rousseau C-E. Evaluation of crack tip fields and fracture parameters in functionally graded materials. PhD Dissertation, Auburn University, 2000.

[18] Weng GJ. Some elastic properties of reinforced solids with special reference to isotropic ones containing spherical inclusions. Int J Eng Sci 1984;22:845–56.

[19] Tippur HV, Krishnaswamy S, Rosakis AJ. A coherent gradient sensor for crack tip deformation measurements: analysis and experimental results. Int J Fract 1991;48:193–204.

[20] Rosakis AJ, Ravi-Chandar K. On crack-tip stress state: an experimental evaluation of three-dimensional effects. Int J Solids Struct 1986;22:121–38.

[21] Sinha JK, Tippur HV, Xu L. An interferometric and finite element investigation of interfacial crack tip fields: role of mode-mixity on 3D stress variation. Int J Solids Struct 1997;34:741–54.

PERGAMON

Engineering Fracture Mechanics 69 (2002) 1695–1711

**Engineering
Fracture
Mechanics**

www.elsevier.com/locate/engfracmech

Investigation of crack growth in functionally graded materials using digital image correlation

Jorge Abanto-Bueno, John Lambros [*]

*Department of Aeronautical and Astronautical Engineering, University of Illinois at Urbana-Champaign,
306 Talbot Lab, MC 236, 104 South Wright Street, Urbana, IL 61801, USA*

Received 1 February 2001; received in revised form 1 July 2001; accepted 10 July 2001

Abstract

The crack growth resistance behavior of functionally graded materials (FGMs) was investigated using the full-field measurement technique of digital image correlation (DIC). Model FGMs were manufactured using ultraviolet irradiation of a photosensitive polyethylene co-polymer. Edge crack FGM fracture specimens were loaded such that stable quasi-static crack growth occurred. A DIC code, generated and tested in-house, was used to obtain the in-plane displacement field surrounding the propagating crack. The results were then used to extract stress intensity factors and generate resistance curves for the FGM. A crack growth resistance behavior of continually increasing toughness with crack extension was observed. The full-field optical technique also allowed for the investigation of regions of K-dominance in the FGM (i.e. areas where the theoretical asymptotic fields well describe the near tip deformation). The results were compared to stress intensity factor values obtained using crack face profile shape measurements and good agreement was found.

Keywords: Functionally graded materials; Crack growth; Digital image correlation; R-curve

1. Introduction

The theoretical study of crack tip stress fields in continuously nonhomogeneous materials (also termed functionally graded materials or FGMs) is well established. Comparable experimental study, however, lags behind because of the inherent difficulties in manufacturing bulk FGMs [1]. Eischen [2] studied the stress and displacement fields around the crack tip in a nonhomogeneous material whose elastic moduli were specified by continuous and generally differentiable functions. Eischen's results show that the asymptotic stress singularity at the crack tip of an FGM is of the same form as that for a crack tip in a homogeneous material, i.e.

$$\sigma_{ij}(r, \theta) = \frac{K_{\mathrm{I}}}{\sqrt{2\pi r}} F_{ij}^{\mathrm{I}}(\theta) + \frac{K_{\mathrm{II}}}{\sqrt{2\pi r}} F_{ij}^{\mathrm{II}}(\theta),
\tag{1}$$

---

[*] Corresponding author. Tel.: +1-217-333-2242; fax: +1-217-244-0720.
 E-mail address: lambros@uiuc.edu (J. Lambros).

0013-7944/02/$ - see front matter © 2002 Elsevier Science Ltd. All rights reserved.
PII: S0013-7944(02)00058-9

where σ_{ij} are the Cartesian stress tensor components. Angular functions $F_{ij}^{I}(\theta)$ and $F_{ij}^{II}(\theta)$ are the same as those in the expression for the case of a homogeneous material [3], with (r, θ) being a polar coordinate system attached to the crack tip and K_{I} and K_{II} the mode I and II stress intensity factors respectively. However, the range of dominance of Eq. (1) (i.e. K-dominance) for an FGM is, in general, unknown and will depend on the specific geometry, loading and material property variation [4].

Although the experimental study of FGMs certainly lags behind in comparison to the theoretical, it is an area of increasing activity. Butcher et al. [5] have manufactured a functionally graded composite by using gravity casting of glass beads in a polymer matrix. Crack tip fields under quasi-static loading, with cracks perpendicular to the direction of elastic gradient, were imaged using optical techniques. The stress intensity factors they obtained were successfully compared with finite element analysis results. Parameswaran and Shukla [6] have succeeded in making FGMs with discrete property variation (i.e. a layered structure) and dynamic fracture in this material was studied using a combination of photoelasticity and high-speed photography. A resistance curve for crack growth in a Ti–TiB FGM was obtained in the work of Carpenter et al. [7], using a boundary load measurement technique.

For a fundamental study of the fracture behavior of large scale FGM specimens, Lambros et al. [8] have developed a low cost method of manufacturing model FGMs through irradiation of a photodegradable polymer with ultraviolet (UV) light. The polymer, polyethylene co-carbon monoxide or ECO, has mechanical properties that are a function of UV exposure time. Li et al. [9] have studied the fracture behavior of such an FGM using a hybrid numerical–experimental method. Using experimentally recorded crack length and applied load, they developed a method to calculate stress intensity factor and energy release rate with the aid of finite elements. In this fashion they were able to obtain crack growth resistance curves for both uniformly irradiated and gradient ECO.

The methods of Li et al. [9] and Carpenter et al. [7] are based on measurement of far-field quantities such as applied load and displacement. In order to most accurately determine the resistance curve of the FGM and obtain an accurate failure criterion based on local stress or displacement values, measurement of *near crack tip* fields is needed. The objective of this study is to experimentally determine crack growth resistance curves from near tip measurements taken in real time as the crack propagates quasi-statically. Transmission optical interferometric methods cannot be used with ECO since it does not fully transmit light. In addition the samples used in the experiments cannot be coated as they are very thin (0.42 mm). This prohibits the use of reflective coating optical interferometric methods. Hence, in this study it was decided to use the full-field method of digital image correlation (DIC) [10–13] to measure the in-plane displacement field surrounding the propagating crack tip. The DIC technique is very easy to set up and operate, is extremely cheap, and is very robust as it does not require precision alignment as in the case of interferometric methods.

2. Experimental method

2.1. Specimen preparation and testing

The FGMs used in this study were fabricated from a polyethylene 1% carbon monoxide co-polymer (ECO). The work of Ivanova et al. [14], Andrady and Nakatsuka [15] and Lambros et al. [8] has shown that mechanical degradation of ECO occurs when it is irradiated with UV light. Lambros et al. [8] showed that for this material the strain to failure decreases by a large amount after a few hours of irradiation and the Young's modulus increases from around 150 MPa to around 275 MPa on irradiating the material for about 140 h. Fig. 1 shows a sequence of tensile stress–strain curves for ECO subjected to varying times of UV irradiation, clearly illustrating the simultaneous stiffening and embrittling effect.

Graded ECO specimens were obtained by slowly removing a UV shield covering the material, while UV irradiation was incident on the specimen. Note that the base material used is in the form of thin (0.42 mm)

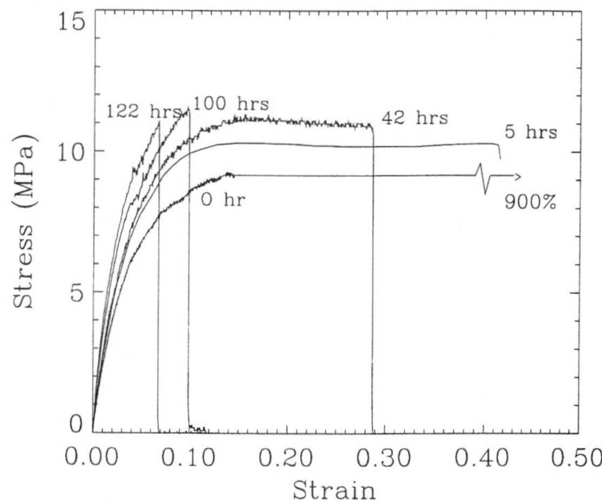

Fig. 1. Effect of UV irradiation on the uniaxial tensile stress–strain response of polyethylene 1% CO co-polymer (ECO).

sheets, so irradiation effects penetrate through the sheet thickness. The resulting graded material exhibits a property variation in the in-plane directions. A simultaneous increase in modulus and decrease in failure strain is attained. The resulting material therefore is a plane stress specimen which in-plane exhibits similar mechanical behavior to some metal–ceramic FGMs, which would also be characterized by an increasing stiffness and decreasing failure strain with distance. For more details on specimen fabrication see [8].

In the analytical study of FGMs (e.g. [16]) most often an exponential variation of Young's modulus and a constant Poisson's ratio are assumed. Typically, a variation of the type

$$E(x) = E_0 e^{\beta x},\tag{2}$$

where E_0 is the modulus at one end of the sample, is used. Quantity β, which has dimensions of $1/L$, is an intrinsic material length scale denoting the spatial extent of nonhomogeneity. When the scale of observation is comparable to $1/\beta$, inhomogeneity effects are expected to be significant. It is easy to see that,

$$\beta = (1/h)\ln(E_1/E_0),\tag{3}$$

where E_0 and E_1 denote the variation of Young's modulus over a length h, which may or may not be the entire specimen length. In our case the length scale of interest is the extent of crack growth, so all values of β have been computed for the variation of E_0 and E_1 over the field of physical observation. In the present work FGMs with β of $-6/m$ and $-30/m$ have been tested.

The mode I fracture behavior of functionally graded ECO was studied using the single edge notch specimen shown in Fig. 2. Note that this loading fixture is somewhat different from that used in [9] as it has rigid arms which provide a uniform extension boundary condition along the entire length of the sample. This was done in order to provide an experimental boundary condition that better approximates the fixed grip condition used in the theoretical analysis of [16]. Specimen dimensions are also shown in Fig. 2. A 30 mm starter notch was cut in each case using a razor blade. The radius of curvature of this starter notch is extremely small since it very easy to cut this material with a sharp blade. The pre-notch was located in the more brittle side (i.e. longer irradiation times) and the crack was grown towards the more ductile side (i.e. shorter irradiation times). This provides for a stable crack growth configuration. The specimen was mounted in an MTS machine and loaded at a constant displacement rate of 0.5 mm/min. Load was re-corded on a digitizing oscilloscope (Tektronix TDS 540). One side of the ECO sheet was sprayed with black paint that would form the basis of the speckle pattern to be used in the DIC technique. During and after

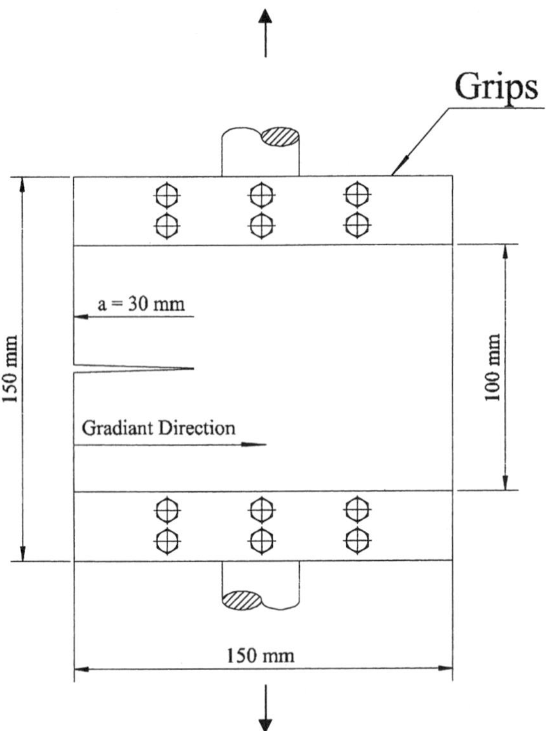

Fig. 2. Specimen dimensions and geometry for single edge crack fracture experiments on homogeneous and graded ECO. For graded specimens material gradient was in the direction of crack-line.

each test it was confirmed that the painted speckle pattern did not debond from the specimen. By conducting tensile experiments on painted and unpainted (unirradiated) samples, it was also verified that the thin paint coating did not affect the mechanical response of the ECO sheet. In both cases identical results were obtained (within the error of the experimental set-up) for both modulus and yield strength.

Fig. 3 shows a selected sequence of CCD images depicting crack growth in a graded ECO sample. Times denoted in the figure are with respect to $t = 0$ s as the instant of load application. Crack initiation corresponds to about $t = 350$ s. A region of dimensions approximately 45 mm length and 35 mm height is recorded in each frame. A 640×480 8-bit (256 gray scale levels) Sony XC-77 CCD camera was used for imaging. In most cases appropriate lenses to produce a spatial resolution of 0.03 mm/pixel were used. The optical method of DIC was used to analyze the various images by comparing them with an image taken prior to deformation. Details of the DIC method are provided in the next section.

2.2. Digital image correlation

The method of digital image correlation (DIC) is well developed and its details are well known. A brief description of the method is given in this section and more details can be found in the literature [10–13]. Consider a planar body undergoing a two-dimensional deformation, such as the one shown in Fig. 4. A material point P_0 at coordinate (x_1, x_2) in the undeformed configuration is shifted to coordinate (y_1, y_2) because of the deformation. If we consider a point P at coordinates $(\tilde{x}_1, \tilde{x}_2)$ in subset S and if the subset S is small enough for the deformation to be locally homogeneous around P_0, then:

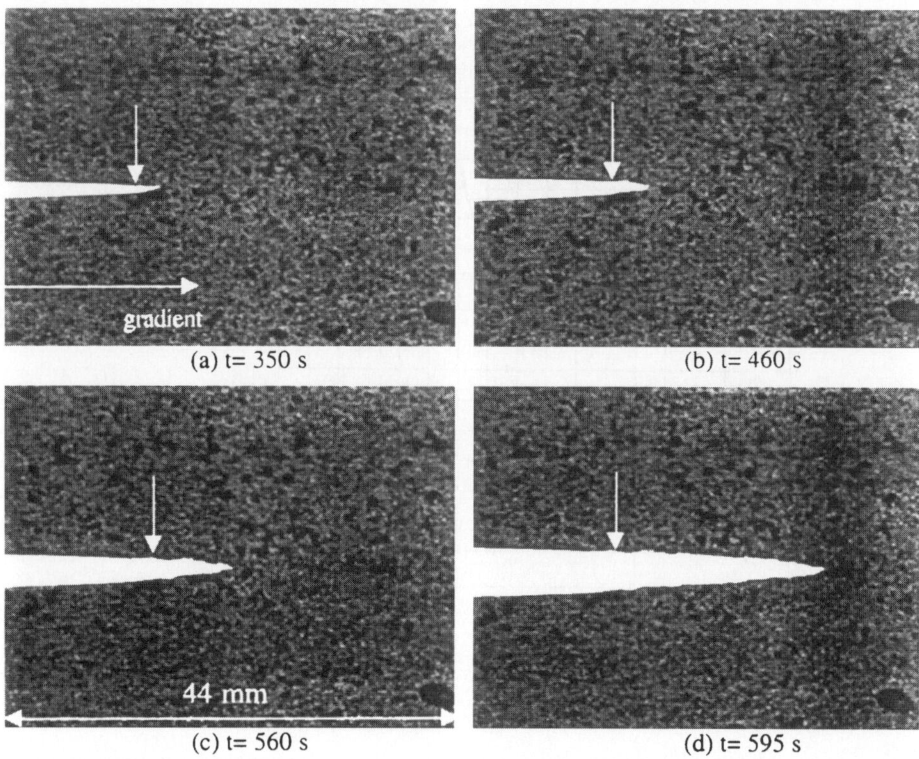

(a) t= 350 s

(b) t= 460 s

(c) t= 560 s

(d) t= 595 s

Fig. 3. Sequence of CCD images at four different times depicting quasi-static crack growth in a graded ECO sample. Note the speckle pattern used in the DIC technique. Vertical arrow indicates the razor-cut initial notch tip. Time $t = 0$ s corresponds to initial application of load.

$$\tilde{y}_1 = \tilde{x}_1 + u_1 + u_{1,1}\Delta x_1 + u_{1,2}\Delta x_2, \tag{4a}$$

$$\tilde{y}_2 = \tilde{x}_2 + u_2 + u_{2,1}\Delta x_1 + u_{2,2}\Delta x_2, \tag{4b}$$

where u_1 and u_2 are the displacements of $P_0(x_1, x_2)$ in the 1 and 2 directions respectively, $u_{\alpha,\beta}$ denotes the differentiation $\partial u_\alpha / \partial x_\beta$ and $\Delta x_\alpha = \tilde{x}_\alpha - x_\alpha$, $\alpha, \beta = 1, 2$.

The DIC method, originally developed in [10], uses two-dimensional correlation of digitized images of an object in the deformed and undeformed configurations to determine the deformation field of the object, i.e. the six parameters u_1, u_2, $u_{1,1}$, $u_{2,2}$, $u_{1,2}$, $u_{2,1}$ in Eqs. (4a) and (4b). Refs. [12,13,17] discuss the application of the DIC method to determine displacements in the cases of bending of a cantilever beam, rotational motion of a disk, vibratory motion of a rod, uniform translation, rotation and uniform finite-strain. McNeill et al. [18] were the first to use the DIC method to determine mode I stress intensity factors in fracture specimens.

The speckle pattern depicted in the frames of Fig. 3 was used in the DIC technique. To determine the displacements of the specimen surface, two images of the object, one prior to deformation and one after the deformation, were compared. In all cases the current deformed image was compared with an image taken at zero load. Since the DIC scheme used in this work is based on a linearized theory, its results become increasingly inaccurate as large strain and rotation effects become significant. It has been shown that using the small strain implementation of DIC accurate measurement of strains up to 20% is possible [19], at a resolution of between 50 and 200 microstrain [17,20]. In our case the maximum recorded strains, i.e. near the crack tip and at the highest loads, are about 2% while the smallest strains, at distances in excess of 15 mm

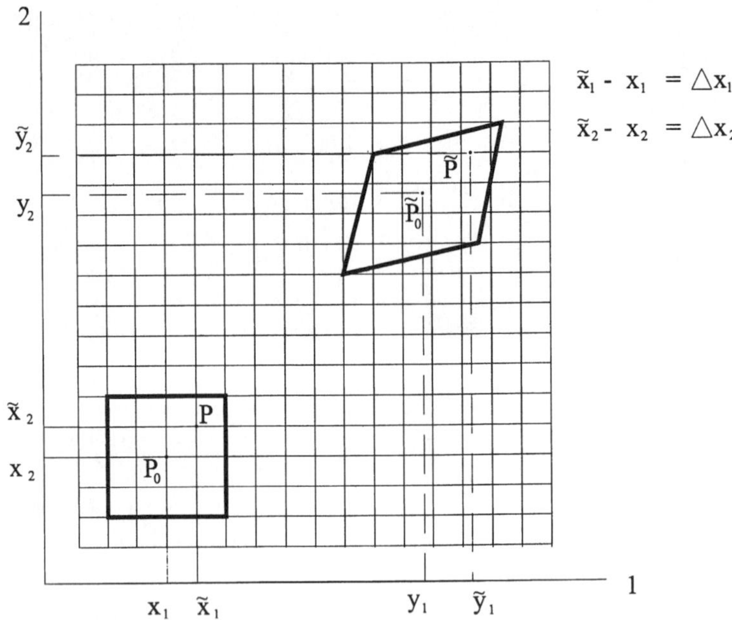

Fig. 4. Body undergoing planar deformation. Undeformed pixel subset is denoted by dark square and is deformed into the region denoted by dark diamond.

from the crack tip are about 500 microstrain, i.e. well within the upper and lower limits of the DIC technique. In addition the rigid body rotations recorded here are of the order of 0.01°, again well within the limits of small deformation DIC which can accurately record up to 10° rotation [13].

The DIC comparison process involves pixel subsets (dark square in Fig. 4) of the first image being compared with subsets (dark diamond) of the second image, to determine the displacements at the center, P_0, of the subset S in the undeformed configuration. In this method, there are two important steps—interpolation and optimization.

Interpolation: Since a pixel in the undeformed configuration may be displaced to a *nonpixel* point in the deformed configuration, it may become necessary to obtain the gray level intensity at a point (y_1, y_2) that may not lie on the rectangular grid of the deformed digitized image. This estimation of intensity is done through interpolation from the known intensity values at the pixel points of the deformed image. The interpolation function used may be either bilinear [11,12,17] or bicubic [21]. Note that this interpolation is needed only on the deformed image. After interpolation, a comparison can be made between the pixel intensity in the undeformed and deformed subsets.

Optimization: Let S be the subset around $P_0(x_1, x_2)$ and its image in the deformed configuration be $\tilde{S}$. (In our analysis, the subset S is a square grid of points shown as the dark square in Fig. 4.) Intensity, $f(P(x_1, x_2))$, of the subset S in the undeformed configuration is then compared to the intensity, $g(\tilde{P}_0(y_1, y_2))$, of the deformed subset $\tilde{S}$. As was discussed in [22], to perform this comparison, we can use the following least square correlation coefficient C,

$$C = \frac{\sum_{P_S \in S}(f(P_S) - \tilde{g}(P_S, \mathbf{V}))^2}{\sum_{P_S \in S} f^2(P_S)}, \tag{5}$$

$f(P)$ is the intensity of the point P and $g(\tilde{P}(P, \mathbf{V}))$ the intensity of point $\tilde{P}$. Point $\tilde{P}$ itself depends on P and the vector $\mathbf{V} = \{u_1 \ u_2 \ u_{1,1} \ u_{2,2} \ u_{1,2} \ u_{2,1}\}$, consisting of the displacement and displacement derivatives, using the notation introduced in [22]. The value of the correlation coefficients will be zero if the six components of the vector **V** are determined exactly. Note that since we deal with discrete images, the expression of the correlation coefficient contains sums of intensity over a grid of points P_S contained in S. Quantity $\tilde{g}(P, \mathbf{V})$ is the *interpolated intensity* value at the coordinate $\tilde{P}$, which depends on P and $\mathbf{V}$, determined either using the bilinear or the bicubic interpolation. Minimization of the above correlation coefficients yields the values of the six parameters, $\mathbf{V} = \{u_1 \; u_2 \; u_{1,1} \; u_{2,2} \; u_{1,2} \; u_{2,1}\}$, on which the deformation is dependent. This minimization can be performed by using either the coarse–fine (CF) method or the Newton–Raphson (NR) method.

The CF scheme [12] is a procedure in which the optimum solution is obtained by searching over a given range for each of the six parameters. The main drawbacks of the CF method are its severe computational requirements and, to some extent, reduced accuracy. To achieve shorter run times, the NR method was developed by Sutton et al. [13]. This method involves *simultaneous* minimization of all six parameters. However, it requires the furnishing of an initial guess for all parameters, to which the method is very sensitive. In all results presented in this work the outcome of an initial CF minimization was used as an initial guess in a subsequent NR minimization. The NR minimization results are the ones quoted. In most cases a 35 pixel subset was used and the analysis with both the CF and the NR methods was done over approximately 1000 correlation grid points.

An in-house computer code was written to implement the DIC technique. The code was tested by duplicating experimental results available in the literature [13,17,21]. Testing consisted of baseline (i.e. zero displacement) tests, uniform translation, uniform rotation and a three point bend edge crack experiment on PMMA (as in [18]). In all cases agreement with results previously published was excellent. More details on the implementation and testing of the DIC code can be found in [23].

2.3. Extraction of stress intensity factors and region of K-dominance

Fig. 5(a) shows a vector plot of the displacement field as obtained experimentally from the DIC scheme for a graded ECO sample. The axes are in millimeters and represent the physical dimensions over which the analysis has been performed. The crack tip is located at the origin and the crack-line is shown in the figure. The length of the arrows plotted is a measure of the magnitude of in-plane displacement, but magnified 20 times with respect to the axes. This displacement field contains contributions from the strain field generated by the presence of the crack and from possible rigid body motion.

It was shown in [2] that the dominant term of the asymptotic expansion for displacements near a crack tip in a FGM was identical to that for homogeneous materials, but evaluated with the local crack tip material properties. Thus, if we assume the contribution of the crack strain field to the displacement as being given only by the most dominant term (i.e. in effect assuming K-dominance) the total displacement, u_2^{tot}, in the x_2-direction can be expressed as

$$u_2^{\text{tot}}(r, \theta) = K_{\text{I}} \frac{r^{1/2}}{\mu_{\text{tip}}\sqrt{2\pi}} \sin\frac{\theta}{2}\left(\frac{2}{1+v} - \cos^2\frac{\theta}{2}\right) + A_1 r \cos\theta + A_2, \tag{6}$$

where μ_{tip} is the shear modulus of the FGM at the current crack tip location and v is the material's Poisson's ratio. In this work Poisson's ratio for all homogeneous and graded ECO samples has been taken as constant at $v = 0.45$, a value which is typical of polyethylene [24]. The first term on the right hand side of Eq. (6) represents the contribution by the crack tip field, while the other two terms represent a rigid body rotation and translation respectively, with A_1 and A_2 being unknown motion amounts. The least square minimization of Eq. (6) and the DIC experimental results for u_2^{tot} yields the values of the three parameters K_{I}, A_1 and A_2. On determining the three coefficients from Eq. (6) we can compute the opening, u_2 displacement induced *only* by the strain field of the crack tip, by subtracting the rigid body motion terms from the full-field displacement computed using the DIC method. A vector plot of the displacement field adjusted

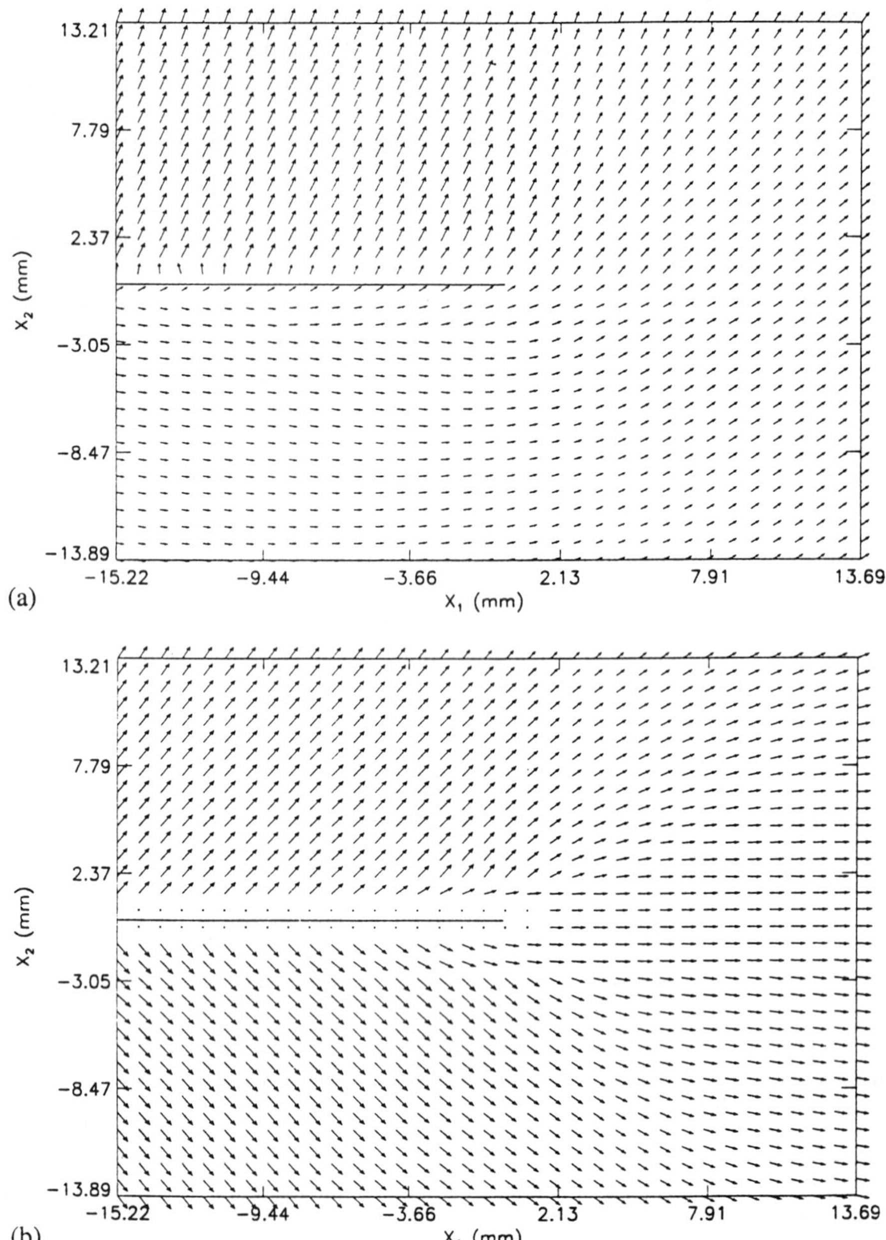

Fig. 5. Displacement field as obtained from the DIC technique for a graded ECO sample (a) without removal of rigid body motion and (b) after removal of x_2-direction rigid body motion.

for rigid body motion in the x_2-direction is shown in Fig. 5(b). As can be seen in this figure the DIC technique captures the displacement field surrounding the crack tip quite well. Since this is an opening type (mode I) experiment, this is especially true for the dominant displacement component, u_2.

Use of Eq. (6) in modeling the observed displacement field implicitly requires two conditions to be met: (a) the material should be well modeled as a linear elastic (inhomogeneous) isotropic solid and (b) data points used in the least squares fit of Eq. (6) be taken from a region of K-dominance. As can be seen in Fig. 1 the material used in this work exhibits a clear nonlinearity at higher stress levels. However, the amount of nonlinearity significantly decreases with increasing amounts of irradiation, i.e. as the material becomes stiffer it also becomes considerably more brittle. The extent of the nonlinear deformation region surrounding the crack tip has been studied in [9]. In that work a full-field nonlinear finite element solution was used to show that in a 60 h homogeneously irradiated ECO sample, nonlinear effects were confined to within an area of about 4 mm surrounding the crack tip (Fig. 9 in Ref. [9]). Keep in mind that the zone of nonlinearity will decrease with increasing irradiation and that the FGMs tested here usually involve crack growth in regions between about 120 and 70 h of irradiation. Thus, if data used for the fit of Eq. (6) are taken *outside* a 4 mm radius surrounding the tip, as was done in all cases here, linear elastic material response will be applicable.

The outer boundary of data points to be fitted to Eq. (6) will be dictated by requirement (b) above. A second finite element analysis conducted in [9] provided regions of K-dominance for a 60 h homogeneously irradiated ECO fracture specimen. The regions of K-dominance predicted in [9] were used for extraction of data points in the present work, when testing both homogeneous and graded samples. However, once the extraction of K_I from (6) has been performed, it is possible to experimentally verify K-dominance by comparing the results of Eq. (6), evaluated using the measured value of K_I, with the experimental data from which the value was extracted. Such a comparison, for the case of a growing crack in a 60 h homogeneously irradiated ECO sample, is shown in the form a contour plot in Fig. 6. The crack tip is located at the origin of the plot and the current crack length is 37 mm, i.e. after 7 mm of growth. Because of symmetry in the results only the upper half field is shown. The solid contours represent experimental data as measured by DIC, and the dashed contours represent the result from Eq. (6) applied to a homogeneous material. Fig. 6 shows that a significant region of K-dominance exists in the area of 90–135° surrounding the crack tip. Note

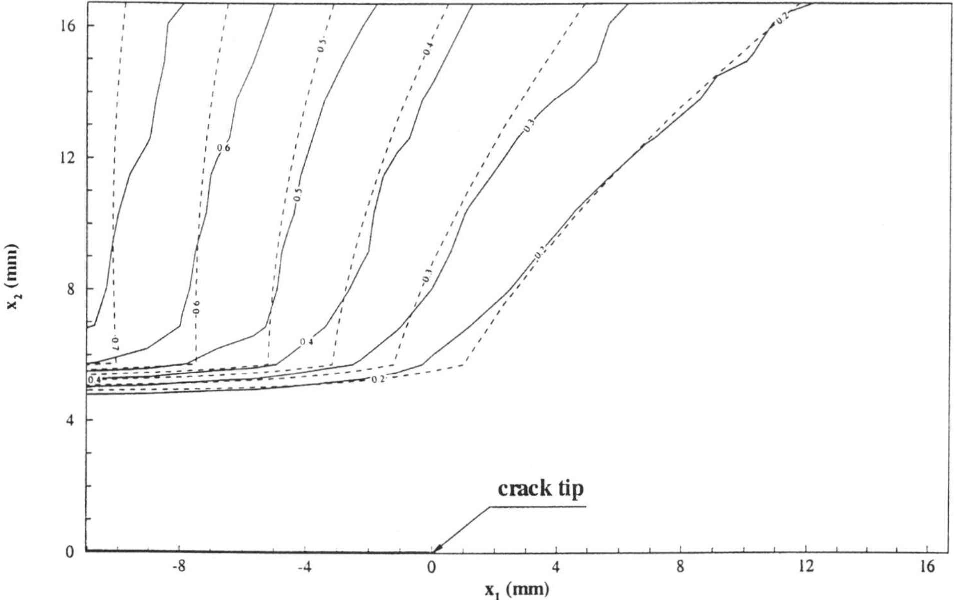

Fig. 6. Contour plot of u_2 displacement illustrating the extent of K-dominance in a 60 h homogeneously irradiated ECO. Solid contours depict raw experimental data extracted from DIC and dashed contours are obtained from the asymptotic displacement field with a value of K_I extracted from fitting the DIC data. Because of symmetry in the results only the upper half field is shown. (Picture corresponds to a crack extension of 7 mm.)

also that a zone of data points spanning 4 mm around the crack tip (i.e. the predicted nonlinear region) have been discarded. The results illustrated in Fig. 6 are in very good agreement with the finite element predictions of [9] (Fig. 10(b) in [9]). This confirms both the appropriate exclusion of points near the crack tip and the extent of K-dominance. In addition it shows the validity of using the small strain approximations in the DIC scheme.

2.4. Crack profile measurements

Values of stress intensity factor K_I were also extracted independently using an alternative method based on the crack opening profile. The crack face profile is digitized at various time instances from the photographs of Fig. 3. The opening profile predicted from the asymptotic theory for FGMs [2] is given by

$$u_2^\pm(r, \pm\pi) = \pm \frac{2K_I r^{1/2}}{\mu_{\text{tip}}(1 + v)\sqrt{2\pi}} \tag{7}$$

i.e. the first term on the right hand side of Eq. (6) evaluated for $\theta = \pm\pi$. The digitized crack face profiles provide experimental measurements of $u_2^\pm$ versus radial distance behind the crack tip. Solving Eq. (7) for K_I yields an equation for the "provisional" stress intensity factor, Y_I, as

$$Y_I(r, \pm\pi) = \frac{(1 + v)\mu_{\text{tip}} u_2^\pm \sqrt{2\pi}}{2r^{1/2}} . \tag{8}$$

Y_I becomes K_I only when it is constant as a function of radius behind the tip. In the present work, values were taken from points in the range $r = 4$–30 mm behind the tip and K-dominance was ensured by verifying in each case that indeed the measured value of Y_I was constant with r. In a manner similar to the evaluation of K-dominance presented in Fig. 6, we can evaluate the accuracy of the crack profile measurements by overplotting the digitized crack face profile from the experiment and the results of Eq. (7) evaluated with the measured K_I. Fig. 7 shows such a plot for the case of an FGM. Experimentally measured and analytically derived crack face profiles are compared for a number of time instances (130, 240, 320 and 410 s)

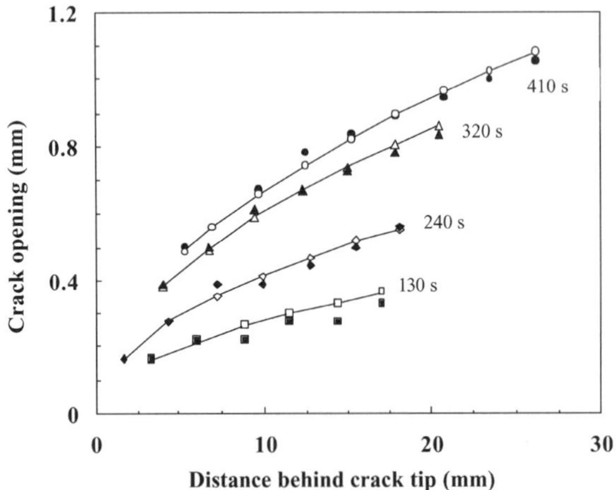

Fig. 7. Comparison of crack face profile between experimentally measured and analytically calculated (through Eq. (7)) results at four times during crack growth in an FGM fracture experiment. Time is denoted in seconds after initial application of load. Crack initiation occurs at approximately 130 s. Points with solid symbols denote actual experimental measurements of crack face profile and points with open symbols are the results of Eq. (7).

during crack growth. Time $t = 0$ s corresponds to initial application of load, and crack initiation occurred at approximately $t = 130$ s. As can be seen the agreement is quite good, thus justifying the use of Eq. (7) to describe the crack profile in the region of 4–30 mm behind the tip, and for a large duration of the experiment. Note that in comparison to traditional brittle materials, the crack opening displacements seen here are quite large. This is a direct consequence of the very low elastic modulus of the material. However, as was discussed earlier, strains are still small, thus allowing for accurate use of small strain techniques.

3. Results and discussion: homogeneously irradiated eco

As a basis for comparison with the subsequent graded material experiments, the mode I fracture behavior of homogeneously irradiated ECO was studied using the single edge notch specimen shown in Fig. 5. A homogeneously irradiated case of 60 h irradiation was tested. The Young's modulus for the 60 h material is 200 MPa (measured in a uniaxial tension test). To obtain resistance curves for crack growth in this material, the data analysis procedure described in Sections 2.2 and 2.3 was applied to images during crack extension. Each current image was correlated with the image from the completely unloaded state. It is worth pointing out again that excessive yielding or large strains can cause problems in both the linear elastic material assumption used for obtaining K_I and the small strain assumption implicit in the DIC scheme used here. The limits of these effects were discussed in Sections 2.2 and 2.3. Keep in mind, however, that the material used in this study becomes increasingly brittle when subjected to UV irradiation, and that most results presented here, even for the FGM case, are for longer irradiation times (>50 h).

Fig. 8(a) shows experimentally obtained resistance curves (i.e. K_I versus crack extension) extracted using the far-field measurements of [9] for three cases of homogeneously irradiated ECO. Comparing the resistance curves for the 5, 60 and 106 h irradiation clearly illustrates the changing nature of the material fracture response when subjected to UV irradiation. The material becomes considerably more brittle, because of the decrease in failure strain, and the resistance curves become flatter. Fig. 8(b) shows resistance curve results for a 60 h irradiated sample as obtained by three different techniques: the near tip measurement of DIC, the crack face profile measurement and the far-field measurements of [9]. As can be seen the near tip measurements compare well with the far-field hybrid experimental/numerical method used in [9], although the far-field measurements consistently overestimate K, indicating the need for accurate near tip measurements. Note that in all cases of near tip measurements using either DIC or crack face profile an investigation of K-dominance was made, as illustrated in Figs. 6 and 7, and the quoted results are those extracted only from K-dominant regions. This fact, which cannot be experimentally verified when far-field measurement are used, as in [9], increases the confidence of the near tip measurements considerably.

4. Results and discussion: functionally graded material

The FGM specimens were tested as described in Section 2. In all FGM fracture experiments reported here the 30 mm pre-notch was set parallel to the material gradient direction and was located in the more brittle side (i.e. longer irradiation times). Fig. 9(a) shows the resistance curves of an FGM computed using the far-field measurements of [9]. This FGM had a continuous variation of Young's modulus over 150 mm of length with a $\beta = -6/m$. The continually rising R-curve in Fig. 9(a), which does not reach a final plateau as those in Fig. 8, is characteristic of an FGM's inherent resistance to crack growth and clearly illustrates the necessity of a continually increasing driving force in order to sustain crack growth in the FGM. Such a "built-in" crack growth resistance has been theoretically postulated [25] and was first experimentally recorded in [7] and [9]. In addition, it is worth noting that the convex curvature of the observed resistance

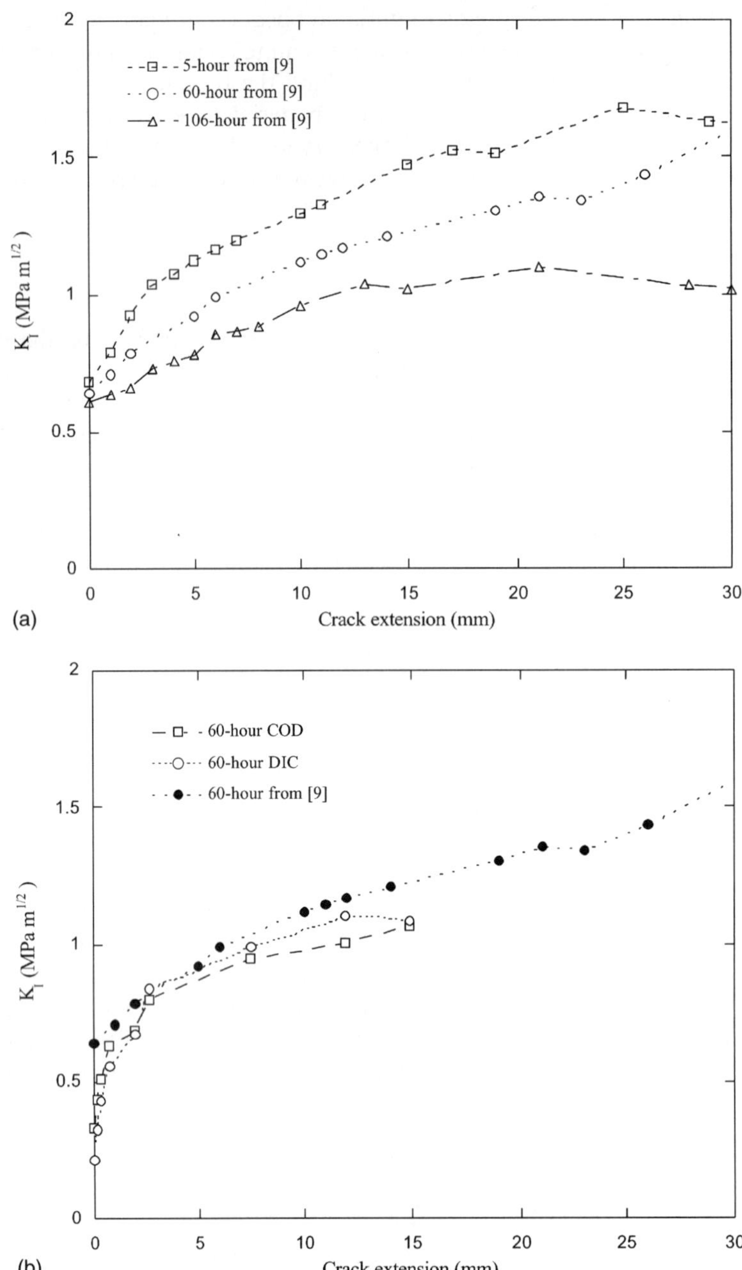

Fig. 8. Resistance curves for homogeneously irradiated ECO samples. (a) Effect of UV irradiation on fracture resistance (from [9]) and (b) comparison between DIC, crack face profile and hybrid experimental/numerical results (from [9]) for a 60 h homogeneously irradiated ECO sample.

curve suggests a cohesive type failure, since it resembles the theoretically calculated resistance curve of Jin and Batra [26] when they assumed a crack bridging law in their model. An investigation into the details of such a cohesive failure law is currently underway.

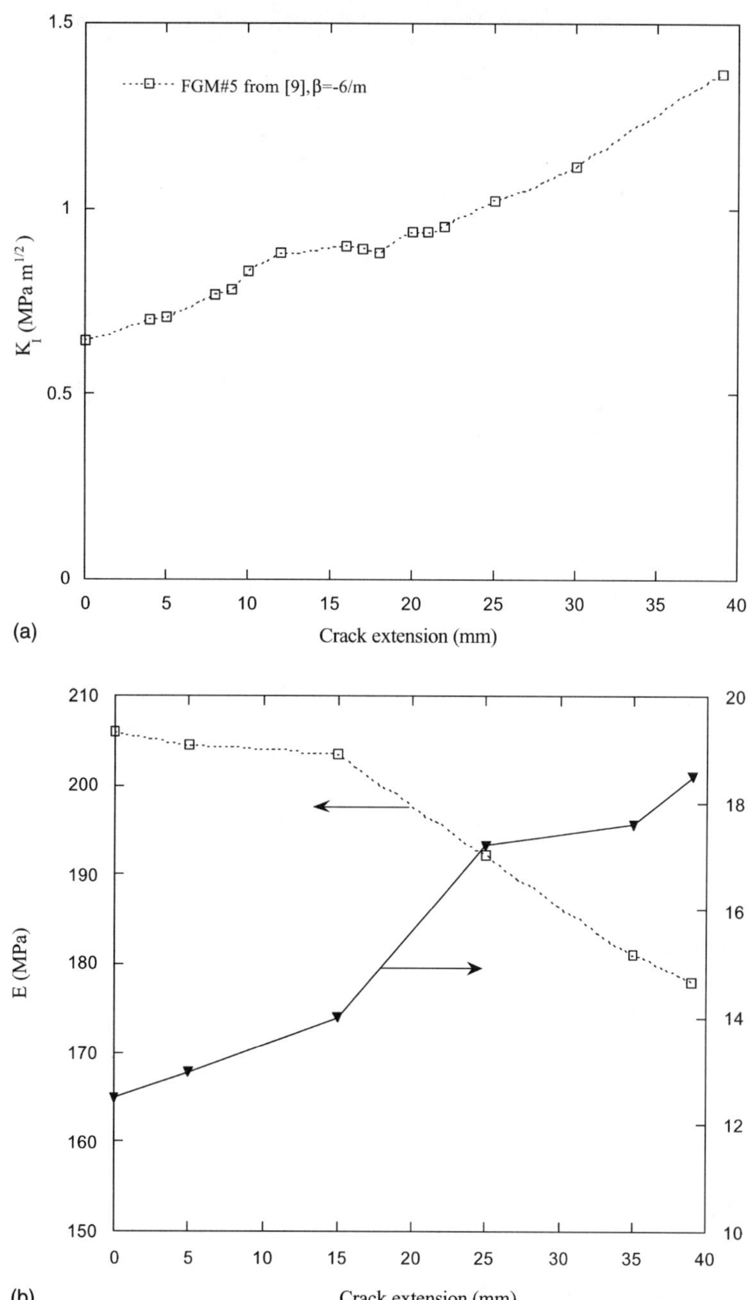

Fig. 9. (a) Resistance curve for crack growth in FGMs obtained using a hybrid experimental/numerical method (from [9]), (b) variation of Young's modulus and failure strain for this FGM.

Since in the current method of preparing the FGM it is not possible to separate the change of modulus and failure strain, Fig. 9(b) shows both these quantities (as measured experimentally through uniaxial tension tests) for the FGM corresponding to the results of Fig. 9(a). It is clear that a simultaneous decrease

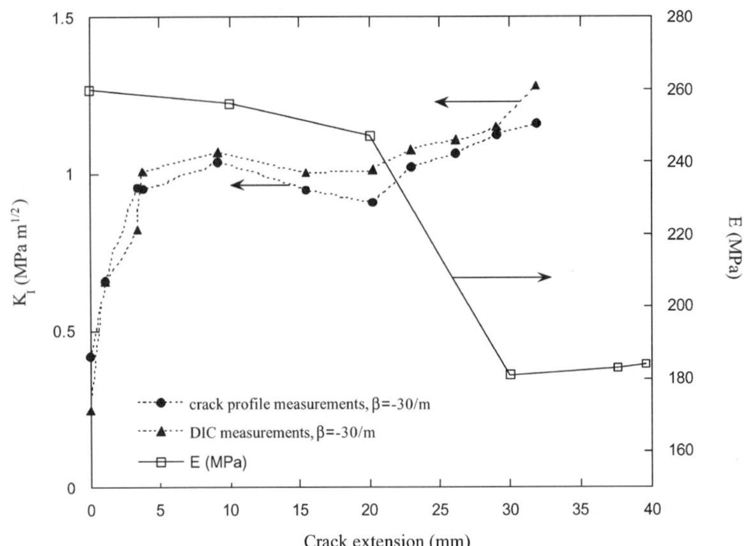

Fig. 10. Resistance curve as obtained in the present study using DIC and crack face profile measurements. Variation of Young's modulus is also shown.

of Young's modulus and increase in failure strain occur in the direction of crack growth. The change in modulus affects the development of the stress field surrounding the crack and therefore continuously affects the *available* crack driving force. The change in failure strain provides an increased spatial toughness and thus affects the *required* driving force. Unfortunately at this stage it is very difficult to separate these two competing effects, i.e. the simultaneous change in available *and* required driving force.

Fig. 10 shows a crack growth resistance curve for an FGM with a much steeper gradient variation ($\beta = -30$/m) in the vicinity of the crack tip. An overplot of the modulus variation for this FGM is also shown in the figure. As can be seen, the DIC and crack face profile results agree very well for the entire duration of crack growth. In the region of the sample which is almost homogeneous (i.e. up to about 20 mm of crack extension) we see a concave resistance curve which reaches a plateau value, much like in homogeneously irradiated ECO (Fig. 8). However, when the crack enters the graded region (after about 20 mm of growth) the resistance curve becomes convex, as in the results of Fig. 9(a), illustrating more clearly the differing effect of the FGM on crack growth resistance, i.e. the convex, continually increasing, resistance curve.

The advantage of using a full-field measurement technique over the far-field techniques used in [7,9] is that, in addition to obtaining values of stress intensity factor—a global quantity—we have detailed information about the structure of the displacement field near the crack tip. Such information can be used to determine the extent of validity of the asymptotic equations for the case of an FGM. To our knowledge this has not been done in the past. Fig. 11 shows a contour plot of u_2 displacement (i.e. opening) illustrating the extent of K-dominance in an FGM ($\beta = -30$/m) during crack growth (corresponding to a crack extension of 23 mm, i.e. total length of 53 mm). The solid lines in this figure are raw experimental data extracted from DIC, after removal of rigid body motion. The superimposed dashed contours are the results obtained from the asymptotic displacement field for the FGM (i.e. the first term on the right hand side of Eq. (6)) using the value of K_I extracted from fitting the DIC data for this particular picture. As can be seen, the agreement is quite good, denoting a substantial region of K-dominance.

It is expected that the extent of K-dominance will be very sensitive to the relation of three parameters, β (which as may be recalled is the intrinsic FGM length scale), the crack ligament and the scale of obser-

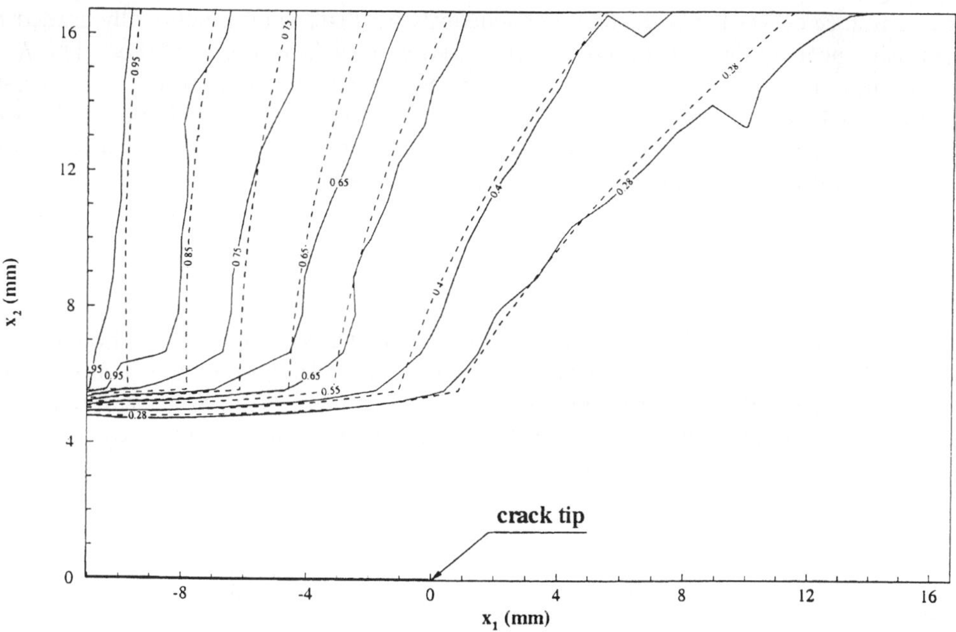

Fig. 11. Contour plot of u_2 displacement illustrating the extent of K-dominance in an FGM. Solid contours depict raw experimental data extracted from DIC and dashed contours are obtained from the asymptotic displacement field for the FGM with a value of K_I extracted from fitting the DIC data. Because of symmetry in the results only the upper half field is shown. (Picture corresponds to a crack extension of 23 mm.)

vation. In a separate work, [4], we have conducted a thorough investigation of dependence of the extent of K-dominance on these three parameters and for this loading geometry. In cases when βx (i.e. scale of observation) is very small, then the results of the present study are valid, since locally, even in the FGM, the results for a homogeneous material can be used (a direct consequence of asymptotics, [2]). If βx is very large, however, then it may become necessary to use the asymptotic field of Erdogan [25] in lieu of Eq. (1). This field is given by

$$\sigma_{ij}(r,\theta) = e^{\beta \cos\theta r}\left[\frac{K_I}{\sqrt{2\pi r}}F_{ij}^I(\theta) + \frac{K_{II}}{\sqrt{2\pi r}}F_{ij}^{II}(\theta)\right],$$

where the quantities in the square brackets as well as β have been previously defined. Clearly for small r this reduces to the result of Eischen [2]. More discussion on this point can be found in [4].

5. Conclusions

In this work, we have investigated the fracture of FGMs using the full-field measurement technique of DIC. The DIC method was implemented using bilinear and bicubic interpolation, and CF and NR optimization algorithms. After extensive testing of the DIC code, it was used to investigate fracture of uniformly irradiated and graded polyethylene-CO co-polymer (ECO) samples. The crack opening displacement fields were well captured using DIC. K-dominance in ECO was investigated and it was found that the regions beyond a radial distance of 4 mm and at angles $\beta = \pm 90°$ to $\pm 135°$ were K-dominant.

The crack resistance curves for both the homogeneously irradiated and functionally graded ECO were also obtained using near tip data of the DIC method and a crack face profile method. The R-curves obtained from the near tip methods were in general agreement with previous results using far-field measurements, although better agreement was always obtained between DIC and the crack face opening methods. Using a full-field technique also allowed for investigation of K-dominance regions in the FGM, something which is not possible using only boundary measurements.

Acknowledgements

The authors wish to acknowledge the generous support of the National Science Foundation through grant number CMS-9712831. We also wish to thank Hi-Cone Division of Illinois Tools Work Co., and especially Mr. C. Arends V.P. of Engineering and Packaging and Mr. W.N. Weaver Manager of Product Development, for providing the testing material, ECO, free of charge for our research.

References

[1] Rabin BH, Shiota I. Functionally graded materials. Mater Res Soc Bull 1995;20(3):427–31.

[2] Eischen JW. Fracture of nonhomogeneous materials. Int J Fract 1987;34:3–22.

[3] Anderson TL. Fracture mechanics: fundamentals and applications. Boca Raton, FL: CRC Press Inc.; 1994.

[4] Anlas G, Santare MH, Lambros J. K-dominance effects in cracked inhomogeneous plates. Int J Fract, in press.

[5] Butcher RJ, Rousseau C-E, Tippur HV. A functionally graded particulate composite: preparation, measurements and failure analysis. Acta Mater 1998;47(1):259–68.

[6] Parameswaran V, Shukla A. Dynamic fracture of a functionally gradient material having discrete property variation. J Mater Sci 1998;33:3303–11.

[7] Carpenter RD, Paulino GH, Munir ZA, Gibeling JC. A novel technique to generate sharp cracks in metallic/ceramic functionally graded materials by reverse 4-point bending. Scripta Mater 2000;43:547–52.

[8] Lambros J, Santare MH, Li H, Sapna III G. A novel technique for the fabrication of laboratory scale functionally graded materials. Exp Mech 1999;39(3):183–9.

[9] Li H, Lambros J, Cheeseman BA, Santare MH. Experimental investigation of the quasi-static fracture of functionally graded materials. Int J Solids Struct 2000;37(27):3715–32.

[10] Peters WH, Ranson WF. Digital imaging techniques in experimental stress analysis. Opt Eng 1982;21(3):427–31.

[11] Peters WH, Ranson WF, Sutton MA, Chu TC, Anderson J. Application of digital image correlation methods to rigid body mechanics. Opt Eng 1983;22(6):738–42.

[12] Sutton MA, Wolters WJ, Peters WH, Ranson WF, McNeill SR. Determination of displacements using an improved digital image correlation method. Image Vision Comput 1983;1(3):133–9.

[13] Sutton MA, Cheng M, Peters WH, Chao YJ, McNeill SR. Application of an optimized digital image correlation method to planar deformation analysis. Image Vision Comput 1986;4:143–50.

[14] Ivanova E, Chudnovsky A, Wu S, Sehanobish K, Bosnyak CP. A new experimental technique for modeling of a micro-heterogeneous media. Exp Tech 1996;20(6):11–3.

[15] Andrady AL, Nakatsuka S. Studies on enhanced degradable plastics III. The effect of weathering of polyethylene and ethylene-carbon monoxide copolymers on moisture and carbon dioxide permeability. J Environ Polym Degrad 1994;2(2):161–7.

[16] Erdogan F, Wu BH. The surface crack problem for a plate with functionally graded properties. J Appl Mech 1997;64:449–56.

[17] Chu TC, Ranson WF, Sutton MA, Peters WH. Application of digital image correlation techniques to experimental mechanics. Exp Mech 1985;25(3):232–45.

[18] McNeill SR, Peters WH, Sutton MA. Estimation of stress intensity factor by digital image correlation. Eng Fract Mech 1987;28(6):101–12.

[19] Tong W. Strain characterization of propagative deformation bands. J Mech Phys Solids 1998;46(10):2087–102.

[20] Smith BW, Li X, Tong W. Error assessment for strain mapping by digital image correlation. Exp Tech 1998;22(4):19–21.

[21] Bruck HA, McNeill SR, Sutton MA, Peters WH. Digital image correlation using Newton–Raphson method of partial differential corrections. Exp Mech 1989;46:261–7.

[22] Vendroux G, Knauss WG. Submicron deformation field measurements: Part 2. Improved digital image correlation. Exp Mech 1988;38(2):86–92.

[23] Narayanaswamy A. Use of digital image correlation in the investigation of quasi-static fracture of functionally graded materials. MS thesis. University of Delaware, 1999.
[24] Domininghaus H. Plastics for engineers: materials, properties and applications. Munich, Germany: Hanser Publications; 1993.
[25] Erdogan F. Fracture mechanics of functionally graded materials. Compos Eng 1995;5(7):753–70.
[26] Jin ZH, Batra RC. Some basic fracture mechanics concepts in functionally graded materials. J Mech Phys Solids 1996;44(8): 1221–35.

PERGAMON

Engineering Fracture Mechanics 69 (2002) 1713–1728

Engineering Fracture Mechanics

www.elsevier.com/locate/engfracmech

Thermal fracture behavior of metal/ceramic functionally graded materials

A. Kawasaki *, R. Watanabe

Department of Materials Processing, Graduate School of Engineering, Tohoku University, 02 Aoba Aramaki Aobaku Sendai 980-8579, Japan

Received 26 February 2001; received in revised form 21 November 2001; accepted 18 January 2002

Abstract

Thermal fracture behavior of metal/ceramic functionally graded materials (FGMs) was evaluated by a well controlled burner heating method using a H_2/O_2 combustion flame, which simulated real environment.

Partially stabilized zirconia (PSZ)/IN100 FGMs having finely mixed microstructures and PSZ/Inco718 FGMs having rather coarse microstructures were prepared by a slurry dipping and HIP sintering process. Also, three types of functionally graded thermal barrier coatings (TBCs) as well as duplex coatings, each designed to have the same thermal resistance, were fabricated by an air plasma spraying process. The fracture mechanism has been discussed on the basis of the crack morphology, the analysis of acoustic emissions and the variation of effective thermal conductivity. The thermal shock fracture behavior is discussed on PSZ/In100 FGMs and PSZ/Inco718 FGMs, while the cyclic fracture behavior is discussed on plasma sprayed coatings. The cyclic fracture behavior is found to be: orthogonal crack formation on the top surface during cooling, then transverse crack formation in the graded layer during heating, and subsequent growth of transverse cracks and their coalescence which eventually causes the ceramic coat to spall. Compared to duplex coatings, it has been revealed that functionally graded TBCs possess the desirable effect for improvement of spallation life under cyclic thermal loads. The dependence of spallation life on composition profile in functionally graded coatings has been discussed.
© 2002 Elsevier Science Ltd. All rights reserved.

Keywords: TBC; FGM; Sintered FGM; Plasma spray coating; Spallation life; Cyclic thermal fracture; Fracture mechanism; Acoustic emission

1. Introduction

The concept of a functionally graded material (FGM) is now accepted worldwide [1]. One of the application of FGMs is to thermal barrier coatings (TBCs). For example, the use of thermal barrier-type FGM coatings in advanced gas turbines offers a reduction in fuel consumption by allowing much less flow of coolant or higher turbine inlet temperature, while improving durability by decreasing base metal

---

* Corresponding author. Tel./fax: +81-22-217-7314.
E-mail address: kawasaki@material.tohoku.ac.jp (A. Kawasaki).

temperature [2,3]. The major problem in the use of a TBC on a metal substrate is the spalling of the ceramic coating due to the large thermal stresses produced during the thermal cycling in an oxidation environment [4,5]. The thermal stresses are induced mainly by the thermal expansion mismatch between a ceramic coating and a metal bond coat as well as the temperature gradient within the coating and the substrate [6–8]. The introduction of a functionally graded structure into the coating improves the resistance of the TBC to thermal shock cracking, because the structure, so-called FGMs [9,10], is capable of effectively reducing the thermal stresses, which are generated at the interface between a ceramic coating and a metal substrate. On the basis of powder metallurgical processes the authors have been fabricating functionally graded TBCs for thermal stress relief, along with the increase of bonding strength between the coating and substrate. The graded coating must be designed and processed in a manner that prevents its delamination and spallation during thermal loading in service conditions. However, the design criteria have not been well established at present, because of insufficient understanding of the fracture mechanisms of the FGM coating when subjected to actual thermal environments. For appropriate design of functionally graded TBCs to meet the above requirements, therefore, the failure mechanisms of the coating must be understood.

This paper describes the thermal shock fracture behavior, the cyclic thermal fracture behavior and the evaluation of spallation life of functionally graded TBCs. Two kinds of FGM coatings, PSZ/superalloy FGMs and plasma sprayed FGMs, were fabricated. The thermal shock fracture behavior is discussed on PSZ/In100 FGMs and PSZ/Inco718 FGMs, while the cyclic fracture behavior is discussed on plasma sprayed FGM coatings. The effect of compositional profile in the FGM coatings on spallation life has been discussed.

2. Experimental procedures

2.1. Specimen preparation

Ceramic/metal functionally graded coatings were fabricated through the route of slurry dipping method and plasma spray process.

For the slurry dipping method [11], the raw material powders are commercially available partially stabilized zirconia (3 mol%Y_2O_3PSZ) and superalloys of IN100 and Inco718 with mean particle sizes of 0.07, 24.9 and 9.6 μm, respectively. The chemical compositions of the superalloys are given in Table 1. Fig. 1 shows a flow-chart of the fabrication process by the slurry dipping method. Disk-shaped green compacts of the superalloy powders, 14 mm in diameter and 5 mm in thickness, were prepared by die compaction and cold isostatic pressing (CIP) as substrates for slurry dipping. The metal and/or ceramic powders were suspended in ethanol and milled by tumbler ball mill to get a slurry having an appropriate viscosity for dipping. A substrate was dipped in the slurry, then withdrawn and dried. After drying, the coated substrate was CIPed again to settle the intended green layer. This process was repeated with slurries of different compositions to get a graded layer. The formed compacts were densified by hot isostatic pressing (HIP). Two types of FGM coatings, having the same graded structure with different microstructures, were fabricated. PSZ/IN100 FGM has a finely mixed microstructure and PSZ/Inco718 FGM has a rather coarse microstructure, as shown in Fig. 2. PSZ/IN100 and PSZ/Inco718 FGMs were brazed on copper holders so

Table 1
Chemical compositions of IN100 and Inconel718 powders in wt.%

Material	Ni	Al	Cr	Fe	Co	Mn	Mo	Nb	V	Si	Ti	C	S
IN100	Bal	0.54	18.5	18.5	–	0.10	3.1	5.4	–	0.19	0.99	0.040	0.006
Inconel718	Bal	5.00	12.38	0.10	18.45	0.01	3.16	–	0.76	0.04	4.34	0.065	0.003

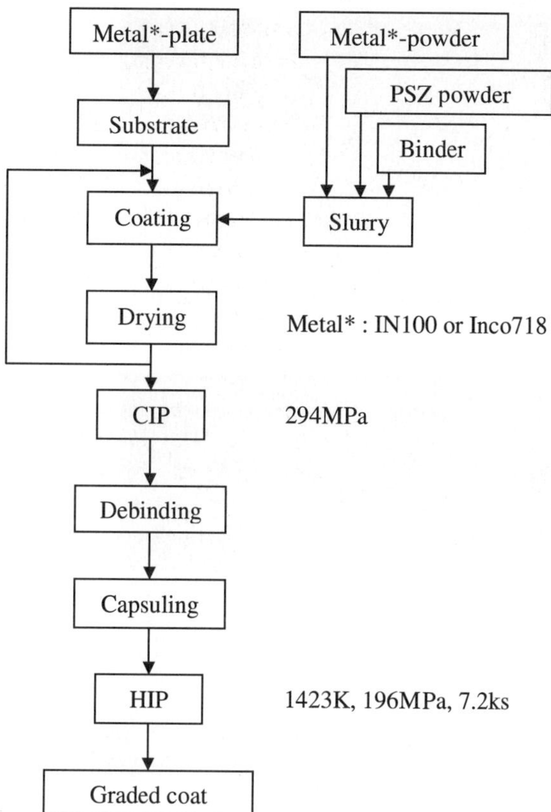

Fig. 1. Flow-chart of slurry dipping process for PSZ/superalloy FGM coatings on metal substrate.

as to place them under fully constrained mechanical boundary conditions. For these FGMs fracture toughness at different compositions was determined by a conventional vickers indentation method on non-FGM specimens which were prepared under the same fabricating condition as in the case of FGMs. The specimens were polished to a mirror finish on one of the faces. The test was conducted at an indent load of 196 N according to the standard indentation fracture procedure, JIS R1607 [12]. Crack lengths were measured by optical microscopy to evaluate fracture toughness values.

For the plasma spray process, the raw material powders were commercially available NiCrAlY powder for bond coat (Ni–22wt.%Cr–10Al–1.0Y) and partially stabilized zirconia (3 mol%Y_2O_3PSZ) for top coat. AISI type 304 stainless steel was used for the substrate material, because the thermo-mechanical properties are close to those of Ni base superalloys. The graded structure comprises six layers, where the PSZ content is changed stepwise from 0 vol% at NiCrAlY bond coat to 100 vol%PSZ in the ceramic top coat. Four different kinds of coatings were fabricated in order to study the effect of functionally graded structure on spallation life under cyclic thermal load. The graded structures and thickness of every layer are given in Table 2. Three types of FGM coating having different compositional profiles and a conventional duplex coating, composed of NiCrAlY bond coat followed by PSZ top coat, were prepared by the air plasma spray process. The typical microstructure of graded TBCs are shown in Fig. 3. The plasma sprayed coatings are designed to have the same thermal resistance to ensure the desired thermal performance of the coatings; that is, heat flux of 0.48 MW/m^2 yields temperature difference of about 150 K in the coating.

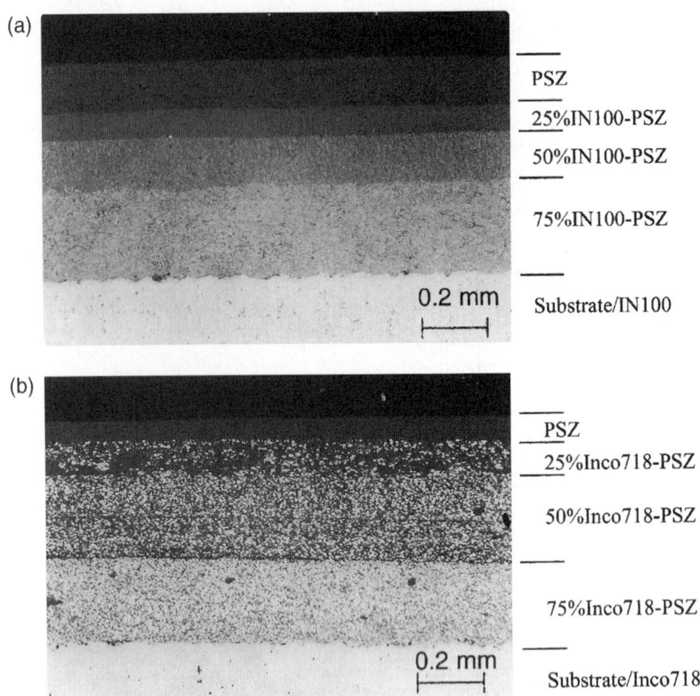

Fig. 2. Microstructures of PSZ/superalloy FGMs prepared by slurry dipping process. The dark part shows PSZ and the light part shows the metal phase. (a) Specimen A: PSZ/IN100-FGM, (b) specimen B: PSZ/Inco718-FGM.

Table 2
Graded structure and size of coatings (mm)

	Type-1	Type-2	Type-3	Duplex
100 vol%PSZ	0.31	0.19	0.11	0.35
NiCrAlY–80 vol%PSZ	0.19	0.19	0.12	
NiCrAlY–60 vol%PSZ	0.09	0.19	0.15	
NiCrAlY–40 vol%PSZ	0.05	0.19	0.21	
NiCrAlY–20 vol%PSZ		0.24	0.96	
100 vol%NiCrAlY	0.1	0.1	0.1	0.1

2.2. Burner heating method

Fig. 4 shows a schematic illustration of a burner heating test system. The surface of the specimen is heated by combustion flame of hydrogen–oxygen gas mixture, and the bottom side of the holder is cooled by water flow. The use of a mass-flow regulator allows precise control of the flow rate as well as the mixing ratio of the gases. A heat shielding board is set at the same level as the specimen surface so as to prevent the flame from impinging on the side surface of the specimen. A shutter, which cuts off the flame makes possible the rapid heating and cooling. The surface temperature (T_s) is monitored by an emission thermometer in which the absorption band of infrared rays from the burning of the mixed gas is cut-off. An emissivity of 0.74 and/or 0.24 was used [13,14]. Three thermocouples spaced 3 mm apart behind the specimen permit the determination of the heat flux and the estimation of the bottom surface temperature (T_b). Temperature difference (ΔT) is defined as the difference between the top and bottom surface temperature of the specimen.

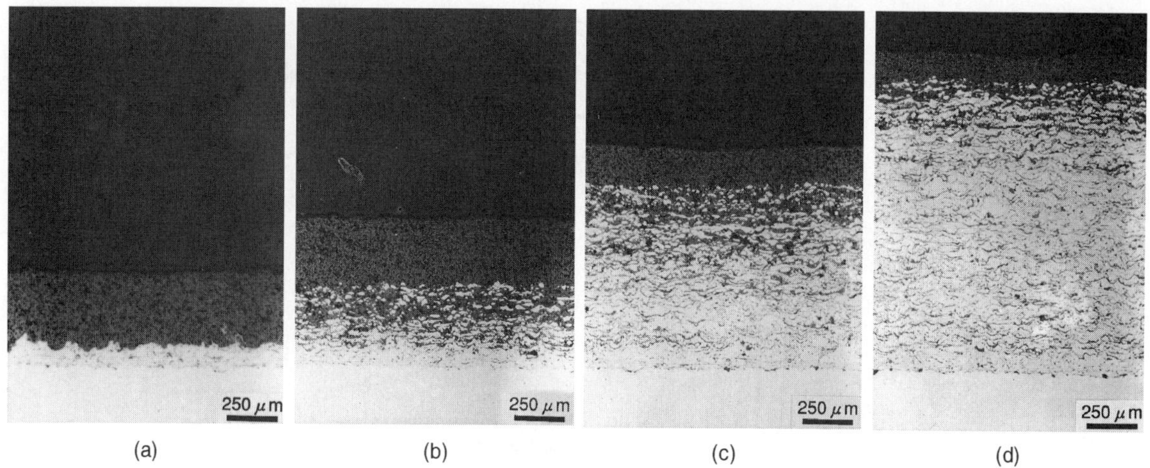

Fig. 3. Microstructure of duplex coating and three types of functionally graded TBCs prepared by plasma spray process. The dark part shows PSZ and the light part shows the metal phase. (a) Duplex TBC, (b) type-1 FGM, (c) type-2 FGM, (d) type-3 FGM.

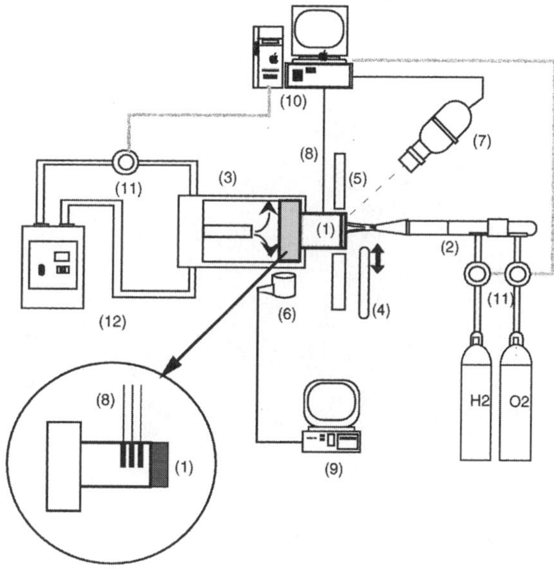

Fig. 4. Test system and sample setting configuration for burner heating test. (1) Test samples, (2) torch burner, (3) cooling chamber, (4) shutter, (5) protect plate, (6) AE sensor, (7) emission pyrometer, (8) thermocouple, (9) AE apparatus, (10) monitor, (11) regulate valve, (12) cooling water.

An AE sensor is mounted on the body of a cooling chamber to detect the onset of cracking by using Physical Acoustics Corporation (PAC)'s acoustic emission analyzer. The sequence of the burner heating test is: (1) heating up; (2) holding for 2 min; (3) cooling down. Fig. 5 shows typical temperature variation during a thermal cycle. About 20–60 s were needed to reach a steady-state heating at a given temperature.

For thermal shock test using PSZ/IN100 and PSZ/Inco718 FGMs, the sequence was performed with increasing the power output until cracking was detected. For cyclic burner heating test using plasma sprayed FGM coatings, the sequence was repeated with some constant power output until the spallation of a coating was detected. Fig. 6 shows the cyclic burner heating test system during testing.

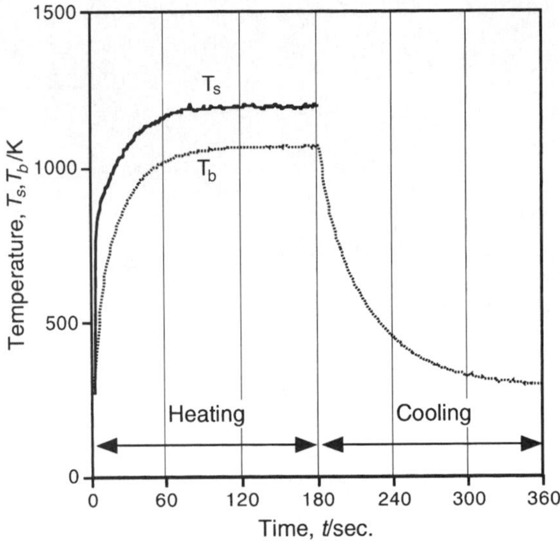

Fig. 5. Typical temperature variation during one thermal cycle. T_s surface temperature, T_b bottom surface temparature of the coating.

Fig. 6. Cyclic burner heating test system. (1) Test sample, (2) cooling chamber, (3) protect plate, (4) torch burner, (5) shutter, (6) AE sensor, (7) cooling water.

After testing, specimens were cut perpendicular to the surface and polished to observe the cracking morphology in the cross-section of every specimen by means of SEM and optical microscopy.

3. Results and discussions

3.1. PSZ/superalloy FGMs by slurry dipping process [15]

3.1.1. Microstructure and fracture toughness

As shown in Fig. 2, the microstructure of specimen A is quite different from that of specimen B. Specimen A has a finely mixed microstructure of PSZ and metal powders of IN100, while specimen B has a rather coarse microstructure where large particles of Inco718 are dispersed in the PSZ matrix. Specimen A shows a typical microstructural transition known in FGMs fabricated by powder metallurgy, where the microstructure is characterized by the gradual replacement from metal matrix to ceramic matrix with increase in the fraction of ceramic phase. In the metal rich side, ceramic particles are dispersed in the metal matrix. With increase in ceramic volume fraction, clusters of the ceramic phase are formed and their further growth results in the formation of a network structure. Then, the network of metal phase is gradually diminished and turns into isolated metal particles dispersed in the ceramic matrix in the ceramic rich side. Thus, in view of the differences in microstructure, it is likely that the fracture toughnesses in specimens A and B are also different.

The results of indentation fracture test are given in Table 3. The fracture toughness of PSZ/25 vol%IN100 is about 5.4 MPa m$^{1/2}$ which is slightly larger than that of monolithic PSZ. However, PSZ/50 vol%Inco718 shows a lower fracture toughness value of 2.7 MPa m$^{1/2}$. In PSZ/IN100 system, the fracture toughness increases with increase in the metal phase content, while in PSZ/Inco718 system, there is no improvement in fracture toughness. The values are considerably lower than those of PSZ/In100 system. Fig. 7 shows typical indentation cracks induced. The crack propagates in the PSZ matrix or along interfaces between the PSZ matrix and the metal particle. This indicates that the crack is deflected and/or arrested by the metal particles. For this reason, in specimen A having finely mixed microstructure, the fracture toughness increases with increase in metal phase content because of the high possibility of crack arrest. On the contrary, in specimen B having rather coarse microstructure, a crack propagates in the PSZ matrix with comparative ease and connectivity of metal particle is lower, resulting in the less possibility of arresting cracks. Thus, in this case, less improvement in fracture toughness results, in spite that the metal phase content increases.

The primary toughening mechanisms of particulate-reinforced ceramics includes: (1) crack front bowing by interaction between the crack front and particles, (2) crack deflection by the particulates ahead of a propagating crack, (3) particulate bridging by ductile particulates and (4) residual stress field due to the mismatch between the coefficients of thermal expansion (CTEs) of the ceramic matrix and particulates. It has been known that the fracture toughness, K_c, of particulate-reinforced ceramic–matrix composites increases with the volume fraction of the dispersed particles. This increase in the fracture toughness seems to be attributed to crack front deflection [16]. The crack deflection observed is considered to be a dominant toughening mechanism in ceramic rich side, while particle bridging toughening by ductile metal particles will contribute to increase the fracture toughness in metal rich side in the present material systems, because the connectivity of metal particles increases with the increase of metal phase.

3.1.2. Thermal shock cracking

The temperature of the first crack formation was defined as the critical surface temperature. The cracks monitored by AE were almost always generated during cooling. The critical temperatures of both specimens A and B were found to be around 1300 K. The crack formation is similar to the case of PSZ/stainless steel FGMs [13], where PSZ/metal FGMs show the critical surface temperature of around 1300 K regardless of specimen size and compositional profile, indicating intrinsic dependence of this temperature on

Table 3
Fracture toughness determined by IF method

Materials	MPa m$^{1/2}$
PSZ	5.0
PSZ/25 vol%IN100	5.4
PSZ/50 vol%IN100	11.4
PSZ/50 vol% Inconel718	2.7

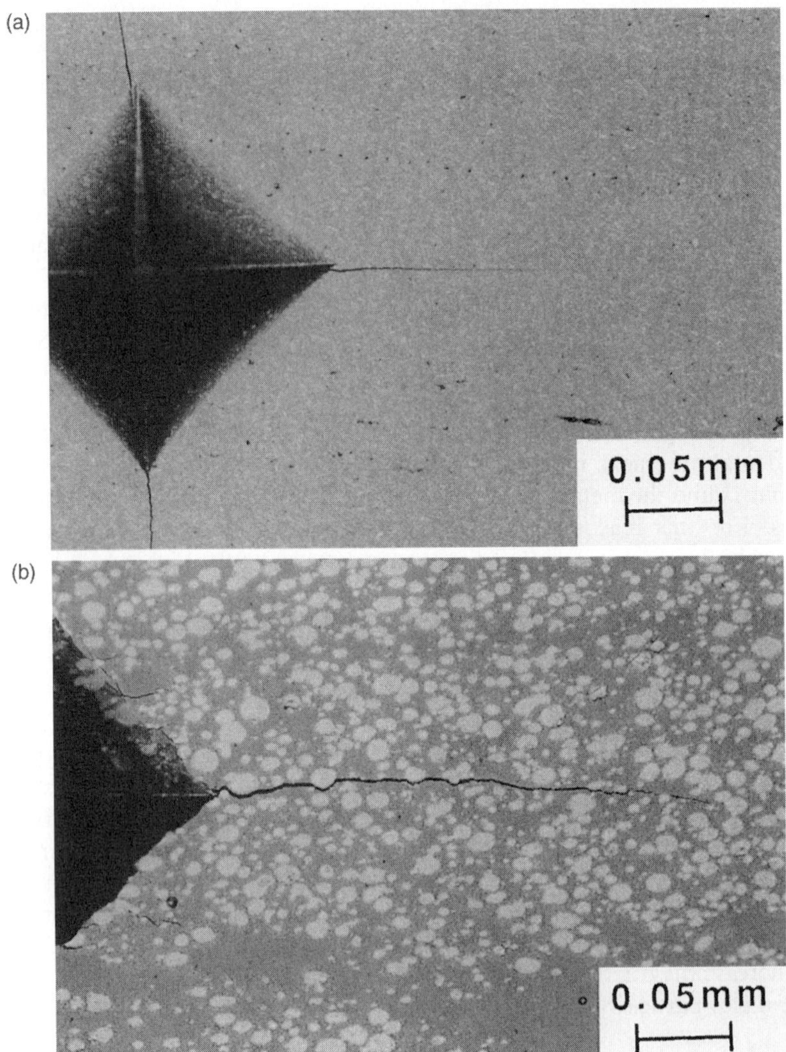

Fig. 7. Typical indentation cracks induced in sintered PSZ/superalloy composites. (a) 25%IN100-PSZ, (b) 50%Inco718-PSZ.

material species. It is reported that PSZ is thermally stable at elevated temperature as high as 1500 K. Sintered PSZ compacts exhibit brittle fracture at room temperature. Although the fracture strength decreases with temperature, brittle fracture still occurs up to 1273 K. However, non-linear deformation begins at about 1373 K. Above 1500 K, fully plastic deformation is observed [17]. The deformation behavior of PSZ turns from brittle to ductile around 1300 K in the narrow temperature range [13,17]. The critical surface temperature mentioned is close to the temperature of the brittle to ductile transition of PSZ.

Fig. 8 shows typical damages observed in specimens A and B after burner heating test. The optical observation is carried out on the top surfaces as well as in the cross-sections of them. Fig. 8 reveals that the cracks in specimen A are generated vertically in the top surface layer of PSZ. They deflect and propagate in the boundary between the PSZ and PSZ-25 vol%IN100 layer in the direction parallel to the graded plane. The deflection of crack propagation causes a surface segment of PSZ to be spalled out by link-up of the

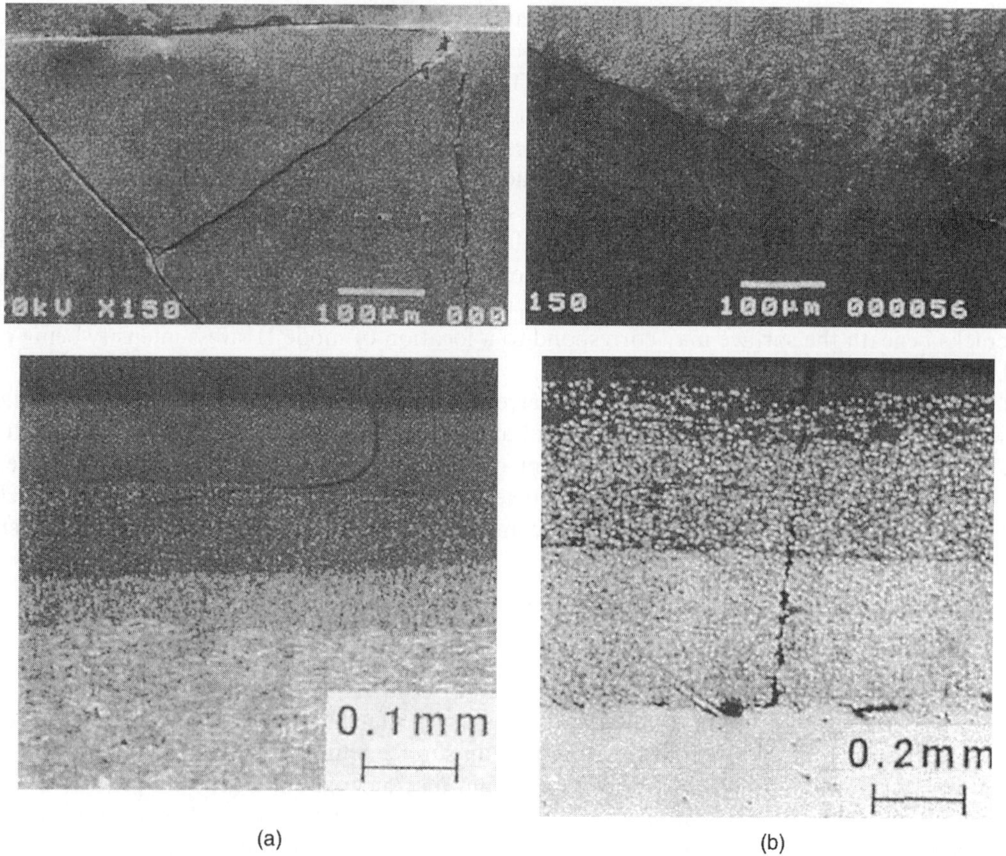

(a)   (b)

Fig. 8. Top views and cross-sections of PSZ/superalloy FGMs after burner heating test. (a) Specimen A: PSZ/IN100-FGM, (b) specimen B: PSZ/Inco718-FGM.

cracks together. In contrast, in specimen B, the cracks generated vertically propagated through the whole graded layer without deflection, reaching the boundary between the graded coating layer and the substrate. In this case, no spalling will occur.

It has been shown that stresses in the center part of the top surface during heating are bi-axial compressive stresses whose values decrease inversely with the radial distance towards the periphery of the specimen [13]. Also, it has been shown that the compressive stress, which is maximum at the top surface, decreases towards the bottom side along the center axis. The axial stress pertinent to delamination is rather small. Thus, bi-axial compressive stress in the surface is the most important in regard to contribution of failure.

From the context, the mechanism of the vertical crack formation has been elucidated as the following sequence [15]. During heating, the top surface of an FGM is in a large bi-axial compressive stress state. The stress causes non-linear deformation when the top surface is heated above the transition temperature of PSZ. During cooling, the resulting strain causes the radial stress to become tensile. The change from compression to tension whose magnitude is large enough to exceed the fracture strength of PSZ causes the vertical crack. The large tensile stress is limited to a shallow surface layer, which follows from the fact that the temperature decreases abruptly with increasing distance from the top surface and non-linear deformation is limited only to the shallow surface layer [13].

Drory [18] reported the fracture behavior of bi-material consisting of the thin film on a substrate. In the case that the thin film is in residual tension, vertical cracks initiate from the film surface and extend into a split in the substrate perpendicular to the interface. Then the crack extend into a trajectory parallel to the surface. It has been reported that cracks tend to branch and deflect into a trajectory with the mode II stress intensity being equal to zero. In the case that the residual tensile stress is generated in a shallow surface layer, mode I stress intensity factor for vertical crack decreases as the crack extends into the material, indicating the feature that the vertical crack may arrest. The initiated vertical cracks in PSZ/Inco718 FGMs having low fracture toughness extend into the interface of FGM/substrate and arrest without deflection. Although vertical cracks in PSZ/IN100 FGMs tend to be arrested in the FGM coating with the extension of the cracks into the graded layer, they deflect toward the direction parallel to the surface. The depth of the parallel cracks beneath the surface may correspond to a location of mode II stress intensity being equal to zero [18].

It has been thought that there is a possibility to arrest vertical cracks and the capability of keeping them stable without deflection by microstructural control and by optimizing the $K_{II} = 0$ location which will be achieved by controlling the compositional gradient of the FGM. Vertical cracks are rather effective in reducing thermal stresses without changing the heat resistance of FGMs and so the design of functionally graded TBCs must include the crack propagation control to prevent the coating from causing to spall out during accrual thermal loading.

3.2. PSZ/NiCrAlY functionally graded TBCs by plasma spray process [14]

3.2.1. Cyclic thermal fracture behavior

Fig. 9 shows the evolution of cyclic thermal damage in specimens of type-1 at desired thermal cycles, tested at the heat flux of 0.6 MW/m^2 which cause the top surface temperature to reach around 1400 K for the temperature difference of about 350 K. It can be seen that orthogonal cracks are generated in the top surface layer after five cycles. The number and the length of the orthogonal cracks increased with further thermal cycling. After 30 cycles, small cracks in the direction parallel to the surface are generated in the graded layer, chiefly in the layer of 60–80 vol%PSZ. The sites of the transverse cracks are shown to be both in PSZ matrix and NiCrAlY particle/PSZ matrix interface. The number of the transverse cracks increases with the increase of thermal cycles and some link-up of the cracks are observed after 57 cycles, which may lead to the delamination crack formation. In the case of duplex coatings, as shown in Fig. 10, orthogonal cracks were generated from the coating surface in the early life, that is the same result as in the case of functionally graded TBCs. However, in contrast to the FGM coatings, a small number of transverse cracks were initiated only at a convex part of the top coat/bond coat interface and grew toward concave part. With further thermal cycling, it is noted that a main crack is formed by coalescence of the transverse cracks at the bond coat/top coat boundary.

Fig. 11 shows AE amplitude and variation of T_s and T_b recorded in the thermal cycling test of the corresponding specimens. Two types of AE are observed. The first type of AE is detected upon cooling from the early life of the specimens and the second type of AE is obtained during heating after 30 cycles. In order to study the difference of the two types of AE source the spectrum analysis was conducted. The typical results showed the spectrum of AE during heating includes rather low frequency, while the spectrum of AE during cooling includes the wide frequency band, indicating thus the origins of the AE sources are different. It is thought that the first type of AE is identified to the orthogonal crack formation during cooling and the second type is in relation to the transverse crack formation and its linking-up. Therefore, the transverse cracks are to be generated during the heating phase. The orthogonal crack formation is considered to be due to a large tensile stress generated upon cooling because of inelastic compressive strain formed during heating in the top surface materials [13,19]. Generally, a thin top layer of a coating is subjected to an in-plane compressive stress during heating [20]. However, FEM analysis showed the pos-

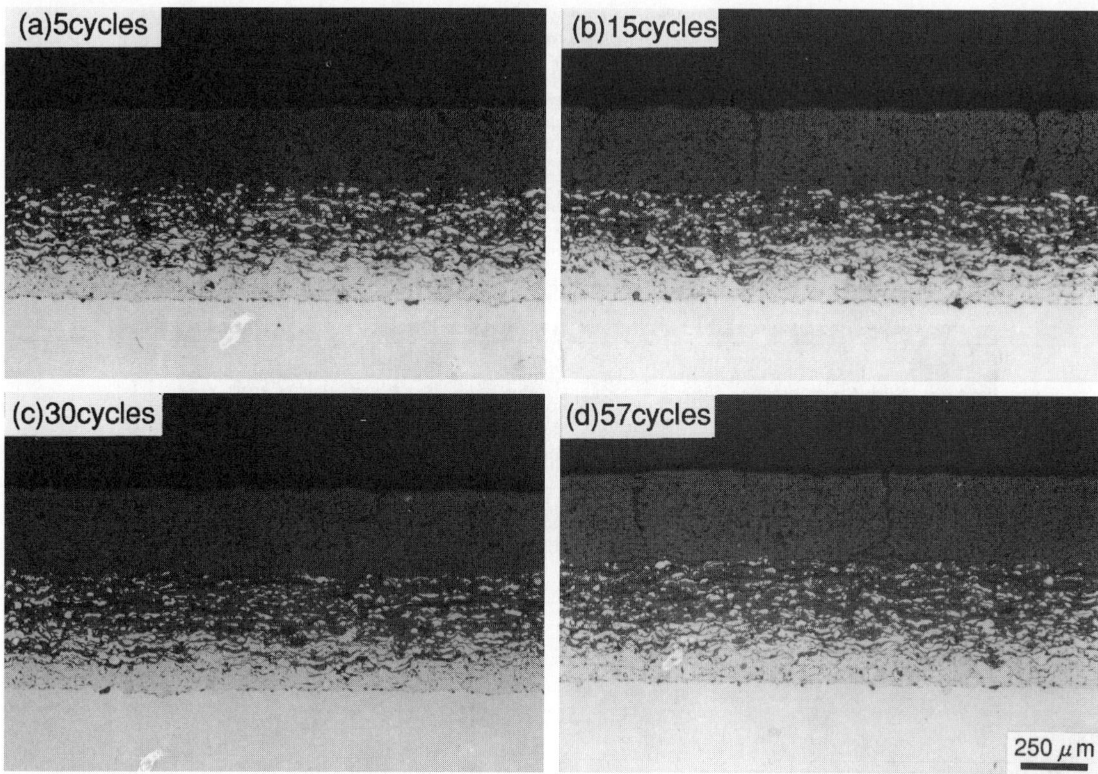

Fig. 9. Cyclic thermal damage in specimens of type-1 at desired thermal cycles, tested at the heat flux of 0.6 W/m², the surface temperature of 1400 K, the temperature difference of about 350 K.

sibility of tensile stress in the direction of transverse cracking occurred owing to an uneven interface between the bond coat and top coat [7]. The relatively large local tensile stresses that form in the presence of boundary waviness like metal/ceramic interface, may exceed the critical fracture stress and permit the formation of transverse cracks. From the context, the sequence of spalling behavior is found to be orthogonal crack formation on the top surface during cooling, then transverse crack formation in a graded layer during heating, followed by the growth of transverse crack to delamination crack and their coalescence which leads eventually the coating to spall.

Fig. 12 shows variation of AE hits during heating and cooling as a function of thermal cycles for the corresponding specimens of type-1. The first increase of AE hits during cooling indicates mainly orthogonal cracks formation and growth. After about 20 thermal cycles AE hits during heating appear to be detected which may be in relation to the formation of small transverse cracks in the graded layers. It should be noted that the final rapid increase of AE hits might be in relation to the formation of a large delamination crack by linking up of the small transverse cracks. The variation of effective thermal conductivity is also plotted in Fig. 12. The effective thermal conductivity of a specimen is defined as

$$\lambda_{\text{eff}} = q/[(T_s - T_b)/t] \tag{1}$$

where q is the heat flux loaded to the specimen, t the thickness of the specimen, T_s the averaged top surface temperature and T_b is the bottom surface temperature of the specimen. The effective thermal conductivity decreases moderately with the increase in the number of thermal cycles and suddenly drops

Fig. 10. Typical thermal cracks in duplex coatings after (a) 20 cycles and (b) 100 cycles.

about 50 thermal cycles. It is known that the change of the effective thermal conductivity is useful in examining thermal fatigue characteristics of a coating, because cracking in a coating generally cause the effective thermal conductivity to decrease. The transverse crack, since it is oriented in the normal direction to the heat flow, is more effective in decreasing the effective thermal conductivity. Thus, the sudden drop of the effective thermal conductivity may well indicate the delamination crack formation. Consequently, the spallation life of the coating can be defined as the critical number of thermal cycle, evaluated from either the rapid increase of AE hits during heating or the sudden drop of the effective thermal conductivity.

3.2.2. Evaluation of spallation life

The number of cycles to failure by a given temperature difference pertinent to an applied heat flux, is used to compare the spallation life of the TBCs under thermal cycling. The tests were completed when the large increase of AE hits during heating or the sudden drop of the effective thermal conductivity was detected. Fig. 13 shows the relation between the temperature difference (ΔT) in a coating and the logarithmic number of cycles to failure of the TBCs. It is noted that the failure life increases linearly with the decrease of the applied thermal load. This relation is similar to that of the fatigue behavior of a metallic material ; that is, the fatigue life is dependent on the mean applied stress, which is well known as an S–N curve. It is clear from this figure that the FGM coatings show much longer failure life than conventional duplex coatings. For example, at a given temperature difference of 300 K, FGM type-1 survives more than 200 cycles, which is ten times as long as that of the duplex coating. Compared to the duplex TBCs, the results have shown that functionally graded TBCs possess the desirable effect in the improvement of failure life under cyclic thermal loads. On the other hand, it is important to note that the failure life of functionally graded TBCs depends on compositional profile of gradient. Among the functionally graded TBCs, FGM type-2 may have good advantage to give a longer failure life. At a given temperature difference of 400 K, FGM type-2 is

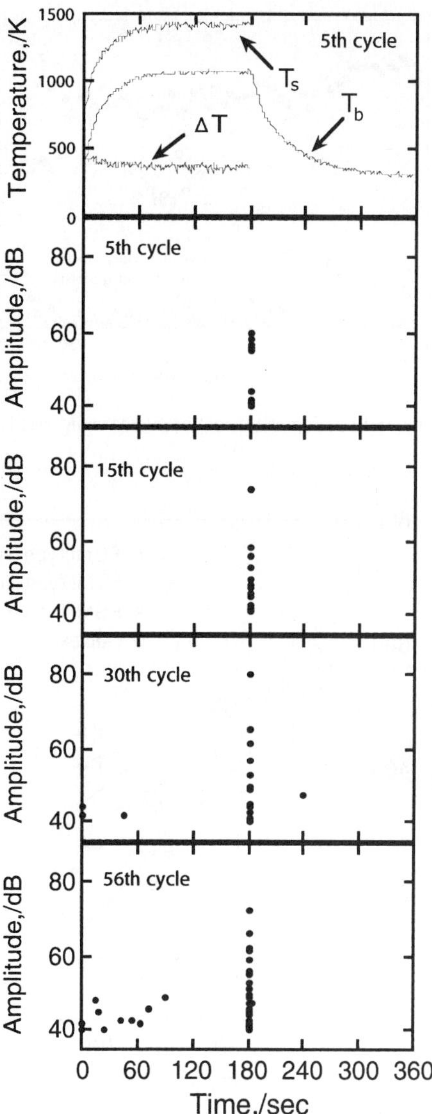

Fig. 11. AE characteristics and variation of T_s and T_b recorded in the thermal cycling test of the corresponding specimens in Fig. 9.

expected to survive ten times more cycles than FGM type-1. Thus, FGM type-2 with linear compositional profile is the one to improve a failure life under thermal cycling in this study.

As mentioned, transverse cracking in a duplex coating is limited to bond coat/top coat interface, while in a functionally graded TBCs shown in Fig. 14, a number of small transverse cracks are generated in the graded layer. Thus, the advantage of the graded coating is attributed to the formation of a large number of the small transverse cracks. This fact is beneficial in reducing the thermal stress effectively, since the presence of cracks decreases the overall stiffness of the coating layer. Consequently, the retardation of the growth of a main delamination crack in the coating is expected.

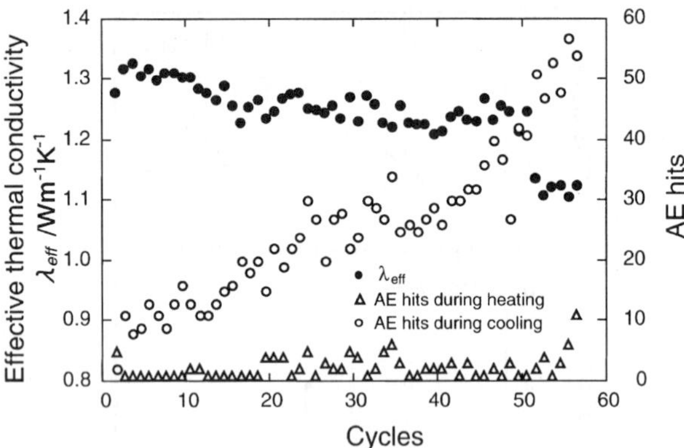

Fig. 12. Variation of AE hits during heating and cooling as a function of thermalcycles for the corresponding specimen of type-1.

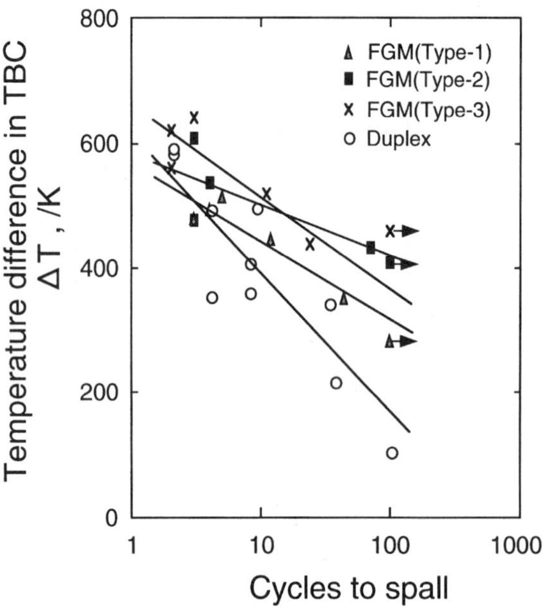

Fig. 13. Spallation life of the TBCs under thermal cycling.

4. Conclusion

The well controlled burner-heating-test was conducted in order to study the thermal fracture behavior, the cyclic thermal fracture behavior and spallation life of metal/ceramic functionally graded TBCs.

In PSZ/IN100 FGMs, the fracture toughness increased with increase in the metal phase content, while in PSZ/Inco718 FGMs it was fairly lower than that of PSZ/In100 FGMs, owing to roughly dispersed metal particles. The crack formation at the top surface during cooling was observed, which was shown to be vertical to the sample surface. On the consideration of the fracture toughness the initiated vertical cracks in PSZ/Inco718 FGMs were considered to extend into the interface of FGM/substrate without deflection. This crack extension behavior was confirmed by observing the cross-section of the tested samples. Although

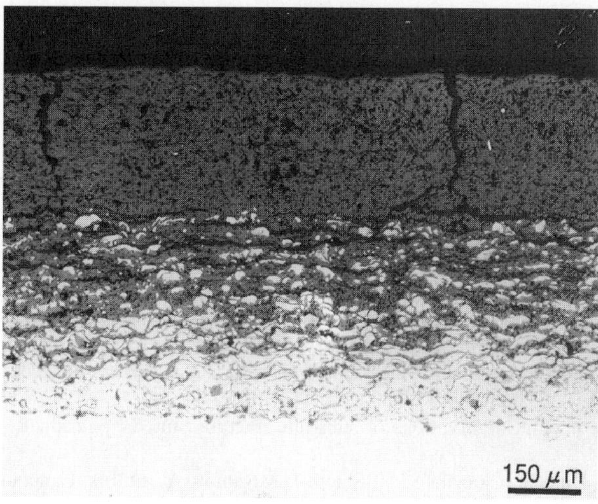

Fig. 14. Cross-section of specimen type-1 after 57 cycles, tested at the heat flux of 0.6 MW/m², the average surface temperature of 1400 K, the temperature difference of 350 K.

vertical cracks in PSZ/IN100 FGMs tend to be arrested in the FGM coating with the extension of the cracks into the graded layer, they deflected toward the direction parallel to the surface.

In plasma sprayed FGMs, the sequence of spalling behavior has been found to be: orthogonal crack formation on the top surface during cooling, then transverse crack formation in the graded layer during heating, and subsequent growth of transverse cracks and their coalescence which leads eventually the ceramic coat to spall. It is shown that the spallation life of the coating can be defined as the critical number of thermal cycles, evaluated from either the rapid increase of AE hits during heating or the sudden drop of the effective thermal conductivity. Functionally graded TBCs possess the desirable effect for improvement of spallation life under cyclic thermal loads. In addition, the dependence of spallation life on the composition profile in functionally graded coatings has been shown.

References

[1] Cherradi N, Kawasaki A, Gasik M. Worldwide trends in functional gradient materials research and development. Compos Eng 1999;4:883–94.

[2] Demasi-Marcin JT, Sheffler KD, Bose S. Mechanism of degradation and failure in a plasma-deposited thermal barrier coatings. ASME J Eng Gas Turbines Power 1992;112:521.

[3] Berndt CC, Herman H. Failure during thermal cycling of plasma-sprayed thermal barrier coatings. Thin Solid Films 1983;108:427–37.

[4] Miller RA, Lowell CE. Failure mechanisms of thermal barrier coatings exposed to elevated temperatures. Thin Solid Films 1982;95:265–73.

[5] Miller RA, Berndt CC. Performance of thermal barrier coatings in high heat flux environments. Thin Solid Films 1984;119:195–202.

[6] Gill SC, Clyne TW. Stress distributions and material response in thermal spraying of metallic and ceramic deposits. Metall Trans 1990;21B:377–85.

[7] Evans AG, Crumley GB, Demaray RE. On the mechanical behavior of brittle coatings and layers. Oxidat Metals 1983;20:193–216.

[8] Ferrari M, Lutterotti L. Thermal stresses in Bi-coated structures. J Eng Mech 1992;118:1928–38.

[9] Kawasaki A, Watanabe R. Finite element analysis of thermal stress of the metal/ceramic multi-layer composites with controlled compositional gradients. J Japan Inst Metals 1987;51(6):525–9.

[10] Khor KA, Gu YW. Thermal properties of plasma-sprayed functionally graded thermal barrier coatings. Thin Solid Films 2000;372:104–13.

[11] Yamaoka H, Yuki M, Tahara K, Irisawa T, Kawasaki A, Watanabe R. Fabrication of functionally gradient material by slurry stacking and sintering process. In: Holt JB, Koizumi M, Hirai T, Munir ZA, editors. Ceramic Trans (Functionally Gradient Materials), vol. 34. 1993. p. 165–72.

[12] Testing method for fracture toughness of high performance ceramics. JIS R 1607, 1990.

[13] Kawasaki A, Watanabe R. Evaluation of thermomechanical performance for thermal barrier type of sintered functionally graded materials. Compos Part B 1997;28B:29–35.

[14] Kawasaki A, Watanabe R. Cyclic thermal fracture behavior and spallation life of PSZ/NiCrAlY functionally graded thermal barrier coatings. Mater Sci Forum 1999;308-311:402–9.

[15] Kawasaki A, Watanabe R, Yuki M, Nakanishi Y, Onabe H. Effect of microstructure on thermal shock cracking of functionally graded thermal barrier coatings studied by burner heating test. Mater Trans JIM 1996;37(4):788–95.

[16] Faber KT, Evans AG. Crack deflection processes—I. Theory Acta Metall 1983;31:565–76.

[17] Jiang W, Li J-F, Kawasaki A, Watanabe R. High temperature deformation and fracture behavior in sintered Mo/PSZ composites as evaluated by small punch test. J Japan Inst Metals 1995;59(10):1055–60.

[18] Drory MD, Evans AG. Experimental observations of substrate fracture caused by residually stressed films. J Am Ceram Soc 1990;73:634–8.

[19] Kawamura M, Matsuzaki Y, Hino H, Okazaki S. In: Shiota I, Miyamoto Y, editors. Functionally Graded Materials 96, 1997. p. 417.

[20] Kokini K, Takeuchi YR, Choules BD. Surface thermal cracking of thermal barrier coatings owing to stress relaxation: zirconia vs. mullite. Surf Coat Technol 1996;82:77–82.

PERGAMON

Engineering Fracture Mechanics 69 (2002) 1729–1751

Engineering Fracture Mechanics

www.elsevier.com/locate/engfracmech

A surface crack in a graded medium loaded by a sliding rigid stamp

Serkan Dag, Fazil Erdogan *

Department of Mechanical Engineering and Mechanics, Lehigh University, 19 Memorial Drive West, Bethlehem, PA 18015 3085, USA

Received 21 February 2001; received in revised form 16 July 2001; accepted 20 July 2001

Abstract

In this article the coupled problem of crack/contact mechanics in a nonhomogeneous medium is considered. The underlying physical problem to which the results might be applicable is the initiation and the subcritical growth of surface cracks in a graded medium loaded by a sliding rigid stamp in the presence of friction. The dimensions of the graded medium are assumed to be very large in comparison with the local length parameters of the crack/contact region. Thus in formulating the problem the graded medium is assumed to be semi-infinite. The objective of the study is to determine, in addition to contact stresses, the in-plane component of the surface stress and the stress intensity factors at the crack tip. These are the primary load factors controlling the initiation and the subsequent growth of surface cracks in the graded medium. The coupled mixed boundary value problem is solved and the results are presented for various combinations of friction coefficient, material nonhomogeneity constant and crack/contact length parameters. © 2002 Published by Elsevier Science Ltd.

Keywords: Functionally graded materials; Sliding contact/crack problems; Stress intensity factors

1. Introduction

Graded materials, also known as functionally graded materials (FGMs) are multiphase composites with continuously varying volume fractions and, as a result, thermomechanical properties. Used as coatings and interfacial zones they reduce the residual and thermal stresses resulting from the material property mismatch, increase the bonding strength, improve surface properties and provide protection against severe thermal and chemical environments. Many of the present and potential applications of FGMs involve contact problems. These are mostly load transfer problems in deformable solids, generally in the presence of friction as in, for example, bearings, gears, cams, machine tools and abradable seals in gas turbines. In such applications the concept of material property grading appears to be ideally suited to improve the surface properties and wear resistance of the components that are in contact.

* Corresponding author. Tel.: +1-610-758-4099; fax: +1-610-758-6224.
E-mail address: fe00@lehigh.edu (F. Erdogan).

0013-7944/02/$ - see front matter © 2002 Published by Elsevier Science Ltd.
PII: S0013-7944(02)00053-X

From the standpoint of failure mechanics an important aspect of contact problems is the surface cracking which is caused by friction forces and which invariably leads to fretting fatigue [1]. In most applications material property grading near the surface is used as a substitute for homogeneous ceramic coatings. In both homogeneous and graded coatings the surface of the composite medium consists of 100% ceramic which generally is a brittle solid. Hence, the "maximum tensile stress" criterion may be used for crack initiation on the surface. Once the crack is initiated, its subcritical growth under repeated loading by a sliding stamp is controlled by stress intensity factors at the crack tip. The main objective of this study is, therefore, the evaluation of peak tensile stresses on the surface for the purpose of studying crack initiation and the stress intensity factors for modeling the subcritical crack growth. Specifically, the objective is the examination of the influence of friction coefficient and material nonhomogeneity parameters on the peak surface stresses and stress intensity factors. The problem is considered under the assumptions of plane strain, Coulomb friction and linear nonhomogeneous elasticity.

The model used in this study is that of continuum solid mechanics, whereas in reality the medium under consideration is generally a multiphase composite with highly complex microstructure. Therefore, certain observations need to be made with regard to the limitations of the model. In dealing with the damage mechanics of such materials, bulk homogeneous or graded, first it should be pointed out that the particular length scale of interest play a major role in the thermomechanical modeling of the medium. If the local flaw size is of the same order as the microstructural length parameters (e.g., grain size, inclusion size, interparticle distance), then it would be necessary to use a micromechanical approach to study the initial phase of the failure process, assuming the constituents of the medium to be isotropic or anisotropic homogeneous materials. On the other hand if the composite medium is bulk-homogeneous and the size of the dominant flaw, generally a planar crack, is large in comparison with the microstructural length parameters of the medium, then the approach would be by using some kind of a homogenization technique to characterize the medium [2] and by applying the method of continuum solid mechanics to solve the related damage or fracture mechanics problem. The material characterization may further be complicated due to the fact that the distributions of size, shape and spacing of the constituents in the composite are generally random.

Second and somewhat more difficult question is how to assess the influence of internal stresses induced by material processing on the damage mechanics of the composite medium. These stresses, which are also known as microscopic, intrinsic or quench stresses [3,4], are unknown and are very highly process dependent. Similar stresses arise under macroscopic thermal or mechanical loading due to local material inhomogeneity [3]. The latter stresses act more like stress concentrations and disappear as the external load is removed, whereas the former are similar to residual stresses and are self-equilibrating. If the medium is analyzed by using a micromechanics approach the effect of these internal stresses can be taken into account by following well-established elastic or inelastic superposition techniques.

The third question is the determination and analytical modeling of the effective thermomechanical properties of bulk-homogeneous or graded composites. As in structural composites, in graded materials, too, it is necessary to express such properties as the mechanical, thermal and piezo-electric parameters analytically in order to study the damage mechanics of the medium. Generally, given the properties of the constituents, volume fractions and microstructures, the effective thermomechanical properties of the composite may be obtained within a satisfactory degree of accuracy [2,3]. In graded materials the problem would reduce essentially to a curve-fitting process. These questions are, off course, not unique to graded materials and have been studied extensively in connection with composites. An excellent up-to-date review of the subject may be found in [3].

Studies in contact mechanics in elastic solids were originated by Hertz [5]. The technical literature on the subject is very extensive. A thorough description and review of the underlying solid mechanics problems in homogeneous materials may be found, for example, in [6,7]. Some sample solutions for contact problems in a semi-infinite graded medium are given in [8–12]. Details of the analysis of homogeneous substrates with FGM coatings having positive or negative curvatures and extensive results regarding the stress distribution

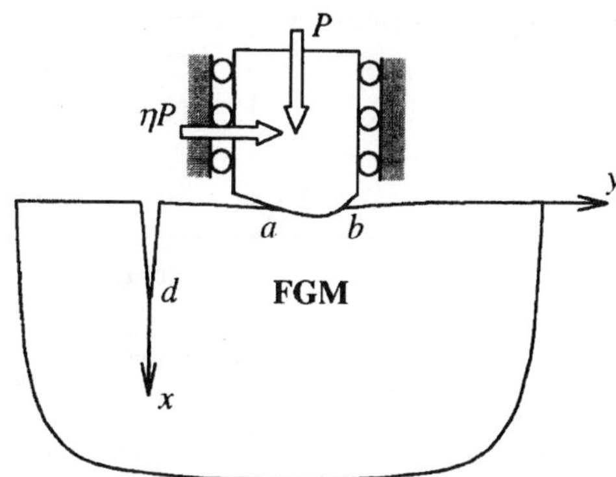

Fig. 1. The general description of the crack/contact problem in a graded medium.

under plane strain conditions and sliding contact are discussed in [13]. The crack/contact problem described in Fig. 1 has also been considered in [14] for a homogeneous half-plane by using the conformal mapping technique. Even though the square root singularity at the crack tip is embedded in the technique, the procedure used in [14] does not account for the singularities at the end points of the contact region a and b, particularly when $a = 0$ (see, Fig. 2). Also, for some reason in [14] the mode II stress intensity factor k_2 is calculated to be negative, implying that under the loading by a sliding stamp the surface crack would tend to curve forward in the direction of the moving stamp, whereas the experimental results (e.g., [15]) and the results found in this study show the opposite.

2. Formulation of the general crack/contact problem

The coupled crack/contact problem for a nonhomogeneous half plane considered in this study is described in Fig. 1. This is a sample problem representing an idealized version of a wide variety of physical applications in brittle solids under sliding contact conditions. Scratch tests performed on glass do indeed indicate that herringbone surface cracks develop and arc backward [3,15]. However, similar experiments performed on a graded alumina–glass composite with 100% glass on the surface and increasing stiffness in depth direction also show that in the graded medium such surface cracking may occur only under very high contact loads, the herringbone cracks are largely absent and, instead a more "graceful" microcrack pattern develops on the surface [3]. In this study we consider the crack/contact problem for a graded medium with an increasing stiffness in depth direction which cover most of the graded ceramic/metal composites used in practice. In formulating the problem it is assumed that the shear modulus of the graded medium may be approximated by a two-parameter exponential function and the Poisson's ratio is constant. Thus, referring to Fig. 1, the elastic parameters of the graded medium may be expressed as [1]

$$\mu(x) = \mu_0 \exp(\gamma x), \quad \kappa = \text{constant} \tag{1}$$

where μ is the shear modulus, $\kappa = 3 - 4v$ for plane strain and $\kappa = (3 - v)/(1 + v)$ for the generalized plane stress conditions, v being the Poisson's ratio. In this continuum solid mechanics formulation of the problem it is assumed that μ and v are known empirically in terms of the variable x (μ monotonically increases with x)

---

[1] In previous studies of crack problems in graded materials it was shown that the stress intensity factors are not significantly influenced by the variations in the Poisson's ratio [16], hence the assumption of $\kappa = $ constant.

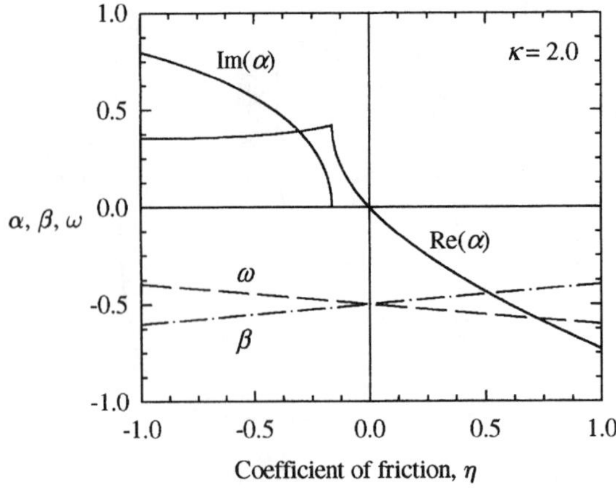

Fig. 2. Variation of the exponents α, β and ω with friction coefficient η.

and are independent of y. The material nonhomogeneity parameters μ_0 and γ and the constant κ may then be determined from the empirical data by using a "best fit" to the model given by (1).

In physical problems the medium is usually finite. However, in this study it is further assumed that the dimensions of the medium, particularly its thickness h in x-direction, are large in comparison with the local dimensions d, a and b (Fig. 1). Thus, the problem is approximated by that of an elasticity for a semi-infinite nonhomogeneous isotropic medium subjected to unbalanced resultant normal and tangential contact forces P and ηP, η being the coefficient of friction. By using the Hooke's law

$$\sigma_{xx}(x, y) = \frac{\mu(x)}{\kappa - 1}\left\{(\kappa + 1)\frac{\partial u}{\partial x} + (3 - \kappa)\frac{\partial v}{\partial y}\right\}, \tag{2a}$$

$$\sigma_{yy}(x, y) = \frac{\mu(x)}{\kappa - 1}\left\{(\kappa + 1)\frac{\partial v}{\partial y} + (3 - \kappa)\frac{\partial u}{\partial x}\right\}, \tag{2b}$$

$$\sigma_{xy}(x, y) = \mu(x)\left\{\frac{\partial u}{\partial y} + \frac{\partial v}{\partial x}\right\}, \tag{2c}$$

the equilibrium equations $\sigma_{ij,j} = 0$ become

$$(\kappa + 1)\frac{\partial^2 u}{\partial x^2} + (\kappa - 1)\frac{\partial^2 u}{\partial y^2} + 2\frac{\partial^2 v}{\partial x \partial y} + \gamma(\kappa + 1)\frac{\partial u}{\partial x} + \gamma(3 - \kappa)\frac{\partial v}{\partial y} = 0, \tag{3a}$$

$$(\kappa + 1)\frac{\partial^2 v}{\partial y^2} + (\kappa - 1)\frac{\partial^2 v}{\partial x^2} + 2\frac{\partial^2 u}{\partial x \partial y} + \gamma(\kappa - 1)\frac{\partial v}{\partial x} + \gamma(\kappa - 1)\frac{\partial u}{\partial y} = 0. \tag{3b}$$

Eqs. (3a) and (3b) must be solved under the following conditions:

$$\sigma_{xx}(0, y) = 0, \qquad \sigma_{xy}(0, y) = 0, \quad -\infty < y < a, \ b < y < \infty, \tag{4}$$

$$\sigma_{xy}(0, y) = \eta \sigma_{xx}(0, y), \qquad \frac{4\mu_0}{\kappa + 1}\frac{\partial}{\partial y}u(0, y) = f(y) \quad a < y < b, \tag{5}$$

$$\sigma_{yy}(x, 0) = 0, \qquad \sigma_{xy}(x, 0) = 0, \quad 0 < x < d, \tag{6}$$

$$\int_a^b \sigma_{xx}(0,y)\,\mathrm{d}y = -P, \tag{7}$$

where the known function $f(y)$ is defined by the derivative of the stamp profile. Note that, in addition to $f(y)$, the external loads are described by the resultant force P and the crack surface tractions given by (6). We also observe that the unknown functions of the problem may be identified as follows (Fig. 1):

$$\frac{2\mu_0}{\kappa+1}\frac{\partial}{\partial x}(v(x,0^+) - v(x,0^-)) = f_1(x), \quad 0 < x < d, \tag{8a}$$

$$\frac{2\mu_0}{\kappa+1}\frac{\partial}{\partial x}(u(x,0^+) - u(x,0^-)) = f_2(x), \quad 0 < x < d, \tag{8b}$$

$$\sigma_{xx}(0,y) = f_3(y), \quad a < y < b. \tag{9}$$

By using the Fourier transforms, the mixed boundary value problem under consideration may be reduced to a system of integral equations of the following form:

$$\sigma_{yy}(x,0) = \lim_{y\to 0}\int_0^d [k_{11}(x,y,t)f_1(t) + k_{12}(x,y,t)f_2(t)]\,\mathrm{d}t + \lim_{y\to 0}\int_a^b k_{13}(x,y,t)f_3(t)\,\mathrm{d}t = 0, \quad 0 < x < d, \tag{10a}$$

$$\sigma_{xy}(x,0) = \lim_{y\to 0}\int_0^d [k_{21}(x,y,t)f_1(t) + k_{22}(x,y,t)f_2(t)]\,\mathrm{d}t + \lim_{y\to 0}\int_a^b k_{23}(x,y,t)f_3(t)\,\mathrm{d}t = 0, \quad 0 < x < d, \tag{10b}$$

$$\frac{4\mu_0}{\kappa+1}\frac{\partial}{\partial y}u(0,y) = \lim_{x\to 0}\int_0^d [k_{31}(x,y,t)f_1(t) + k_{32}(x,y,t)f_2(t)]\,\mathrm{d}t + \lim_{x\to 0}\int_a^b k_{33}(x,y,t)f_3(t)\,\mathrm{d}t = f(y), \quad a < y < b, \tag{10c}$$

where the kernels k_{ij} may be expressed in terms of infinite integrals with known integrands by using the symbolic manipulator MAPLE and the residue theorem. For example, consider the kernels k_{j3}, $j = 1, 2, 3$ that arise from the contact problem, that is, in Fig. 1 and (10a)–(10c) by letting $d = 0$. Using Fourier transforms, from (3a) and (3b) it can be shown that

$$u_3(x,y) = \frac{1}{2\pi}\int_{-\infty}^{\infty}\sum_{j=1}^{2} M_j \exp(s_j x + i\rho y)\,\mathrm{d}\rho, \tag{11a}$$

$$v_3(x,y) = \frac{1}{2\pi}\int_{-\infty}^{\infty}\sum_{j=1}^{2} M_j N_j \exp(s_j x + i\rho y)\,\mathrm{d}\rho, \tag{11b}$$

where

$$N_j(\rho) = \frac{i\big((\kappa+1)s_j^2 + \gamma(\kappa+1)s_j + \rho^2(1-\kappa)\big)}{\rho(2s_j + \gamma(3-\kappa))}, \quad (j=1,2) \tag{12}$$

and s_1, s_2 are the roots of characteristic equation with negative real parts. The characteristic equation and its roots are given by

$$(s^2 + \gamma s - \rho^2 - i|\rho|\delta)(s^2 + \gamma s - \rho^2 + i|\rho|\delta) = 0, \qquad \delta = |\gamma|\sqrt{\frac{3-\kappa}{\kappa+1}}, \tag{13}$$

$$s_1 = -\frac{1}{2}\gamma - \frac{1}{2}\sqrt{\gamma^2 + 4\rho^2 + 4i|\rho|\delta}, \quad \Re(s_1) < 0, \tag{14a}$$

$$s_2 = -\frac{1}{2}\gamma - \frac{1}{2}\sqrt{\gamma^2 + 4\rho^2 - 4i|\rho|\delta}, \quad \Re(s_2) < 0, \tag{14b}$$

$$s_3 = -\frac{1}{2}\gamma + \frac{1}{2}\sqrt{\gamma^2 + 4\rho^2 + 4i|\rho|\delta}, \quad \Re(s_3) > 0, \tag{14c}$$

$$s_4 = -\frac{1}{2}\gamma + \frac{1}{2}\sqrt{\gamma^2 + 4\rho^2 - 4i|\rho|\delta}, \quad \Re(s_4) > 0. \tag{14d}$$

The functions $M_j(\rho)$ may be expressed as

$$M_j(\rho) = \frac{1}{\mu_0}\psi_j(\rho)\int_a^b f_3(t)\exp(-i\rho t)\,dt \quad (j = 1, 2), \tag{15}$$

$$\sum_{j=1}^{2}(s_j(\kappa+1) + i\rho N_j(3-\kappa))\psi_j(\rho) = \kappa - 1, \tag{16a}$$

$$\sum_{j=1}^{2}(i\rho + N_j s_j)\psi_j(\rho) = \eta. \tag{16b}$$

In (11a) and (11b), the subscript 3 stands for the displacements due to stamp loading only. After determining u_3 and v_3, referring to (10a)–(10c) and using the stress–strain relations the kernels may be obtained as follows:

$$k_{13}(x,y,t) = \frac{1}{2\pi}\frac{\exp(\gamma x)}{\kappa-1}\int_{-\infty}^{\infty}\phi_{13}(\rho,x)\exp(i\rho(y-t))\,d\rho, \tag{17a}$$

$$k_{23}(x,y,t) = \frac{\exp(\gamma x)}{2\pi}\int_{-\infty}^{\infty}\phi_{23}(\rho,x)\exp(i\rho(y-t))\,d\rho, \tag{17b}$$

$$k_{33}(x,y,t) = \frac{1}{\pi}\frac{2}{\kappa+1}\int_{-\infty}^{\infty}\phi_{33}(\rho,x)\exp(i\rho(y-t))\,d\rho, \tag{17c}$$

$$\phi_{13}(\rho,x) = \sum_{j=1}^{2}(i\rho N_j(\kappa+1) + s_j(3-\kappa))\psi_j(\rho)\exp(s_j x), \tag{17d}$$

$$\phi_{23}(\rho,x) = \sum_{j=1}^{2}(i\rho + N_j s_j)\psi_j(\rho)\exp(s_j x), \tag{17e}$$

$$\phi_{33}(\rho,x) = i\rho\sum_{j=1}^{2}\psi_j(\rho)\exp(s_j x). \tag{17f}$$

For future reference, one may also express the in-plane component of the stress on the surface due to the stamp loading as

$$\sigma_{yy3}(0,y) = \lim_{x \to 0} \int_a^b k_{y3}(x,y,t)f_3(t)\,dt,\tag{18}$$

$$k_{y3}(x,y,t) = \frac{1}{2\pi}\frac{\exp(\gamma x)}{\kappa - 1}\int_{-\infty}^{\infty}\phi_{y3}(\rho,x)\exp(i\rho(y-t))\,d\rho,\tag{19a}$$

$$\phi_{y3}(\rho,x) = \sum_{j=1}^{2}(s_j(3-\kappa)+i\rho N_j(\kappa+1))\psi_j(\rho)\exp(s_j x).\tag{19b}$$

The remaining kernels k_{ij} $(i = 1, 2, 3; j = 1, 2)$ are obtained by following a procedure which is somewhat more complicated than but essentially quite similar to that followed for the kernels, k_{j3}, $(j = 1, 2, 3)$. The symmetry conditions require that $\sigma_{yy2}(x,0) = 0$ for $f_1(t) = 0, f_2(t) \neq 0$, $f_3(t) = 0$, and $\sigma_{xy1}(x,0) = 0$ for $f_1(t) \neq 0$, $f_2(t) = 0$, $f_3(t) = 0$, from which it follows that

$$k_{12}(x,0,t) = 0, \quad k_{21}(x,0,t) = 0.\tag{20}$$

The details of the analysis and derivation of the remaining kernels, k_{11}, k_{22}, k_{31} and k_{32} may be found in [17]. After obtaining the unknown functions f_1, f_2 and f_3 the stresses and displacements in the graded medium may be determined from Eqs. (2a)–(2c), (11a) and (11b), and (15) and their equivalents for nonvanishing f_1 and f_2.

3. Singular behavior of the solution

Singular nature of the integral equations (10a)–(10c) and, consequently, the singular behavior of the unknown functions f_1, f_2 and f_3 are obtained by examining the singularities of the kernels k_{ij}. In the problem under consideration the kernels have the form, (see, for example, (17a)–(17f))

$$\lim_{(x\,\text{or}\,y)\to 0} k_{ij}(x,y,t) = \hat{k}_{ij}(x^*,t) = \int_{-\infty}^{\infty} K_{ij}(x^*,t,\rho)\,d\rho, \quad (i,j=1,2,3),\tag{21}$$

where $x^* = x$ for $i = 1, 2$ and $x^* = y$ for $i = 3$. It can be shown that [17] with the exception of one case that will be discussed below, the integrands K_{ij} in (21) are bounded and continuous for $(\rho) < \infty$ and integrable at $\rho = 0$. The singular nature of the kernels $\hat{k}_{ij}$ is, therefore, determined by examining the asymptotic behavior $K_{ij}(x^*,t,\rho)$ as (ρ) tends to infinity. Designating the asymptotic values of K_{ij} by K_{ij}^{∞} and by evaluating the integrals

$$\int_{-\infty}^{\infty} K_{ij}^{\infty}(x^*,t,\rho)\,d\rho,\tag{22}$$

in closed form, the system of integral equations (10a)–(10c) may be expressed as follows:

$$\sigma_{yy}(x,0)\exp(-\gamma x) = \int_0^d \left[\frac{1}{\pi}\frac{1}{t-x} + h_{11s}(x,t) + h_{11f}(x,t)\right]f_1(t)\,dt$$

$$+ \int_a^b \left[h_{13s}(x,t) + h_{13f}(x,t)\right]f_3(t)\,dt = 0, \quad 0 < x < d,\tag{23a}$$

$$\sigma_{xy}(x,0)\exp(-\gamma x) = \int_0^d \left[\frac{1}{\pi}\frac{1}{t-x} + h_{22s}(x,t) + h_{22f}(x,t)\right] f_2(t)\,\mathrm{d}t$$

$$+ \int_a^b \left[h_{23s}(x,t) + h_{23f}(x,t)\right] f_3(t)\,\mathrm{d}t = 0, \quad 0 < x < d, \tag{23b}$$

$$\frac{4\mu_0}{1+\kappa}\frac{\partial}{\partial y}u(0,y) = \int_0^d \left[h_{31s}(y,t) + h_{31f}(y,t)\right]f_1(t)\,\mathrm{d}t + \int_0^d \left[h_{32s}(x,t) + h_{32f}(x,t)\right]f_2(t)\,\mathrm{d}t$$

$$-\frac{1}{\pi}\int_a^b \frac{f_3(t)}{t-y}\,\mathrm{d}t - \eta\frac{\kappa-1}{\kappa+1}f_3(y) + \int_a^b h_{33f}(y,t)f_3(t)\,\mathrm{d}t = f(y), \quad a < y < b, \tag{23c}$$

where $h_{ijs}(x^*,t)$ are additional singular kernels (of the order $1/t$) that become unbounded as the arguments x^* and t tend to the end point zero simultaneously and are evaluated from (22) whereas $h_{ijf}(x^*,t)$ are bounded in their respective closed intervals and are evaluated (numerically) from the integrals

$$\int_{-\infty}^{\infty}\left[K_{ij}(x^*,t,\rho) - K_{ij}^{\infty}(x^*,t,\rho)\right]\mathrm{d}\rho. \tag{24}$$

Evaluating the integrals (22), the singular kernels are found to be

$$h_{11s}(x,t) = h_{22s}(x,t) = \frac{1}{\pi}\left(\frac{1}{t+x} + \frac{2t}{(t+x)^2} - \frac{4t^2}{(t+x)^3}\right), \quad 0 < (t,x) < d, \tag{25a}$$

$$h_{13s}(x,t) = \frac{1}{\pi}\left(\frac{2xt^2}{(t^2+x^2)^2} - \eta\frac{2t^3}{(t^2+x^2)^2}\right), \quad a < t < b, \ 0 < x < d, \tag{25b}$$

$$h_{23s}(x,t) = \frac{1}{\pi}\left(\eta\frac{2xt^2}{(t^2+x^2)^2} - \frac{2tx^2}{(t^2+x^2)^2}\right), \quad a < t < b, \ 0 < x < d, \tag{25c}$$

$$h_{31s}(y,t) = -\frac{1}{\pi}\frac{4t^2 y}{(t^2+y^2)^2}, \quad 0 < t < d, \ a < y < b, \tag{25d}$$

$$h_{32s}(y,t) = \frac{1}{\pi}\frac{4t^3}{(t^2+y^2)^2}, \quad 0 < t < d, \ a < y < b. \tag{25e}$$

Note that the singular kernels h_{11s} and h_{22s} given by (25a) are the standard expressions found for edge cracks in homogeneous materials [17,18].

The singular behavior of the unknown functions f_1, f_2 and f_3 at the end points of their respective intervals may be determined by carrying out a function theoretic analysis [19,20]. The two cases, $a > 0$ and $a = 0$ are considered separately.

$a > 0$: In this case we observe that the kernels (25b)–(25e) are bounded in their corresponding closed intervals and would not contribute to the singularities of the functions f_1, f_2 and f_3. Expressing f_i by

$$f_1(x) = x^{\theta_1}(d-x)^{\lambda_1}F_1(x), \quad 0 < x < d, \tag{26a}$$

$$f_2(x) = x^{\theta_2}(d-x)^{\lambda_2}F_2(x), \quad 0 < x < d, \tag{26b}$$

$$f_3(x) = (y-a)^{\omega}(b-y)^{\beta}F_3(y), \quad a < y < b, \tag{26c}$$

and substituting in (23a)–(23c), the conditions of boundedness of $\sigma_{yy}(x,0)$ and $\sigma_{xy}(x,0)$ in $0 < x < d$ and $f(y)$ in $a < y < b$ would give the following equations to determine θ_1, λ_1, θ_2, λ_2, ω and β:

$$\theta_1 = 0, \qquad \theta_2 = 0, \tag{27}$$

$$\cot(\pi\lambda_1) = 0, \qquad \cot(\pi\lambda_2) = 0, \quad (\lambda_1 = \lambda_2 = -0.5), \tag{28}$$

$$\cot(\pi\omega) = \eta\frac{\kappa - 1}{\kappa + 1}, \qquad \cot(\pi\beta) = -\eta\frac{\kappa - 1}{\kappa + 1}, \tag{29}$$

where the acceptable roots are $\lambda_1 = -0.5$, $\lambda_2 = -0.5$, $\Re(\omega) < 0$ if a is known and is a sharp corner, $\Re(\omega) > 0$ if a is unknown and is a point of smooth contact. Similarly, $\Re(\beta) < 0$ if b is a known sharp corner and $\Re(\beta) > 0$ if b is unknown and the contact is smooth.

$a = 0$: In this case all kernels $h_{ijs}(x^*, t)$ become unbounded as $x^* \to 0$ and $t \to 0$ simultaneously and contribute to the singularity of the unknown functions. Again, we express the unknown functions in the following form:

$$f_1(x) = x^\alpha(d - x)^{\lambda_1} G_1(x), \quad 0 < x < d, \tag{30a}$$

$$f_2(x) = x^\alpha(d - x)^{\lambda_2} G_2(x), \quad 0 < x < d, \tag{30b}$$

$$f_3(y) = y^\alpha(b - y)^\beta G_3(y), \quad 0 < y < b. \tag{30c}$$

Now, by substituting (30a)–(30c) into (23a)–(23c), evaluating the singular integrals by using the complex function theory [19], multiplying both sides of the equations first by $(d - x)^{-\lambda_1}$ then by $(d - x)^{-\lambda_2}$ and letting $x \to d$, and then by $(b - y)^{-\beta}$ and letting $y \to b$ we find

$$\cot(\pi\lambda_1) = 0, \qquad \cot(\pi\lambda_2) = 0, \quad (\lambda_1 = \lambda_2 = -0.5), \tag{31}$$

$$\cot(\pi\beta) = -\eta\frac{k - 1}{k + 1}. \tag{32}$$

Similarly, after integrating all singular kernels if we multiply both sides of (23a)–(23c) by $x^{-\alpha}$, $x^{-\alpha}$ and $y^{-\alpha}$, respectively, and let $x \to 0$, $y \to 0$, we obtain the following system of equations to determine α:

$$\begin{bmatrix} a_{11}(\alpha) & 0 & a_{13}(\alpha) \\ 0 & a_{21}(\alpha) & a_{23}(\alpha) \\ a_{31}(\alpha) & a_{32}(\alpha) & a_{33}(\alpha) \end{bmatrix} \begin{bmatrix} \sqrt{d}\,G_1(0) \\ \sqrt{d}\,G_2(0) \\ b^\beta G_3(0) \end{bmatrix} = \begin{bmatrix} 0 \\ 0 \\ 0 \end{bmatrix}, \tag{33}$$

where the coefficient matrix $a_{ij}(\alpha)$ is known in closed form. Since $G_i(0) \neq 0$, for a nontrivial solution the characteristic equation to determine α is obtained by expressing $|a_{ij}(\alpha)| = 0$, giving

$$\frac{2\alpha^2 + 4\alpha + 1 - \cos(\pi\alpha)}{(\kappa + 1)\sin^2(\pi\alpha)}(\eta(4\alpha^2 + 10\alpha + 5 + (\kappa - 1)\cos(\pi\alpha) + \kappa(2\alpha + 3)) + (\kappa + 1)\sin(\pi\alpha)) = 0. \tag{34}$$

The eigenvalue problem (33) yields the following additional compatibility equations which relate $G_1(0)$, $G_2(0)$ and $G_3(0)$ and which must be taken into consideration in the numerical solution of the integral equations (23a)–(23c):

$$G_1(0)\sqrt{d} = -\frac{\eta\cos(\pi\alpha/2)(2 + \alpha) + \sin(\pi\alpha/2)(1 + \alpha)}{2\alpha^2 + 4\alpha + 1 - \cos(\pi\alpha)}G_3(0)b^\beta, \tag{35a}$$

$$G_2(0)\sqrt{d} = -\frac{\eta\sin(\pi\alpha/2)(1 + \alpha) - \cos(\pi\alpha/2)\alpha}{2\alpha^2 + 4\alpha + 1 - \cos(\pi\alpha)}G_3(0)b^\beta. \tag{35b}$$

Note that the main results of the asymptotic analysis expressed by (27)–(29) and (31)–(33) are independent of μ_0 and the material nonhomogeneity parameter γ and are dependent on the Poisson's ratio (through κ) and the coefficient of friction η only, meaning that in the coupled problem considered the stress singularities for graded and homogeneous materials are identical.

The characteristic equation (34) has been verified independently by considering the 90° elastic wedge under appropriate boundary conditions and by using Mellin transforms (see Appendix A).

For a constant Poisson's ratio $v = 0.25$ Fig. 2 shows the singularities associated with a rigid flat stamp. For $a > 0$ the contact stresses are given by $\sigma_{xx}(0, y) = f_3(y)$, $\sigma_{xy}(0, y) = \eta f_3(y)$ and referring to (26c), the singularities ω and β are obtained from (29). Note that in the case of sliding contact the physically relevant problem is related to $a \geqslant 0$ and $\eta > 0$, otherwise the contact stresses would tend to close the crack. Thus, for $\eta > 0$ from Fig. 2 it is seen that the trailing end of the stamp has a stronger and the leading end a weaker singularity than the frictionless stamp problem, that is $-\omega > 0.5$ and $-\beta < 0.5$. Similarly, in the smooth contact case $\omega > 0$, $\beta > 0$, $\beta > 1/2 > \omega$ and again the contact stresses are concentrated near the trailing end of the stamp. Fig. 2 also shows the singularity α at $a = 0$ for the flat stamp obtained from (34). α is real and negative for $\eta > 0$ and $\alpha = 0$ for $\eta = 0$. Also for $\eta < 0$, $\Re(\alpha) > 0$ and for $\eta \lesssim -0.16$ α is complex.

4. The contact problem for decreasing stiffness ($\gamma < 0$)

Consider the sliding contact problem for a graded medium without a surface crack. The half-plane is thus subjected to a pair of unbalanced resultant forces P and ηP. In formulating the general problem in the previous sections it was stated that the integrands $K_{ij}(x^*, t, \rho)$ in (21) are integrable around $\rho = 0$, with one exception that being K_{33} for $\gamma < 0$. All other integrands appear to be well-behaved at $\rho = 0$ for all values of γ. From (17a)–(17f), and (21) the infinite integral giving k_{33} may be expressed as

$$k_{33}(x, y, t) = \frac{1}{\pi} \frac{2}{\kappa + 1} \int_0^\infty [H_1(\rho, x) \sin(\rho(y - t)) + \eta H_2(\rho, x) \cos(\rho(y - t))]\, d\rho. \tag{36}$$

For the homogeneous medium $\gamma = 0$, $K_{33} = K_{33}^\infty$ and the kernel is evaluated in closed form as shown in (23c). For $\gamma \neq 0$ the asymptotic expansion of $H_1(\rho, 0)$ and $H_2(\rho, 0)$ near $\rho = 0$ gives

$$H_1(\rho, 0) = a_1\rho + a_3\rho^3 + O(\rho^5) \tag{37a}$$

$$a_1 = \frac{2(\kappa - 1)}{\gamma(\kappa + 1)}, \tag{37b}$$

$$a_3 = \frac{8(\kappa - 3)(\kappa(1 + \mathrm{sign}(\gamma)) - 4)}{\gamma^3(\kappa + 1)^2(1 + \mathrm{sign}(\gamma))}, \tag{37c}$$

$$H_2(\rho, 0) = b_2\rho^2 + b_4\rho^4 + O(\rho^6), \tag{38a}$$

$$b_2 = \frac{4(2 - \kappa)}{\gamma^2(1 + \kappa)}, \tag{38b}$$

$$b_4 = \frac{-16(\kappa^2(1 + \mathrm{sign}(\gamma)) - \kappa(7 + 9\,\mathrm{sign}(\gamma)) + 10 + 16\,\mathrm{sign}(\gamma))}{\gamma^4(\kappa + 1)^2(1 + \mathrm{sign}(\gamma))}. \tag{38c}$$

Observing that $\mathrm{sign}(\gamma) = 1$ for $\gamma > 0$, $\mathrm{sign}(\gamma) = -1$ for $\gamma < 0$, from (37a)–(37c) and (38a)–(38c) it is seen that H_1, and H_2 are well-behaved near $\rho = 0$ for $\gamma > 0$. However, for $\gamma < 0$ coefficients a_3 and b_4 become unbounded and as a result k_{33} as expressed by (36) also becomes unbounded. Consequently, it is seen that for a graded medium with an exponentially decaying stiffness the contact problem has no solution. Physically the problem that is analogous to $\gamma < 0$ case is a homogeneous infinite strip of finite thickness under an unbalanced transverse load P (in thickness direction) which has no solution (see [21] for the explanation). Thus for graded half-planes with or without a crack, if $\gamma < 0$ the contact problem has no solution.

5. On the solution of integral equations

Once the exponents λ_1, λ_2, ω, β and α are determined, from (26a)–(26c) and (30a)–(30c) the weight functions w_i and the general form of the solution of the integral equations may be obtained by normalizing the intervals $(0, d)$ and (a, b) or $(0, b)$ to be $(-1, 1)$, e.g. by defining

$$f_i(t) = \phi_i(r), \quad i = 1, 2, 3, \quad -1 < r < 1, \tag{39}$$

$$t = \frac{d}{2}(1 + r), \quad 0 < t < d, \quad -1 < r < 1, \tag{40a}$$

$$t = \frac{b - a}{2}r + \frac{b + a}{2}, \quad a < t < b, \quad -1 < r < 1, \ (a > 0), \tag{40b}$$

$$t = \frac{b}{2}(1 + r), \quad 0 < t < b, \quad -1 < r < 1, \ (a = 0), \tag{40c}$$

The integral equations (23a)–(23c) with generalized Cauchy kernels would then have the form

$$\sum_{j=1}^{3} \int_{-1}^{1} m_{ij}(s^*, r)\phi_j(r)\,\mathrm{d}r = g_i(s^*), \quad -1 < s^* < 1, \ i = 1, 2, 3, \tag{41}$$

where

$$x = d(1 + s^*)/2 \quad \text{for } i = 1, 2,$$
$$y = (b - a)s^*/2 + (b + a)/2, \quad (a > 0),$$
$$y = b(1 + s^*)/2, \quad (a = 0), \quad \text{for } i = 3.$$

From (26a)–(26c)–(29) it is seen that for $a > 0$ the solution of (41) may be expressed as

$$\phi_1(r) = w_1(r)\sum_{n=0}^{\infty} A_{1n}P_n^{(-1/2,0)}(r), \quad w_1(r) = (1 - r)^{-1/2}, \tag{42a}$$

$$\phi_2(r) = w_2(r)\sum_{n=0}^{\infty} A_{2n}P_n^{(-1/2,0)}(r), \quad w_2(r) = (1 - r)^{-1/2}, \tag{42b}$$

$$\phi_3(r) = w_3(r)\sum_{n=0}^{\infty} A_{3n}P_n^{(\beta,\omega)}(r), \quad w_3(r) = (1 - r)^{\beta}(1 + r)^{\omega}. \tag{42c}$$

Similarly for $a = 0$, from (30a)–(30c)–(32) and (34) we obtain

$$\phi_1(r) = w_1(r)\sum_{n=0}^{\infty} B_{1n}P_n^{(-1/2,\alpha)}(r), \quad w_1(r) = (1 - r)^{-1/2}(1 + r)^{\alpha}, \tag{43a}$$

$$\phi_2(r) = w_2(r) \sum_{n=0}^{\infty} B_{2n} P_n^{(-1/2,\alpha)}(r), \quad w_2(r) = (1-r)^{-1/2}(1+r)^{\alpha}, \tag{43b}$$

$$\phi_3(r) = w_3(r) \sum_{n=0}^{\infty} B_{3n} P_n^{(\beta,\alpha)}(r), \quad w_3(r) = (1-r)^{\beta}(1+r)^{\alpha}, \tag{43c}$$

where $P_n^{(\alpha_1,\alpha_2)}(r)$, $-1 < r < 1$, are Jacobi polynomials. The integral equation (41) are solved numerically by using (42a)–(42c) or (43a)–(43c) and an appropriate collocation technique. In this solution the equilibrium condition (7) and the compatibility relations (35a) and (35b) are used as additional conditions. Also, the following property of Jacobi polynomials is used to regularize the singular parts of the integral equations (23a)–(23c), that is the terms involving the Cauchy kernels and the free term

$$\frac{1}{\pi} \int_{-1}^{1} (1-r)^{\beta}(1+r)^{\alpha} P_n^{(\beta,\alpha)} \frac{dr}{r-s} = \cot(\pi\beta)(1-s)^{\beta}(1+s)^{\alpha} P_n^{(\beta,\alpha)}(s) - \frac{2^{(\alpha+\beta)} \Gamma(\beta)\Gamma(n+\alpha+1)}{\pi\Gamma(n+\alpha+\beta+1)}$$
$$\times F(n+1, -n-\beta-\alpha; 1-\beta; (1-s)/2), \tag{44}$$

where $\Gamma(\ )$ is the gamma function and $F(n+1, -n-\beta-\alpha; 1-\beta; (1-s)/2)$ is the hypergeometric function.

After solving the integral equations for f_1, f_2 and f_3, the contact stresses $\sigma_{xx}(0,y) = f_3(y) = \phi_3(s^*)$, $\sigma_{xy}(0,y) = \eta f_3(y) = \eta \phi_3(s^*)$, the in-plane component of the surface stress $\sigma_{yy}(0,y)$ (see for example (18) and (19a) and (19b) for the procedure to be followed) and the stress intensity factors at the crack tip $(d,0)$ may be evaluated by using the results. The stress intensity factors are defined by and calculated from

$$k_1 = \lim_{x \to d+0} \sqrt{2(x-d)}\sigma_{yy}(x,0) = -\lim_{x \to d-0} \frac{2\mu(x)}{\kappa+1} \sqrt{2(d-x)} \frac{\partial}{\partial x}(v(x,0^+) - v(x,0^-)), \tag{45a}$$

$$k_2 = \lim_{x \to d+0} \sqrt{2(x-d)}\sigma_{xy}(x,0) = -\lim_{x \to d-0} \frac{2\mu(x)}{\kappa+1} \sqrt{2(d-x)} \frac{\partial}{\partial x}(u(x,0^+) - u(x,0^-)). \tag{45b}$$

From (42a)–(42c), (43a)–(43c), and (45a) and (45b) it then follows that

$$k_1 = -\exp(\gamma d)\sqrt{d} \sum_{n=0}^{\infty} A_{1n} P_n^{(-1/2,0)}(1), \tag{46a}$$

$$k_2 = -\exp(\gamma d)\sqrt{d} \sum_{n=0}^{\infty} A_{2n} P_n^{(-1/2,0)}(1), \tag{46b}$$

for $a > 0$ and

$$k_1 = -2^{\alpha} \exp(\gamma d)\sqrt{d} \sum_{n=0}^{\infty} A_{1n} P_n^{(-1/2,\alpha)}(1), \tag{47a}$$

$$k_2 = -2^{\alpha} \exp(\gamma d)\sqrt{d} \sum_{n=0}^{\infty} A_{2n} P_n^{(-1/2,\alpha)}(1), \tag{47b}$$

for $a = 0$.

The other quantity of physical interest is the in-plane stress $\sigma_{yy}(0,y)$ on the surface which has a bearing on crack initiation and which may be expressed as

$$\sigma_{yy}(0,y) = \sum_{j=1}^{3} \sigma_{yyj}(0,y) = \lim_{x\to 0}\sum_{j=1}^{2}\int_0^d k_{yj}(x,y,t)f_j(t)\,\mathrm{d}t + \lim_{x\to 0}\int_a^b k_{y3}(x,y,t)f_3(t)\,\mathrm{d}t, \quad -\infty < y < \infty,$$

(48)

where $k_{yj}(x,y,t)$, $(j=1,2,3)$ are known kernels corresponding to the in-plane stress components $\sigma_{yyj}(0,y)$, $(j=1,2,3)$, (see, for example (18), (19a) and (19b) for the in-plane stress due to sliding contact). Knowing the stress components σ_{xx}, σ_{yy} and σ_{xy} on the surface, at the critical location (that is, the trailing end $y=a$ of the contact region) the cleavage stress $\sigma_{\theta\theta}(r,\theta)$ may be expressed as

$$\sigma_{\theta\theta}(r,\theta) = \sigma_{xx}\sin^2(\theta) + \sigma_{yy}\cos^2(\theta) - \sigma_{xy}\sin(\theta)\cos(\theta),$$

(49)

where for $\theta = \theta_{cr}$, $\sigma_{\theta\theta}$ is maximum and positive, and it can be shown that $\theta_{cr} = 0$ and $(\sigma_{\theta\theta})_{cr} = \sigma_{yy}(0,a)$, namely, the crack initiation is perpendicular to the surface.

6. Results and discussion

The calculated results in this study mainly consist of the contact stresses $(\sigma_{xx}(0,y),\sigma_{xy}(0,y), a < y < b)$, stress intensity factors $(k_1(d),k_2(d))$, and the in-plane component of the stress on the surface $(\sigma_{yy}(0,y), -\infty < y < \infty)$. The contact stresses are obtained directly by solving the integral equations (23a)–(23c) or (41) as follows:

$$\sigma_{xx}(0,y) = f_3(y) = \phi_3(s), \qquad \sigma_{xy}(0,y) = \eta\phi_3(s), \quad a < y < b, \quad -1 < s < 1.$$

(50)

Some examples showing the surface stresses in a graded medium in the absence of a crack and loaded by a sliding flat stamp are given in Fig. 4 (see Fig. 3, $d=0$). The figure shows the normal component of the contact stress $\sigma_{xx}(0,y)$, $a < y < b$, and the in-plane stress $\sigma_{yy}(0,y)$, $-\infty < y < \infty$, that is parallel to the surface. The shear component of the contact stress is obtained from $\sigma_{xy}(0,y) = \eta\sigma_{xx}(0,y)$. The results are given for various values of friction coefficient η and material nonhomogeneity parameter γ. For $\eta = 0$ the results are symmetric and stresses on the surface (including σ_{yy}) have square-root singularities at $y=a$ and $y=b$. On the other hand for $\eta > 0$ there is a greater stress concentration near the trailing end of the stamp, $y=a$ and $|\omega| > |\beta|$, ω and β being the singularities at $y=a$ and $y=b$, respectively (see Fig. 2). The important conclusion one may draw from Fig. 4 is that at the trailing end of the stamp the in-plane component of the stress $\sigma_{yy}(0,y)$ is unbounded and discontinuous and has a singularity of the order $(a-y)^\omega$, where $-\omega > 1/2$ (see Fig. 2). This implies that $y=a$ is a likely location of surface crack initiation.

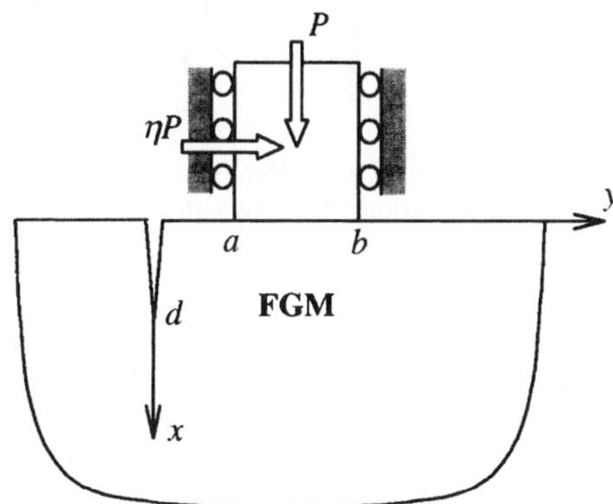

Fig. 3. The geometry of crack/contact problem for a flat stamp.

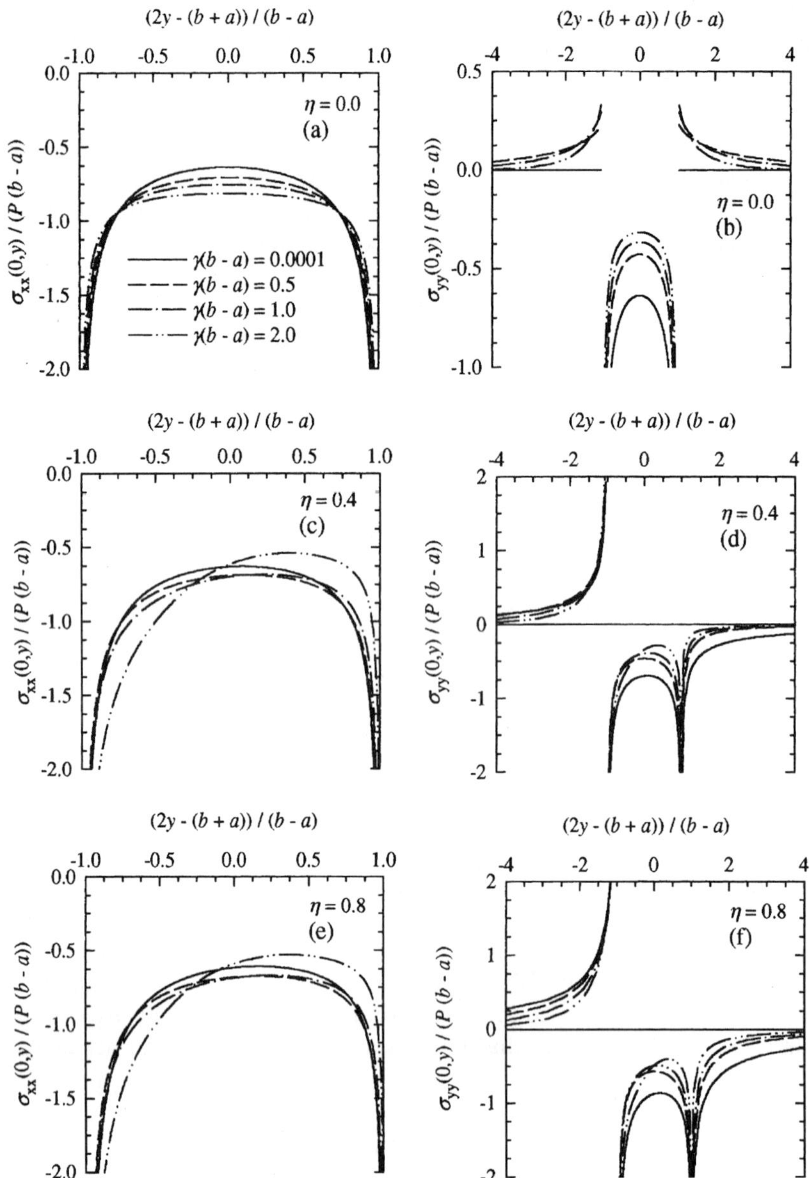

Fig. 4. (a–f) The distribution of the contact stress $\sigma_{xx}(0, y)$ and the in-plane stress $\sigma_{yy}(0, y)$ on the surface in a graded medium without a crack ($d = 0$) loaded by a flat stamp for $\kappa = 2$ and for various values of γ and η.

The contact stress distribution for a graded medium with a surface crack and loaded by a flat stamp is shown in Fig. 5 for $a = 0$. In this case the stress singularities α and β at the end points $a = 0$ and b are given by (32) and (34), respectively. Fig. 2 shows that for relatively small values of (positive) η the stress singularity at b is greater than that at $a = 0$ (i.e., $-\beta > -\alpha$), hence the skewed distribution in Fig. 5a and b where $\eta = 0.4$. On the other hand for relatively large values of η, $|\alpha| > |\beta|$ (see Fig. 2) the trend seems to be reversed and there is a greater stress concentration near the end $a = 0$ (see Fig. 5c and d).

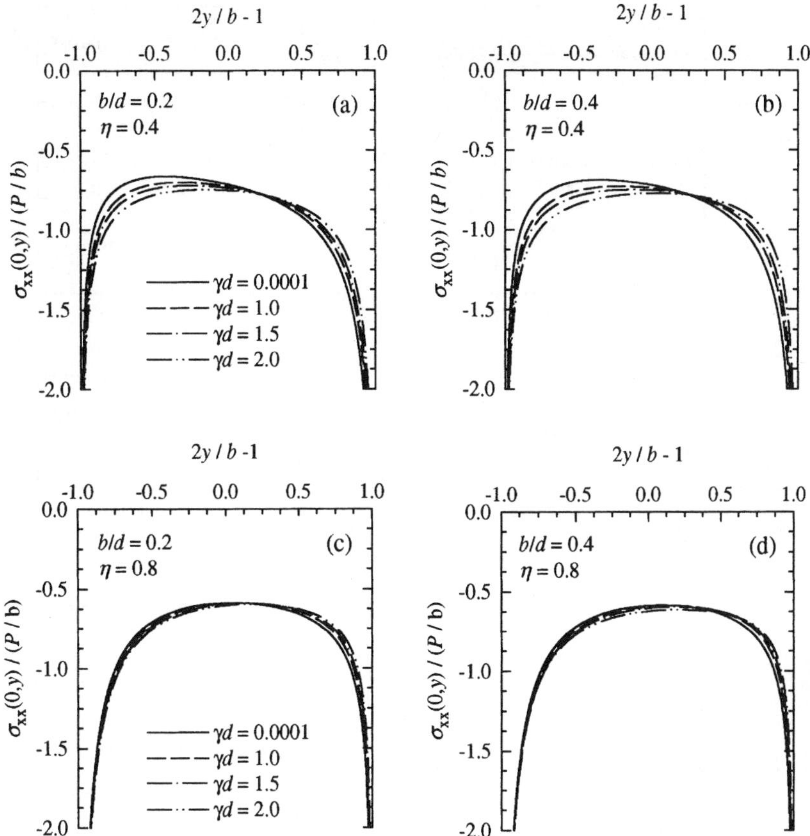

Fig. 5. (a–d) The contact stress distribution for $a = 0$, $\kappa = 2$ and for various values of η, γ and relative stamp dimension b/d. (see Fig. 3).

Some results showing the modes I and II stress intensity factors k_1 and k_2 at the crack tip $x = d$ in a graded medium loaded by a flat stamp (Fig. 3) are given in Figs. 6–8. Fig. 6 shows the results as functions of a/d for $a > 0$, $(b - a)/d = 1$ and various values of γ and η. For a frictionless stamp acting on a homogeneous medium η, γ and the tangential force that would tend to open the crack are zero, the region is mostly under compression and, consequently, k_1 is negative and k_2 is positive. As η increases the resultant friction force ηP also increases and gradually k_1 becomes positive and k_2 negative. Fig. 6 shows that the influence of not only η, which is expected, but also of the material nonhomogeneity parameter γ on the stress intensity factors could be quite significant. Incidentally, the solution presented in Figs. 6–8 corresponding to the values of η, γ and $(b - a)/d$ (or b/d for $a = 0$) for which $k_1 < 0$ is, of course, not valid due to crack closure. But the results can still be applicable and useful in superposition with an uncoupled solution resulting, for example, remote loading [18,22], provided the resultant k_1 is positive. Otherwise, the problem needs to be formulated by taking into account the crack closure and determining the closure distance from the condition of $k_1 = 0$.

Stress intensity factors similar to that shown in Fig. 6 are also given in Fig. 7, the only difference being in the relative stamp size, namely $(b - a)/d = 1$ in Fig. 6 and $(b - a)/d = 0.1$ in Fig. 7.

In the special case of $a = 0$, Fig. 8 shows the stress intensity factors k_1 and k_2 as functions of the relative stamp size b/d (see Fig. 4) for $\kappa = 2$ and for various values of γ and η. For a homogeneous medium the solution is obtained in two different ways: first by assuming $\gamma d = 0.0001$ and using the nonhomogeneous

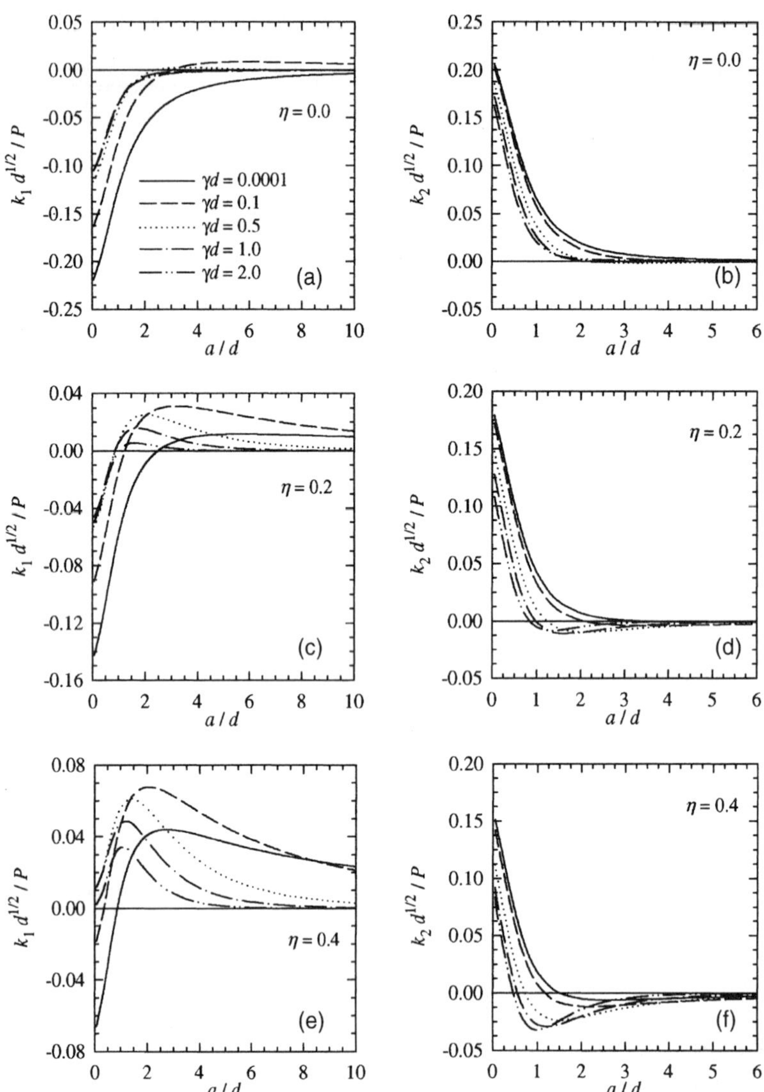

Fig. 6. (a–l) Variation of stress intensity factors with γ, η and stamp location a/d (see Fig. 3) for $\kappa = 2$, $(b - a)/d = 1.0$ and $a > 0$.

material program and then by solving the integral equations for a homogeneous medium (for which all kernels are known in closed form). In Fig. 8, the results of the former solution are given as solid lines and that of the latter as closed circles. Note that for all intents and purposes the two sets of results are identical. It can also be observed that in a homogeneous medium the mode II stress intensity factors are positive for all values of the friction coefficient considered. Consequently, the crack will curve backward in a direction opposite to that of the applied friction force.

7. Some conclusions

In sliding contact problems the trailing end of the contact region is a point of higher (tensile) stress concentration and, consequently, likely location of surface crack initiation.

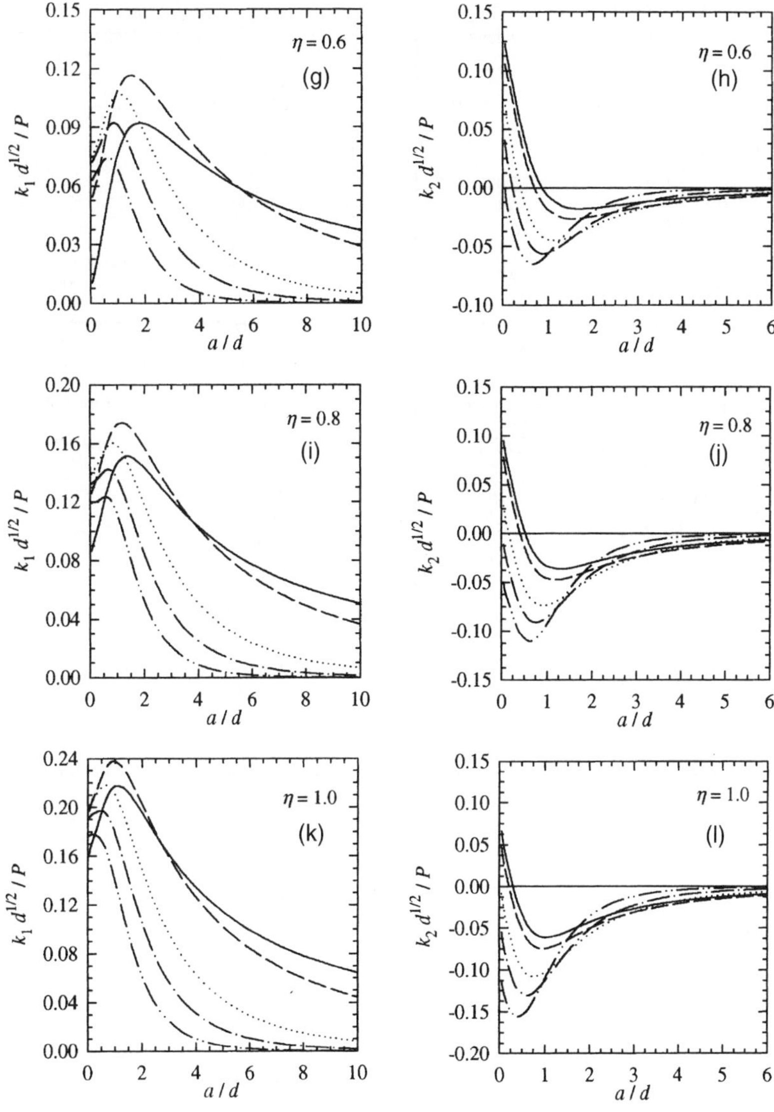

Fig. 6 (*continued*)

In a homogeneous medium ($\gamma = 0$) containing a surface crack and loaded by a sliding rigid stamp, the mixed-mode stress state at the crack tip is such that the crack tend to curve backward, that is, in the opposite direction of the stamp's motion. Immediately upon crack initiation or incremental crack growth (i.e. for small values of a, Fig. 1) k_1 at the crack tip becomes negative. For initiation of another crack or resumption of incremental crack growth of an existing crack the stamp must move further until the peak value of the in-plane surface stress or k_1, k_2 combination reach certain values. This seems to provide an explanation of or a mechanism for periodic cracking in, for example, the scratch tests. In graded materials loaded by a moving stamp, because of the additional factor of material nonhomogeneity constant γ, the relative variations in k_1 and k_2 are much more complicated. However, by carefully following the peak value

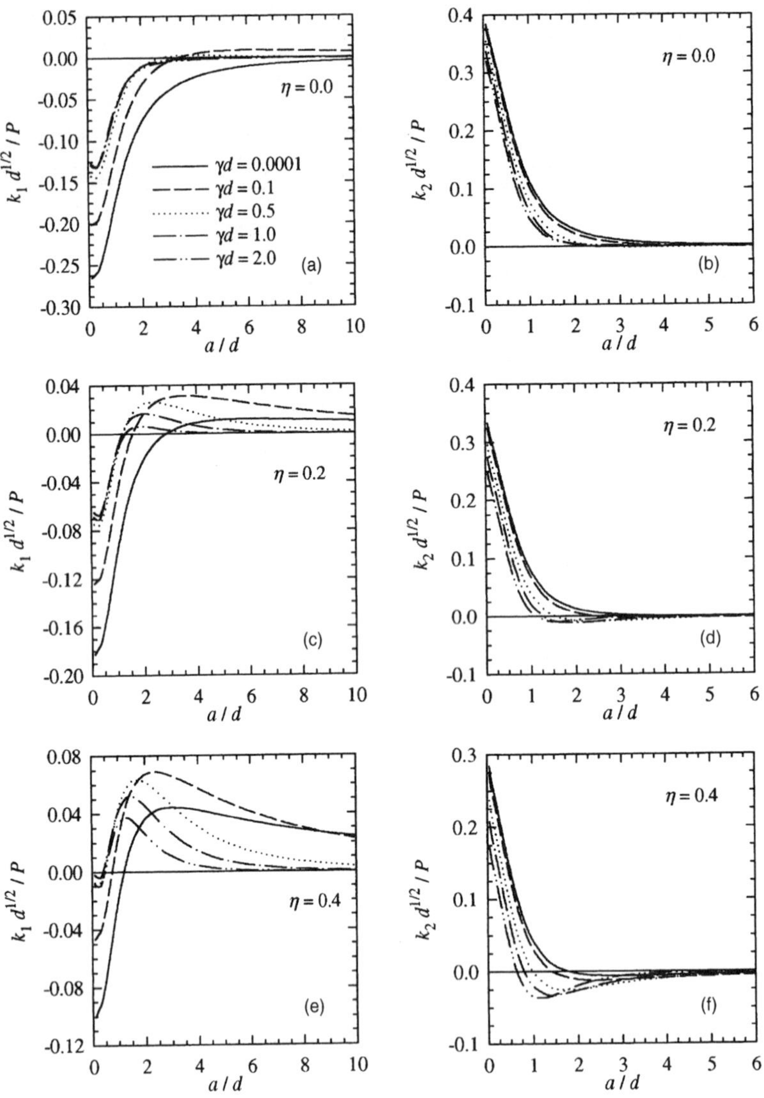

Fig. 7. (a–l) Same as Fig. 6, $(b - a)/d = 0.1$.

of in-plane stresses and k_1, k_2 combinations (signs and magnitudes) it is possible to make arguments similar to homogeneous materials with regard to shielding and crack periodicity.

In the coupled crack/contact problems for a graded medium, the singularities α, β and ω (Fig. 2) are independent of the material nonhomogeneity constants μ_0 and γ (Eq. (1)) and depend only on the friction coefficient η and the surface value of the Poisson's ratio (through the elastic constant κ). Consequently, for the same κ and η, the singularities α, β and ω are the same for a graded and a homogeneous medium.

The contact problem for a graded half plane with exponentially decaying stiffness has no solution.

The results show that in graded materials the influence of not only the friction coefficient η (which is expected) but also the material nonhomogeneity constant γ on the stress intensity factors can be quite significant. An unexpected analytical result of the solution of the crack/contact problem is that, in the case

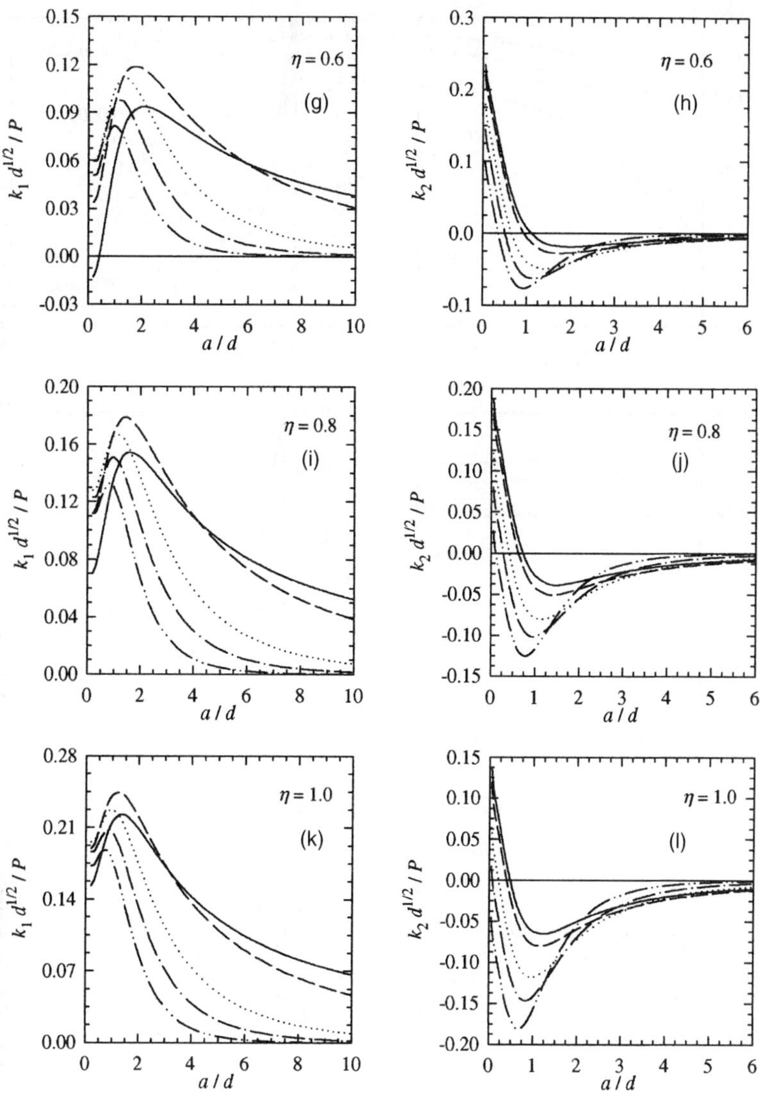

Fig. 7 (*continued*)

of $\eta \neq 0$ and coinciding crack surface and trailing end of the contact region (Fig. 3, $a = 0$, $x = 0$) the intersection $a = 0$, $x = 0$ is a point of singularity, where for sufficiently high values of η the singular power can be greater than 1/2.

Acknowledgements

This work was partially supported by AFOSR under the grant F49620-98-1-0028. The author (F.E.) also gratefully acknowledges the support provided by Alexander von Humboldt Foundation during his stay at the FZK in Karlsruhe, Germany.

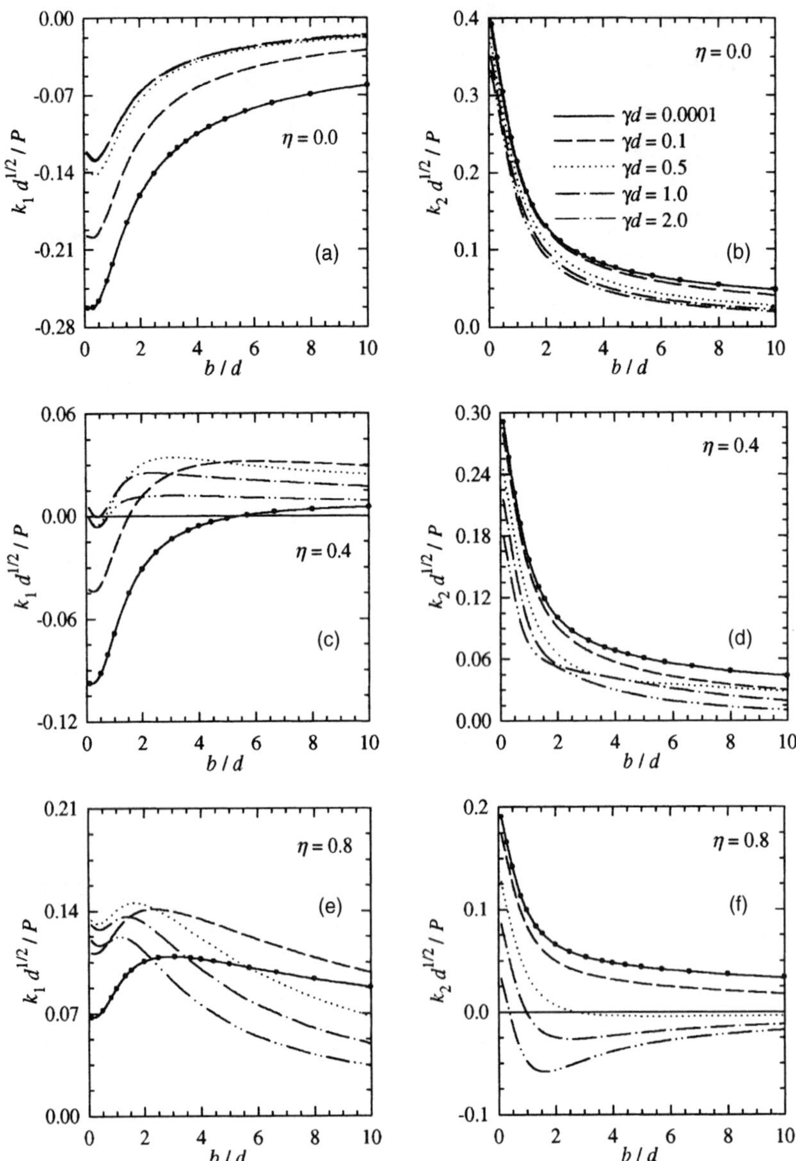

Fig. 8. (a–f) Variation of stress intensity factors with the relative stamp size for $a = 0$, $\kappa = 2$ and for various values of γ and η (see Fig. 3).

Appendix A. Mellin transform method for the derivation of the characteristic equation

The characteristic equation given by (34) can also be obtained by considering a 90° homogeneous elastic wedge as shown in Fig. 9 and using Mellin transformation. The boundary conditions of the problem are

$$\sigma_{r\theta}(r,0) - \eta\sigma_{\theta\theta}(r,0) = 0, \quad 0 < r < \infty, \tag{A.1a}$$

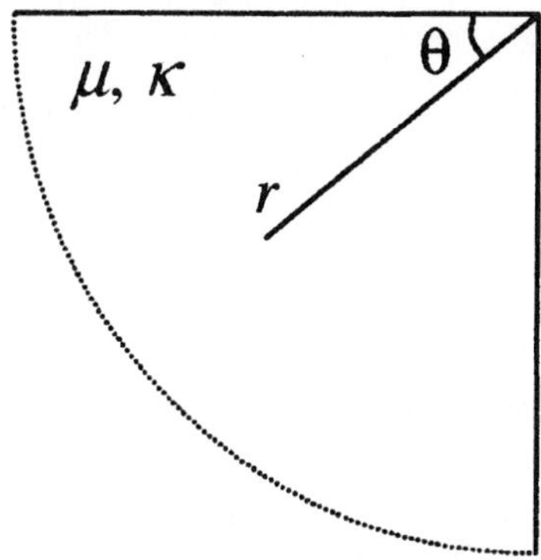

Fig. 9. The geometry of the 90° elastic wedge.

$$\frac{\partial}{\partial r} u_\theta(r,0) = f(r), \quad 0 < r < \infty, \tag{A.1b}$$

$$\sigma_{\theta\theta}(r,\pi/2) = 0, \quad 0 < r < \infty, \tag{A.1c}$$

$$\sigma_{r\theta}(r,\pi/2) = 0, \quad 0 < r < \infty. \tag{A.1d}$$

In polar coordinates, following definition of the stresses in terms of a stress function χ identically satisfies the equilibrium equations

$$\sigma_{rr}(r,\theta) = \frac{1}{r}\frac{\partial\chi(r,\theta)}{\partial r} + \frac{1}{r^2}\frac{\partial^2\chi(r,\theta)}{\partial\theta^2}, \tag{A.2a}$$

$$\sigma_{\theta\theta}(r,\theta) = \frac{\partial^2\chi(r,\theta)}{\partial r^2}, \tag{A.2b}$$

$$\sigma_{r\theta}(r,\theta) = -\frac{\partial}{\partial r}\left(\frac{1}{r}\frac{\partial\chi(r,\theta)}{\partial\theta}\right). \tag{A.2c}$$

Thus, the compatibility condition becomes

$$\nabla^2\nabla^2\chi(r,\theta) = 0, \tag{A.3}$$

where,

$$\nabla^2 = \frac{\partial^2}{\partial r^2} + \frac{1}{r}\frac{\partial}{\partial r} + \frac{1}{r^2}\frac{\partial^2}{\partial\theta^2}. \tag{A.4}$$

It is also known that (see, for example, [23]), displacements can be expressed in terms of the biharmonic stress function χ and another harmonic function ϕ in the following form

$$2\mu u_r(r,\theta) = -\frac{\partial\chi(r,\theta)}{\partial r} + \left(\frac{\kappa+1}{4}\right)r\frac{\partial\phi(r,\theta)}{\partial\theta}, \tag{A.5a}$$

$$2\mu u_\theta(r,\theta) = -\frac{1}{r}\frac{\partial\chi(r,\theta)}{\partial\theta} + \left(\frac{\kappa+1}{4}\right)r^2\frac{\partial\phi(r,\theta)}{\partial r}, \tag{A.5b}$$

where $\chi(r, \theta)$ and $\phi(r, \theta)$ are related by,

$$\nabla^2 \chi(r, \theta) = \frac{\partial}{\partial r}\left(r\frac{\partial \phi(r, \theta)}{\partial \theta}\right), \tag{A.6}$$

and $\phi(r, \theta)$ is a harmonic function,

$$\nabla^2 \phi(r, \theta) = 0. \tag{A.7}$$

Mellin transform of a function $f(r)$ and its inverse are defined by

$$\hat{f}(p) = \int_0^\infty f(r)r^{p-1}\,\mathrm{d}r, \qquad f(r) = \frac{1}{2\pi\mathrm{i}}\int_{c-\mathrm{i}\infty}^{c+\mathrm{i}\infty} \hat{f}(p)r^{-p}\,\mathrm{d}p. \tag{A.8}$$

The transform of the derivative can be written as,

$$\int_0^\infty \frac{\mathrm{d}^n f(r)}{\mathrm{d}r^n}r^{p-1+n}\,\mathrm{d}r = (-1)^n\frac{\Gamma(p+n)}{\Gamma(p)}\hat{f}(p), \tag{A.9}$$

provided,

$$r^{p+m-1}\frac{\mathrm{d}^{m-1}}{\mathrm{d}r^{m-1}}f(r) = 0 \tag{A.10}$$

as $r \to 0$ and $r \to \infty$ for $m = 1, 2, 3, \ldots, n$. Taking the Mellin transform of Eqs. (A.3), (A.6) and (A.7) we obtain,

$$\frac{\mathrm{d}^4\hat{\chi}(\theta, p)}{\mathrm{d}\theta^4} + (p^2 + (p+2)^2)\frac{\mathrm{d}^2\hat{\chi}(\theta, p)}{\mathrm{d}\theta^2} + p^2(p+2)^2\hat{\chi}(\theta, p) = 0, \tag{A.11a}$$

$$\frac{\mathrm{d}^2\hat{\phi}(\theta, p)}{\mathrm{d}\theta^2} + p^2\hat{\phi}(\theta, p) = 0, \tag{A.11b}$$

$$\frac{\mathrm{d}}{\mathrm{d}\theta}\hat{\phi}(\theta, p+2) = -\frac{1}{p+2}\left(p^2\hat{\chi}(\theta, p) + \frac{\mathrm{d}^2\hat{\chi}(\theta, p)}{\mathrm{d}\theta^2}\right). \tag{A.11c}$$

Solving Eqs. (A.11a) and (A.11b) and using (A.11c) displacements and stresses can be expressed as follows:

$$\mu r^2\left(\frac{\partial u_r}{\partial r} + \mathrm{i}\frac{\partial u_\theta}{\partial r}\right) = -\frac{1}{2\pi\mathrm{i}}\int_{c-\mathrm{i}\infty}^{c+\mathrm{i}\infty}(p+1)(Ap\exp(\mathrm{i}p\theta) + B(p+1)\exp(\mathrm{i}(p+2)\theta)$$
$$+ \kappa\bar{B}\exp(-\mathrm{i}(p+2)\theta))r^{-p}\,\mathrm{d}p, \tag{A.12a}$$

$$\frac{r^2}{2\mathrm{i}}(\sigma_{r\theta} + \mathrm{i}\sigma_{\theta\theta}) = \frac{1}{2\pi\mathrm{i}}\int_{c-\mathrm{i}\infty}^{c+\mathrm{i}\infty}(p+1)(Ap\exp(\mathrm{i}p\theta) + B(p+1)\exp(\mathrm{i}(p+2)\theta) + \bar{B}\exp(-\mathrm{i}(p+2)\theta))r^{-p}\,\mathrm{d}p. \tag{A.12b}$$

In Eqs. (A.12a) and (A.12b) A and B are complex constants and $\bar{B}$ is the complex conjugate of B. Using Eqs. (A.12a) and (A.12b) and boundary conditions (A.1a)–(A.1d) following expressions are obtained

$$\frac{\sigma_{\theta\theta}(r, 0)}{4\mu} = \int_0^\infty f(t)\,\mathrm{d}t\,\frac{1}{2\pi\mathrm{i}}\int_{c-\mathrm{i}\infty}^{c+\mathrm{i}\infty}\frac{2\alpha^2 + 4\alpha + 1 - \cos(\pi\alpha)}{D(\alpha)}\frac{r^\alpha}{t^{\alpha+1}}\,\mathrm{d}\alpha, \tag{A.13a}$$

$$\frac{1}{2(\kappa+1)}\frac{\partial}{\partial r}u_\theta(r,\pi/2) = \int_0^\infty f(t)\,\mathrm{d}t\,\frac{1}{2\pi\mathrm{i}}\int_{c-\mathrm{i}\infty}^{c+\mathrm{i}\infty}\frac{\eta\cos(\pi\alpha)(\alpha+2)+\sin(\pi\alpha/2)(\alpha+1)}{D(\alpha)}\frac{r^\alpha}{t^{\alpha+1}}\,\mathrm{d}\alpha, \qquad \text{(A.13b)}$$

where,

$$D(\alpha) = \eta(4\alpha^2 + 10\alpha + 5 + (\kappa-1)\cos(\pi\alpha) + \kappa(2\alpha+3)) + (\kappa+1)\sin(\pi\alpha). \qquad \text{(A.14)}$$

Note that $D(\alpha)$ is the same as the characteristic equation (34). If one performs the inversions in (A.13a) and (A.13b) by using the theory of residues, the negative roots of $D(\alpha)$ give stresses that are singular as r approaches zero. Thus, $D(\alpha) = 0$ is the characteristic equation of the problem.

References

[1] Hills DA, Nowell D, Sackfield A. Mechanics of elastic contacts. London: Butterworth–Heinemann; 1993.

[2] Christensen RM. Mechanics of composite materials. New York: John Wiley; 1979.

[3] Suresh S, Mortensen A. Fundamentals of functionally graded materials—processing and thermomechanical behaviour of graded metals and metal–ceramic composites. IOM Communications Ltd; 1998.

[4] Buryachenko VA. Multiparticle effective field and related methods in micromechanics of composite materials. Appl Mech Rev 2001;54:1–47.

[5] Hertz H. Über die Berührung elastischer Körper. J Reine und Angewandte Mathematik 1882;92:156–71.

[6] Gladwell GM. Contact problems in the classical theory of elasticity. Dordrecht: Kluwer Academic Publishers; 1980.

[7] Johnson KL. Contact mechanics. Cambridge University Press; 1985.

[8] Bakirtas I. The problem of a rigid punch on a nonhomogeneous elastic half space. Int J Eng Sci 1980;18:597–610.

[9] Selvadurai APS, Singh BM, Vrbik J. A Reissner–Sagoci problem for a non-homogeneous elastic solid. J Elas 1986;16:383–91.

[10] Giannakopoulos AE, Suresh S. Indentation of solids with gradients in elastic properties: Part II Axisymmetric indenters. Int J Solids Struct 1997;34:2393–428.

[11] Suresh S, Giannakopoulos AE, Alcala J. Spherical indentation of compositionally graded materials: Theory and experiments. Acta Mater 1997;45:1307–21.

[12] Giannakopoulos AE, Pallot P. Two dimensional contact analysis of elastic graded materials. J Mech Phys Solids 2000;48:1597–631.

[13] Guler MA. Contact mechanics of FGM coatings. PhD dissertation, Lehigh University, Bethlehem, PA, USA, 2000.

[14] Hasebe N, Okumura M, Nakamura T. Frictional punch and crack in plane elasticity. ASCE, J Eng Mech 1989;115:1137–49.

[15] Lawn B. Fracture of brittle solids. Cambridge University Press; 1993.

[16] Delale F, Erdogan F. Crack problem for a nonhomogeneous plane. ASME, J Appl Mech 1983;50:609–14.

[17] Dag S. Crack and contact problems in graded materials. PhD dissertation, Lehigh University, Bethlehem, PA, USA, 2001.

[18] Erdogan F, Wu BH. Surface crack problem for a plate with functionally graded properties. ASME, J Appl Mech 1997;64:449–56.

[19] Erdogan F. Mixed boundary value problems in mechanics. In: Nemat-Nasser S, editor. Mechanics today, vol. 4. Pergamon Press; 1978. p. 1–86.

[20] Muskhelishvili NI. Singular integral equations. Groningen: P. Noordhoff Ltd; 1953.

[21] Ratwani M, Erdogan F. On the plane contact problem for a frictionless elastic layer. Int J Solids Struct 1973;9:921–36.

[22] Kasmalkar M. The surface crack problem for a functionally graded coating bonded to a homogeneous layer. PhD dissertation, Lehigh University, Bethlehem, PA, USA, 1997.

[23] Hein VL, Erdogan F. Stress singularities in a two-material wedge. Int J Fract Mech 1973;7:317–30.

PERGAMON

Engineering Fracture Mechanics 69 (2002) 1753–1768

Engineering Fracture Mechanics

www.elsevier.com/locate/engfracmech

On the dynamic propagation of a finite crack in functionally graded materials

S.A. Meguid [a,*], X.D. Wang [b], L.Y. Jiang [b]

[a] *Department of Mechanical and Industrial Engineering, Engineering Mechanics and Design Laboratory, University of Toronto, 5 King's College Road, Toronto, Ontario, M5S 3G8 Canada*
[b] *Department of Mechanical Engineering, University of Alberta, Edmonton, Alberta, T6G 2G8 Canada*

Received 6 February 2001; received in revised form 13 June 2001; accepted 20 June 2001

Abstract

This article provides a comprehensive theoretical investigation of the singular behaviour of a propagating crack in a functionally graded material (FGM) with spatially varying elastic properties under plane elastic deformation. The analytical formulations are developed in terms of Fourier transforms and the solution of the resulting singular integral equations by using Chebyshev polynomials. In this study, we examine the effect of the gradient of material properties and the speed of crack propagation upon the stress intensity factors, the strain energy release rate and the crack opening displacement. The results reveal that although the singular stress field around the tip of a crack in this FGM is governed by the traditional square root singularity, significant discrepancy in the local stress field exist between it and a homogeneous solid. This indicates that the presence of a gradient in the mechanical properties affects the local stress distribution significantly and, ultimately, the fracture behaviour of the solid.
© 2002 Elsevier Science Ltd. All rights reserved.

Keywords: Functionally graded materials; Dynamic fracture; Stress intensity factor; Strain energy release rate; Crack opening displacement

1. Introduction

The concept of using materials with progressively changing properties, functionally gradient materials (FGMs), for improving material performance has received considerable attention from the research community. FGMs are very attractive for high-temperature applications and wear-protective coatings. The spatial variation of the elastic and physical properties of the material makes FGMs an attractive alternative to composite solids. The fracture behaviour of FGMs is an important design consideration. A crack in a FGM may exhibit complex behaviour. The degree of the complexity of this behaviour is governed by the

---

[*] Corresponding author. Tel.: +1-416-978-5741; fax: +1-416-978-7753.
E-mail address: meguid@mie.utoronto.ca (S.A. Meguid).

function describing the variation in the mechanical properties of the material. For example, the fracture modes of a crack embedded in FGMs are inherently mixed when they are loaded due to the lack of symmetry in the material properties. In the limiting case, where the gradient of the material property of FGMs is very high, the material may behave like an interface with discontinuous material constants. When the gradient of the material property is very low, the FGMs will behave as a homogeneous medium. It should be mentioned that these cases represent two distinct fracture mechanisms with the fracture being governed by a square root singularity and an oscillating singularity, respectively.

Significant efforts have been made in the study of the fracture behaviour of FGMs. These include analytical, numerical and experimental investigations. Delale and Erdogan studied analytically the behaviour of a crack in an infinite FGM plate with the elastic material constants varying in the direction parallel to the crack [1]. Their study indicates that Young's modulus has a significant effect on the stress field, but the effect of Poisson's ratio can be ignored. Delale and Erdogan [2] also considered the problem of two dissimilar homogeneous materials bonded through a non-homogeneous interfacial zone containing a crack parallel to the interfaces. The generalized mixed mode problem for a crack of arbitrary orientation in FGMs was discussed in [3]. Eigenfunction expansion technique [4] was used to study the singular behaviour of the stress field near the crack tip. The result shows that the asymptotic crack tip stress field possesses the same square root singularity as that in homogeneous materials. Similar result was observed in [5] for materials with piecewise differential property variations. The stress intensity factor (SIF) for a semi-infinite crack in a strip of an isotropic FGMs under different loading are discussed in [6].

A semi-infinite crack in an interlayer between two dissimilar materials was studied by Yang and Shih [7]. Erdogan presented solutions of a number of typical fracture problems of FGMs in [8]. Jin and Batra [9] summerized the crack-tip fields in general non-homogeneous materials and obtained the SIF for an edge crack in a strip of FGMs. The mode I fracture behaviour of a crack in orthotropic FGMs was studied by Ozturk and Erdogan [10]. The mixed mode crack problem in an orthotropic FGMs, with the crack plane perpendicular to the direction of the variation of the material property, was considered in [11] to investigate the influence of material orthotropy and gradient of the material properties upon the fracture parameters. Relatively fewer experimental and numerical investigations of the fracture behaviour of FGMs have been conducted. Experimental investigations into the fracture of FGMs are limited due to the high cost and elaborate facilities required for processing FGMs [12–14]. The finite element method has also been used to simulate the fracture behaviuor of cracked FGMs [15,16].

Existing studies which consider the fracture behaviour of FGMs are mainly limited to quasi-static problems. These studies are suitable for cases where rapid crack propagation or impact load are not involved. In this case, inertia effects do not play a role and can be ignored. However, it should be mentioned that most FGMs will be used in critical situations, where significant dynamic loading may be involved. The dynamic fracture behaviour of FGMs has received little attention from the scientific community. Examples of dynamic analysis include the study of the steady state dynamic crack propagation in an interphase with spatially varying elastic properties under antiplane loading conditions reported in [17], and the steady state dynamic fracture of FGMs under in-plane loading with the material properties being assumed to vary along the direction of crack propagation reported in [18]. Experimental studies of the dynamic fracture of a FGMs with discrete property variation using photoelasticity technique were also conducted in [19]. In spite of these efforts, the understanding of the dynamic fracture process of FGMs is still limited.

It is therefore the objective of the current study to provide a theoretical analysis of the dynamic behaviour of a propagating crack in an infinite medium with spatially varying elastic properties perpendicular to the direction of the crack propagation. Young's modulus and mass density of the model are assumed to vary exponentially while Poisson's ratio remains constant. The analytical study is based on the use of Fourier transform technique and the solution of the resulting singular integral equations using a series representation of the field variable. Attention will be focused on the effects of the crack speed and the gradient of the material property upon the dynamic behaviour of the crack. Efforts will also be paid to the coupling that exists between the open and sliding modes of the crack deformation as a result of the material gradient.

This paper is divided into five sections. Following this introduction, the formulation of the problem is given in Section 2. Section 3 provides the analytical solution of the problem, followed by Section 4 which includes a detailed discussion of typical results. Finally, in Section 5 we conclude the paper.

2. Formulation of the problem

The problem envisaged is that of a crack propagating at a constant speed V in an infinite inhomogeneous elastic medium, as shown in Fig. 1. Such a problem in homogeneous medium was first studied by Yoffe [20] in 1951 and further discussed in [21]. The crack is subjected to prescribed normal and shear stresses along its surfaces. The basic equations which govern the plane stress deformation behavior of the medium can be expressed in a fixed cartesian coordinate system (x, y) as,

$$\frac{\partial \sigma_{xx}}{\partial x} + \frac{\partial \sigma_{xy}}{\partial y} = \rho \ddot{u},$$

$$\frac{\partial \sigma_{xy}}{\partial x} + \frac{\partial \sigma_{yy}}{\partial y} = \rho \ddot{v} \tag{1}$$

and

$$\sigma_{xx} = \frac{E}{1 - v^2}\left(\frac{\partial u}{\partial x} + v\frac{\partial v}{\partial y}\right),$$

$$\sigma_{yy} = \frac{E}{1 - v^2}\left(v\frac{\partial u}{\partial x} + \frac{\partial v}{\partial y}\right), \tag{2}$$

$$\sigma_{xy} = \frac{E}{2(1 + v)}\left(\frac{\partial u}{\partial y} + \frac{\partial v}{\partial x}\right),$$

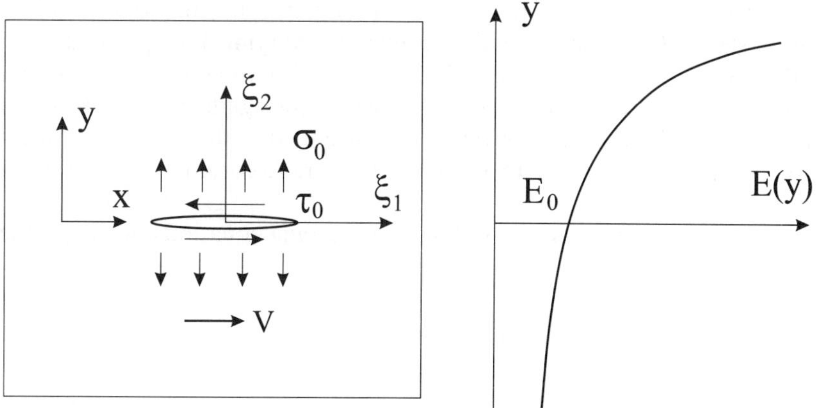

Fig. 1. A finite crack propagating in an infinite inhomogeneous medium with varying properties.

where u and v are the in-plane displacements, σ_{xx}, σ_{yy} and σ_{xy} are the non-zero stress components, E and v are Young's modulus and Poisson's ratio and ρ is the mass density of the material.

For the current problem, it is convenient to introduce the following Galilean transformation

$$\xi_1 = x - Vt, \qquad \xi_2 = y \tag{3}$$

with ξ_1 and ξ_2 being a translating coordinate system, which is attached to the propagating crack. It is, however, assumed that the propagation of the ensuing crack has prevailed for such a long time that the stress distribution around its tip is time invariant in the translating reference frame. Accordingly, the equilibrium equation (1) can be written in terms of the moving coordinate system (ξ_1, ξ_2) as

$$\frac{\partial \sigma_{11}}{\partial \xi_1} + \frac{\partial \sigma_{12}}{\partial \xi_2} = \rho V^2 \frac{\partial^2 u}{\partial \xi_1^2},$$

$$\frac{\partial \sigma_{12}}{\partial \xi_1} + \frac{\partial \sigma_{22}}{\partial \xi_2} = \rho V^2 \frac{\partial^2 v}{\partial \xi_1^2}, \tag{4}$$

which govern the steady state solution of the problem.

For a general inhomogeneous medium with material constants, which are continuous functions of ξ_2, the governing equation of motion can be obtained in the moving coordinate system as

$$\left[\frac{E(\xi_2)}{1 - v^2} - \rho(\xi_2)V^2\right]\frac{\partial^2 u}{\partial \xi_1^2} + \frac{E(\xi_2)}{2(1 + v)}\frac{\partial^2 u}{\partial \xi_2^2} + \left[\frac{vE(\xi_2)}{1 - v^2} + \frac{E(\xi_2)}{2(1 + v)}\right]\frac{\partial^2 u}{\partial \xi_1 \partial \xi_2} + \frac{1}{2(1 + v)}\frac{\partial E(\xi_2)}{\partial \xi_2}\left(\frac{\partial u}{\partial \xi_2} + \frac{\partial v}{\partial \xi_1}\right) = 0,$$

$$\left[\frac{E(\xi_2)}{2(1 + v)} - \rho(\xi_2)V^2\right]\frac{\partial^2 v}{\partial \xi_1^2} + \frac{E(\xi_2)}{1 - v^2}\frac{\partial^2 v}{\partial \xi_2^2} + \left[\frac{E(\xi_2)}{2(1 + v)} + \frac{vE(\xi_2)}{1 - v^2}\right]\frac{\partial^2 u}{\partial \xi_1 \partial \xi_2} + \frac{1}{1 - v^2}\frac{\partial E(\xi_2)}{\partial \xi_2}\left(v\frac{\partial u}{\partial \xi_1} + \frac{\partial v}{\partial \xi_2}\right) = 0. \tag{5}$$

Let us now assume that the variation of Young's modulus E and the mass density ρ in the medium are described by

$$E(\xi_2) = E_0 e^{\alpha \xi_2}, \qquad \rho(\xi_2) = \rho_0 e^{\alpha \xi_2}, \tag{6}$$

where E_0 and ρ_0 are the Young's modulus and the mass density of the medium at the position of the crack and α represents the gradient of the material properties. In this case, the longitudinal and shear wave speeds of the medium are constants. This model includes the effects of varying Young's modulus and mass density, but ignores the effect of the spatial variation of wave speeds. It should be mentioned that this effect is second order in nature if the spatial variation of wave speed is not significant in comparison with that of the Young's modulus and mass density. Earlier work [1] indicates that the solution of the crack problem in inhomogeneous materials is not sensitive to Poisson's ratio v. Accordingly, v is assumed to be constant throughout the entire FGMs medium.

Substituting Eq. (6) into (5) results in the following governing equations of the problem,

$$\beta_1 \frac{\partial^2 u}{\partial \xi_1^2} + \frac{\partial^2 u}{\partial \xi_2^2} + \beta_2 \frac{\partial^2 v}{\partial \xi_1 \partial \xi_2} + \alpha\left(\frac{\partial u}{\partial \xi_2} + \frac{\partial v}{\partial \xi_1}\right) = 0,$$

$$\beta_3 \frac{\partial^2 v}{\partial \xi_2^2} + \frac{\partial^2 v}{\partial \xi_1^2} + \beta_4 \frac{\partial^2 u}{\partial \xi_1 \partial \xi_2} + \beta_3 \alpha\left(v\frac{\partial u}{\partial \xi_1} + \frac{\partial v}{\partial \xi_2}\right) = 0, \tag{7}$$

where

$$\beta_1 = \frac{2}{1-v}\left(1 - \frac{V^2}{c_1^2}\right),$$

$$\beta_2 = \frac{1+v}{1-v},$$

$$\beta_3 = \frac{2}{(1-v)\left(1 - \dfrac{V^2}{c_2^2}\right)},$$

$$\beta_4 = \frac{(1+v)}{(1-v)\left(1 - \dfrac{V^2}{c_2^2}\right)},$$

$$(8)$$

c_1 and c_2 are longitudinal and shear wave speeds of the inhomogeneous medium given by

$$c_1 = \sqrt{\frac{E(\xi_2)}{(1-v^2)\rho(\xi_2)}} = \sqrt{\frac{E_0}{(1-v^2)\rho_0}}, \quad \text{and} \quad c_2 = \sqrt{\frac{E(\xi_2)}{2(1+v)\rho(\xi_2)}} = \sqrt{\frac{E_0}{2(1+v)\rho_0}},$$

respectively.

The general solution of Eq. (7) can be obtained using Fourier transform with respect to ξ_1, which results in

$$\frac{\partial^2 \bar{u}}{\partial \xi_2^2} - \beta_1 s^2 \bar{u} + is\beta_2 \frac{\partial \bar{v}}{\partial \xi_2} + \alpha\left(\frac{\partial \bar{u}}{\partial \xi_2} + is\bar{v}\right) = 0,$$

$$-s^2 \bar{v} + \beta_3 \frac{\partial^2 \bar{v}}{\partial \xi_2^2} + is\beta_4 \frac{\partial \bar{u}}{\partial \xi_2} + \beta_3 \alpha\left(isv\bar{u} + \frac{\partial \bar{v}}{\partial \xi_2}\right) = 0,$$

$$(9)$$

where

$$\bar{u}(s, \xi_2) = \int_{-\infty}^{\infty} u(\xi_1, \xi_2) e^{-is\xi_1}\, d\xi_1$$

and

$$\bar{v}(s, \xi_2) = \int_{-\infty}^{\infty} v(\xi_1, \xi_2) e^{-is\xi_1}\, d\xi_1 \tag{10}$$

are the Fourier transform of the displacement components. The general solution of (9) can be expressed in the following format:

$$\bar{u}(s, \xi_2) = \sum_{1}^{4} C_j(s) e^{\lambda_j \xi_2}, \qquad \bar{v}(s, \xi_2) = \sum_{1}^{4} b_j(s) C_j(s) e^{\lambda_j \xi_2}, \tag{11}$$

where $C_j(s)$ $(j = 1, \ldots, 4)$ are unknown functions to be determined and $\lambda_j(s)$ $(j = 1, \ldots, 4)$ are the roots of the following equation:

$$\lambda^4 + 2\alpha\lambda^3 + \frac{1}{\beta_3}\left[s^2(\beta_2\beta_4 - \beta_1\beta_3 - 1) + \alpha^2\beta_3\right]\lambda^2 + \frac{\alpha s^2}{\beta_3}(\beta_4 + \beta_2\beta_3 v - \beta_1\beta_3 - 1)\lambda + s^2\alpha^2 v + \frac{\beta_1}{\beta_3}s^4 = 0$$

$$(12)$$

and the coefficients $b_j(s)$ $(j = 1, \ldots, 4)$ are given by

$$b_j(s) = \frac{is(\beta_4\lambda_j + \beta_3\alpha v)}{s^2 - \beta_3\lambda_j^2 - \beta_3\alpha\lambda_j}. \tag{13}$$

Eq. (12) has four solutions of λ ($\lambda_1, \lambda_2, \lambda_3, \lambda_4$), with two of them ($\lambda_1$ and λ_3) having positive real parts and two (λ_2 and λ_4) having negative real parts. The solution of $\bar{u}$ and $\bar{v}$ satisfying the regularity conditions at infinity can be written as,

$$\bar{u}(s, \xi_2) = \begin{cases} C_2(s)e^{\lambda_2\xi_2} + C_4(s)e^{\lambda_4\xi_2}, & \xi_2 > 0 \\ C_1(s)e^{\lambda_1\xi_2} + C_3(s)e^{\lambda_3\xi_2}, & \xi_2 < 0 \end{cases}$$
$$\bar{v}(s, \xi_2) = \begin{cases} b_2(s)C_2(s)e^{\lambda_2\xi_2} + b_4(s)C_4(s)e^{\lambda_4\xi_2}, & \xi_2 > 0 \\ b_1(s)C_1(s)e^{\lambda_1\xi_2} + b_3(s)C_3(s)e^{\lambda_3\xi_2}, & \xi_2 < 0 \end{cases} \tag{14}$$

Since the normal and shear stress components must be continuous at $\xi_2 = 0$, then

$$\sigma_{22}(\xi_1, 0^+) = \sigma_{22}(\xi_1, 0^-), \quad \sigma_{12}(\xi_1, 0^+) = \sigma_{12}(\xi_1, 0^-), \quad -\infty < \xi_1 < \infty \tag{15}$$

Accordingly, Eq. (14) can be rewritten as,

$$\bar{u}(s, \xi_2) = \begin{cases} C_2e^{\lambda_2\xi_2} + C_4e^{\lambda_4\xi_2}, & \xi_2 > 0 \\ (e_1C_2 + e_2C_4)e^{\lambda_1\xi_2} + (e_3C_2 + e_4C_4)e^{\lambda_3\xi_2}, & \xi_2 < 0 \end{cases}$$
$$\bar{v}(s, \xi_2) = \begin{cases} b_2C_2e^{\lambda_2\xi_2} + b_4C_4e^{\lambda_4\xi_2}, & \xi_2 > 0 \\ b_1(e_1C_2 + e_2C_4)e^{\lambda_1\xi_2} + b_3(e_3C_2 + e_4C_4)e^{\lambda_3\xi_2}, & \xi_2 < 0 \end{cases} \tag{16}$$

with e_1, e_2, e_3 and e_4 being given in Appendix A.

3. Singular integral equations and solutions

The unknown coefficients $C_2(s)$ and $C_4(s)$ can be determined from the following boundary conditions,

$$\begin{aligned} \sigma_{22}(\xi_1, 0^+) &= -\sigma_0(\xi_1), \quad \sigma_{12}(\xi_1, 0^+) = -\tau_0(\xi_1), \quad -a < \xi_1 < a \\ u(\xi_1, 0^+) &= u(\xi_1, 0^-), \quad v(\xi_1, 0^+) = v(\xi_1, 0^-), \quad a < |\xi_1| < \infty \end{aligned} \tag{17}$$

Substituting the general solution of the displacements (16) into these mixed boundary conditions, a system of dual integral equations in terms of C_2 and C_4 can be obtained. This dual integral is reduced to a system of singular integral equations by introducing the following dislocation density functions

$$\phi_1(\xi_1) = \frac{\partial}{\partial\xi_1}[u(\xi_1, 0^+) - u(\xi_1, 0^-)]$$
$$\phi_2(\xi_1) = \frac{\partial}{\partial\xi_1}[v(\xi_1, 0^+) - v(\xi_1, 0^-)] \tag{18}$$

with '+' and '−' representing the upper and lower surfaces of the crack, respectively.

Applying Fourier transform with respect to ξ_1 on both sides of Eq. (18), $C_2(s)$ and $C_4(s)$ can be expressed in terms of $\phi_j(\xi_1)$ as,

$$C_2(s) = \frac{i}{s\Delta(s)}\left[g_1(s)\bar{\phi}_1(s) - g_2(s)\bar{\phi}_2(s)\right],$$
$$C_4(s) = -\frac{i}{s\Delta(s)}\left[g_3(s)\bar{\phi}_1(s) - g_4(s)\bar{\phi}_2(s)\right], \tag{19}$$

where

$$g_1(s) = b_4 - b_1 e_2 - b_3 e_4, \qquad g_2(s) = 1 - e_2 - e_4$$
$$g_3(s) = b_2 - b_1 e_1 - b_3 e_3, \qquad g_4(s) = 1 - e_1 - e_3 \tag{20}$$
$$\Delta(s) = (e_2 + e_4 - 1)(b_1 e_1 + b_3 e_3 - b_2) - (e_1 + e_3 - 1)(b_1 e_2 + b_3 e_4 - b_4)$$

and

$$\overline{\phi_j}(s) = \int_{-a}^{a} \phi_j(\xi_1) e^{-is\xi_1} \, d\xi_1, \quad (j = 1, 2) \tag{21}$$

is the Fourier transform of ϕ_j.

Using Eqs. (2), (16) and (19), and conducting the inverse Fourier transform, the stress field in the upper half plane $\xi_2 > 0$ can be obtained as,

$$\sigma_{12}(\xi_1, \xi_2) = \frac{E(\xi_2)}{4\pi(1+v)} \int_{-a}^{a} \sum_{j=1}^{2} M_{1j}(\xi_1, \xi_2, \eta) \phi_j(\eta) \, d\eta,$$

$$\sigma_{22}(\xi_1, \xi_2) = \frac{E(\xi_2)}{2\pi(1-v^2)} \int_{-a}^{a} \sum_{j=1}^{2} M_{2j}(\xi_1, \xi_2, \eta) \phi_j(\eta) \, d\eta, \tag{22}$$

where

$$M_{ij}(\xi_1, \xi_2, \eta) = \int_{-\infty}^{\infty} h_{ij}(s, \xi_2) e^{-is(\eta - \xi_1)} \, ds$$

with h_{ij} $(i, j = 1, 2)$ are given in Appendix A.

The stress boundary condition in (17) can be rewritten in terms of ϕ_j as

$$\int_{-a}^{a} \sum_{j=1}^{2} M_{1j}(\xi_1, 0, \eta) \phi_j(\eta) \, d\eta = -\frac{4\pi(1+v)\tau_0(\xi_1)}{E_0},$$

$$\int_{-a}^{a} \sum_{j=1}^{2} M_{2j}(\xi_1, 0, \eta) \phi_j(\eta) \, d\eta = -\frac{2\pi(1-v^2)\sigma_0(\xi_1)}{E_0}. \tag{23}$$

The second equation of (17) is reduced to

$$\phi_1(\xi_1) = 0, \quad \phi_2(\xi_1) = 0, \qquad a < |\xi_1| < \infty,$$

$$\int_{-a}^{a} \phi_1(\xi_1) \, d\xi_1 = 0, \qquad \int_{-a}^{a} \phi_2(\xi_1) \, d\xi_1 = 0. \tag{24}$$

Detailed asymptotic analysis of $h_{ij}(s, \xi_2)$ indicates that h_{11} and h_{22} are odd functions of s and approach constants when s tends to infinity, while h_{12} and h_{21} are even functions of s and tend to zero with increasing s, i.e.

$$\lim_{|s| \to \infty} h_{11}(s, \xi_2) = im_1 \mathrm{Sgn}(s)$$

$$\lim_{|s| \to \infty} h_{22}(s, \xi_2) = im_2 \mathrm{Sgn}(s)$$

$$\lim_{|s| \to \infty} h_{12}(s, \xi_2) = 0 \tag{25}$$

$$\lim_{|s| \to \infty} h_{21}(s, \xi_2) = 0$$

with m_1 and m_2 being two real constants and

$$\text{Sgn}(s) = \begin{cases} 1 & s > 0 \\ -1 & s < 0 \end{cases}$$

being the Heaviside step function.

The asymptotic behaviour of h_{ij} when s tends to infinity governs the singular solution of the problem. After separating the singular parts of the kernels on Eq. (23), the following singular integral equations for the boundary conditions can be obtained,

$$\frac{1}{\pi} \int_{-a}^{a} \left[\frac{m_1 \phi_1(\eta)}{\eta - \xi_1} + N_{11}\phi_1(\eta) + N_{12}\phi_2(\eta) \right] d\eta = -\frac{2(1+v)}{E_0} \tau_0(\xi_1),$$

$$\frac{1}{\pi} \int_{-a}^{a} \left[\frac{m_2 \phi_2(\eta)}{\eta - \xi_1} + N_{21}\phi_1(\eta) + N_{22}\phi_2(\eta) \right] d\eta = -\frac{1-v^2}{E_0} \sigma_0(\xi_1), \quad -a < \xi_1 < a, \tag{26}$$

where,

$$N_{11}(\xi_1, 0, \eta) = \frac{1}{2} \int_{-\infty}^{\infty} (h_{11}(s,0) - im_1 \text{Sgn}(s)) e^{-is(\eta - \xi_1)} ds$$

$$N_{22}(\xi_1, 0, \eta) = \frac{1}{2} \int_{-\infty}^{\infty} (h_{22}(s,0) - im_2 \text{Sgn}(s)) e^{-is(\eta - \xi_1)} ds$$

$$N_{12}(\xi_1, 0, \eta) = \frac{1}{2} \int_{-\infty}^{\infty} h_{12}(s,0) e^{-is(\eta - \xi_1)} ds \tag{27}$$

$$N_{21}(\xi_1, 0, \eta) = \frac{1}{2} \int_{-\infty}^{\infty} h_{21}(s,0) e^{-is(\eta - \xi_1)} ds$$

$\sigma_0(\xi_1)$ and $\tau_0(\xi_1)$ are the applied stress distributed along the crack surfaces.

Eq. (26) can be solved by expanding the dislocation density function $\phi_1(\eta)$ and $\phi_2(\eta)$ using Chebyshev polynomials, as follows:

$$\phi_j(\eta) = \frac{1}{\sqrt{1 - \frac{\eta^2}{a^2}}} \sum_{k=0}^{\infty} A_k^j T_k\left(\frac{\eta}{a}\right), \quad j = 1, 2, \tag{28}$$

where T_k are Chebyshev polynomials of the first kind and A_k^j are unknown coefficients to be determined. From the orthogonality conditions of the Chebyshev polynomials, the boundary condition given by (24) reduces to $A_0^j = 0$, $j = 1, 2$. Substituting (28) into the singular integral equation (26), the following algebraic equations for A_k^j, $j = 1, 2$ are obtained:

$$m_1 \sum_{k=1}^{\infty} A_k^1 U_{k-1}\left(\frac{\xi_1}{a}\right) + \sum_{j=1}^{2} \sum_{k=1}^{\infty} A_k^j g_{1k}^j(\xi_1) = -\frac{2(1+v)}{E_0} \tau_0(\xi_1),$$

$$m_2 \sum_{k=1}^{\infty} A_k^2 U_{k-1}\left(\frac{\xi_1}{a}\right) + \sum_{j=1}^{2} \sum_{k=1}^{\infty} A_k^j g_{2k}^j(\xi_1) = -\frac{1-v^2}{E_0} \sigma_0(\xi_1), \tag{29}$$

where U_k are the Chebyshev polynomials of the second kind and g_{1k}^j, g_{2k}^j $(j = 1, 2)$ are given in Appendix A.

If the Chebyshev polynomials in (28) are truncated to the Nth term and the boundary condition (29) are satisfied at N collocation points, given by

$$\xi_1^l = a \cos\left(\frac{l-1}{N-1}\pi\right), \quad l = 1, 2, \ldots, N \tag{30}$$

then Eq. (29) reduces to the following linear algebraic equations

$$m_1 \sum_{k=1}^{N} A_k^1 \frac{\sin \dfrac{k(l-1)\pi}{N-1}}{\sin \dfrac{(l-1)\pi}{N-1}} + \sum_{j=1}^{2} \sum_{k=1}^{N} A_k^j g_{1k}^j(\xi_1^l) = -\frac{2(1+v)}{E_0} \tau_0(\xi_1^l),$$

$$m_2 \sum_{k=1}^{N} A_k^2 \frac{\sin \dfrac{k(l-1)\pi}{N-1}}{\sin \dfrac{(l-1)\pi}{N-1}} + \sum_{j=1}^{2} \sum_{k=1}^{N} A_k^j g_{2k}^j(\xi_1^l) = -\frac{1-v^2}{E_0} \sigma_0(\xi_1^l), \quad k,l = 1,2,\ldots,N. \tag{31}$$

Once $A_k^j (j=1,2; \ k=1,2,\ldots,N)$ are determined, the fracture parameters, such as SIFs, strain energy release rate and crack opening displacement (COD), can be derived.

The SIFs at the right tip defined by

$$K_{\mathrm{I}}(a) = \lim_{\xi_1 \to a^+} \sqrt{2\pi(\xi_1 - a)}\sigma_{22}(\xi_1, 0^+)$$

$$K_{\mathrm{II}}(a) = \lim_{\xi_1 \to a^+} \sqrt{2\pi(\xi_1 - a)}\sigma_{12}(\xi_1, 0^+) \tag{32}$$

can be obtained as

$$K_{\mathrm{I}}(a) = -\frac{E_0 m_2 \sqrt{\pi a}}{1 - v^2} \sum_{k=1}^{N} A_k^2,$$

$$K_{\mathrm{II}}(a) = -\frac{E_0 m_1 \sqrt{\pi a}}{2(1 + v)} \sum_{k=1}^{N} A_k^1. \tag{33}$$

From (18), the relative displacements between the crack surfaces can be derived as

$$u(\xi_1, 0^+) - u(\xi_1, 0^-) = -a \sum_{k=1}^{\infty} \frac{A_k^1}{k} \sin\left(k \cos^{-1} \frac{\xi_1}{a}\right),$$

$$v(\xi_1, 0^+) - v(\xi_1, 0^-) = -a \sum_{k=1}^{\infty} \frac{A_k^2}{k} \sin\left(k \cos^{-1} \frac{\xi_1}{a}\right). \tag{34}$$

Using the well-known crack closure integral, the strain energy release rate at the right tip of the crack, $\xi_1 = a$, can be obtained in terms of SIFs [22], as

$$G = \frac{1}{E_0}\left(A_{\mathrm{I}} K_{\mathrm{I}}^2(a) + A_{\mathrm{II}} K_{\mathrm{II}}^2(a)\right), \tag{35}$$

where

$$A_{\mathrm{I}} = \frac{V^2 \alpha_1}{(1-v)c_2^2 D}$$

$$A_{\mathrm{II}} = \frac{V^2 \alpha_2}{(1-v)c_2^2 D}$$

with $D = 4\alpha_1 \alpha_2 - (1 + \alpha_2^2)^2$, $\alpha_1 = \sqrt{1 - \dfrac{V^2}{c_1^2}}$ and $\alpha_2 = \sqrt{1 - \dfrac{V^2}{c_2^2}}$.

These fracture parameters will be used to study the effect of the gradient of the properties and crack propagation speed upon the dynamic fracture behaviour of this class of FGMs.

       S.A. Meguid et al. / Engineering Fracture Mechanics 69 (2002) 1753–1768

4. Results and discussion

In this section, attention will be focussed on the effect of the speed of crack propagation and the gradient of the material properties upon the dynamic fracture behaviour of FGMs. The numerical results of the dynamic SIFs and the strain energy release rate are provided to show these effects under different loading conditions.

4.1. Normal loading

First, consider the case where the crack is subjected to a constant normal stress $\sigma_{22} = -\sigma_0$ along its surfaces.

Fig. 2 shows the effect of the gradient of material property on the normalized SIF k_{I} ($k_{\mathrm{I}} = K_{\mathrm{I}}/\sigma_0\sqrt{\pi a}$) for different crack propagation speed (c_2 is the shear wave speed of the medium). It is observed that k_{I} increases with the increase in the material gradient parameter αa. For static problem where $V/c_2 = 0.0$, α shows the weakest effect on k_{I}. With the increase of crack propagation speed, the effect of α increases. Since E and ρ are not symmetric about the crack plane, the opening mode of deformation (mode I) will be coupled with the sliding mode (mode II). Fig. 3 shows the effect of the gradient of the material properties on the normalized stress intensity factor k_{II} ($k_{\mathrm{II}} = K_{\mathrm{II}}/\sigma_0\sqrt{\pi a}$). Significant effect of the gradient of material property αa upon k_{II} can be observed for both the static and the dynamic cases. For example, when $\alpha a = 2$, k_{II} is over 0.5 for static cases. Similar result is observed for the strain energy release rate.

Fig. 4 shows the variation of the normalized SIF k_{I} with the speed of the crack V. For homogeneous materials ($\alpha = 0$), the crack velocity has no effect on k_{I}, corresponding to the well-known Yoffe's results. However, when the material properties are graded, a significant increase in k_{I} is observed with an increase in crack speed and α.

The normalized energy release rate g ($g = GE_0/\pi a\sigma_0^2$) versus the normalized crack speed V/c_2 is depicted in Fig. 5. Similar results as that for k_{I} are observed. However, the energy release rate increases with increasing crack speed even in homogeneous materials. It should be mentioned that both the SIF and the strain energy release rate show rapid increase when the crack speed approach specific values, indicating the possible mode change for high crack speeds.

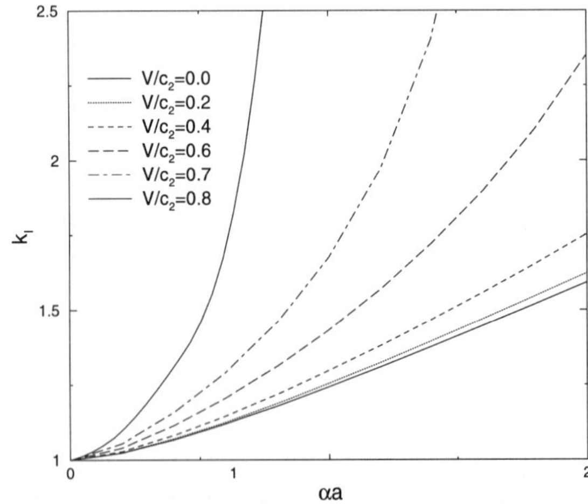

Fig. 2. Variation of the normalized SIF k_{I} with αa under normal loading.

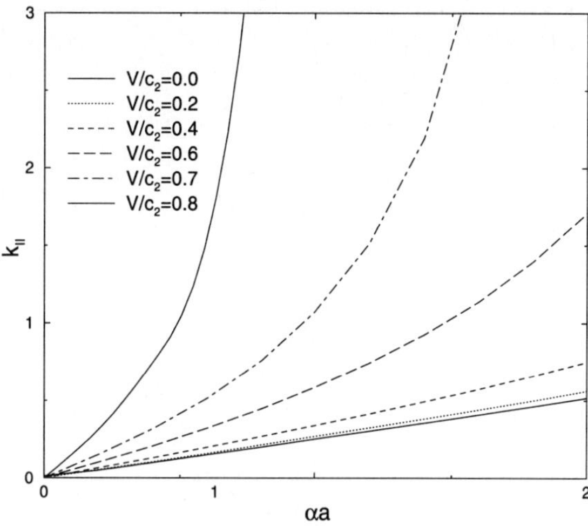

Fig. 3. Variation of the normalized SIF k_{II} with αa under normal loading.

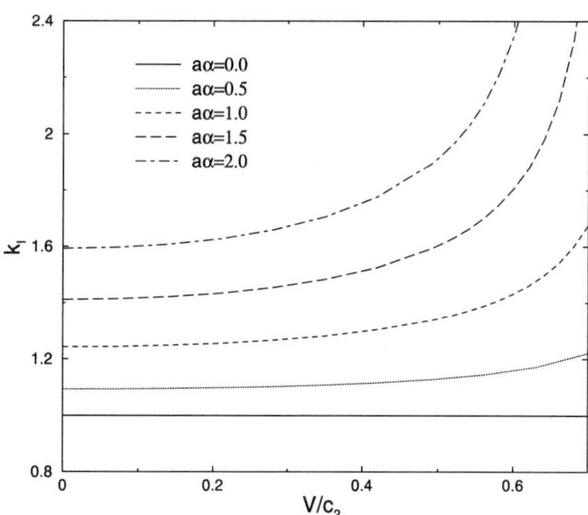

Fig. 4. Variation of the normalized SIF k_{I} with v/c_2 under normal loading.

4.2. Shear loading

Consider now the case where the crack is subjected to a constant shear stress $\sigma_{12} = -\tau_0$ along its surfaces. Fig. 6 shows the effect of the gradient in E and ρ on the normalized stress intensity factor $k_{\mathrm{II}} = K_{\mathrm{II}}/\tau_0\sqrt{\pi a}$. When the crack speed is low, the effects of αa upon the SIF is insignificant. For high crack velocity, k_{II} decreases with the increase of αa. The effect of the gradient of material property upon $k_{\mathrm{I}} = K_{\mathrm{I}}/\tau_0\sqrt{\pi a}$ is depicted in Fig. 7. It can be seen that k_{I} increases with the increase in αa. This effect of α in this case is much more significant than that on k_{II}. This study was further extended to examine the COD and the crack closure phenomenon as a result of the applied shear stress. The work reveals that under shear loading, crack closure exists. The crack closure in this case is determined by the relative displacement

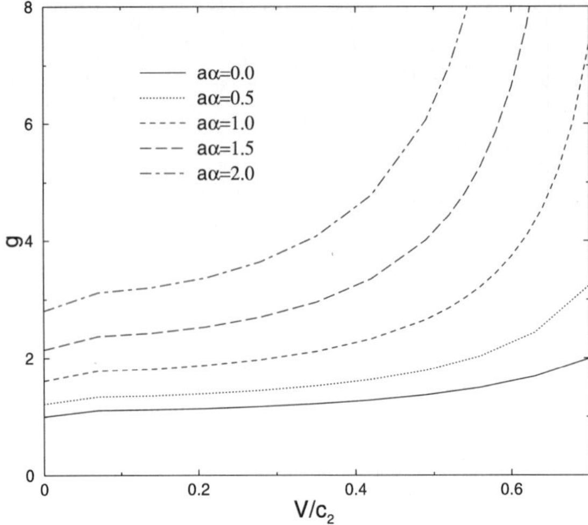

Fig. 5. Variation of the normalized strain energy release rate with v/c_2 under normal loading.

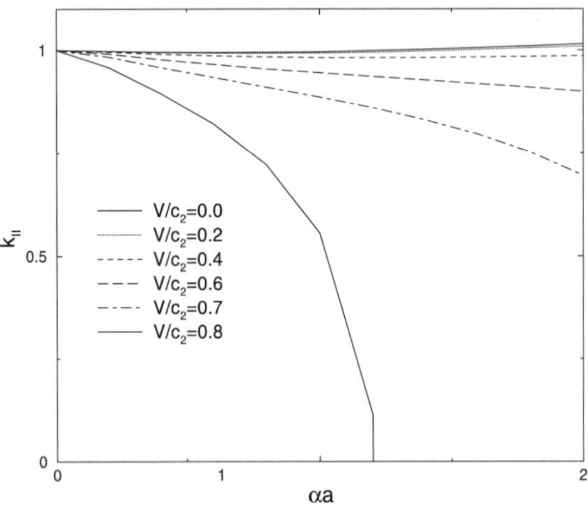

Fig. 6. Variation of the normalized SIF k_{II} with αa under shear loading.

between the crack surfaces under specific shear loads examined. Depending on the direction of the applied shear stress, crack surface contact may occur at different positions along the crack surface.

The deformation of the crack surfaces for $\tau_{12} = -\tau_0 > 0$ is presented in Fig. 8. Crack surface contact at the left part of the crack is observed, while the right part keeps open. It is worth noting that the crack closure phenomenon will influence the calculation of SIFs. Further work is needed in order to account for the effects of crack closure, which will lead to decreasing the SIF and, as the result, improve the integrity of the component.

Fig. 9 shows the effect of the gradient of the material upon the strain energy release rate. When the crack propagation speed is low, the effects of α can be ignored. Significant change of g with α, however, takes place when V approaches higher values. Fig. 10 provides the effect of V upon k_I for different gradient of

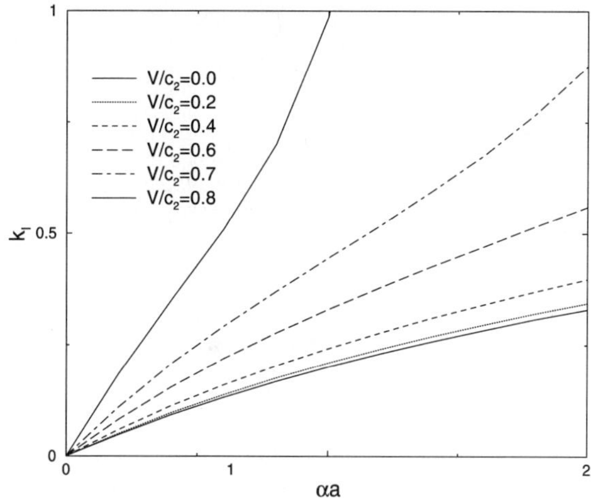

Fig. 7. Variation of the normalized SIF k_I with αa under shear loading.

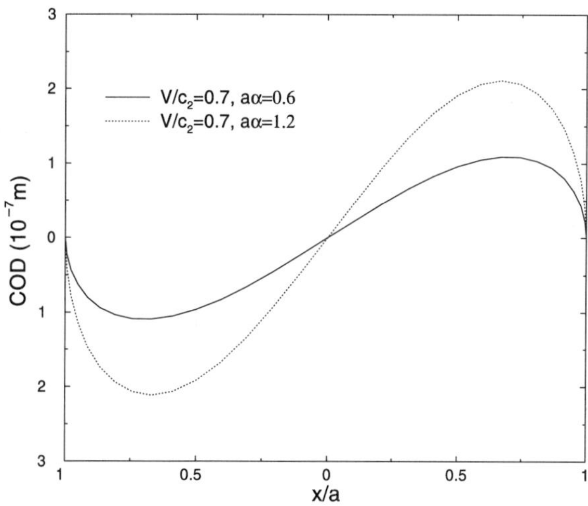

Fig. 8. Crack surface displacement under shear loading.

material property. It can be seen that k_I increases with the increase of V for the inhomogeneous materials. For the homogeneous materials, the crack propagation speed does not have effect on k_I, but with the increase of the gradient of material property, this effect increases.

5. Conclusions

An encompassing analytical model has been developed to study the singularity behaviour of a crack propagating in an infinite medium with spatially varying properties under in-plane loading. The current mixed boundary value problem is formulated in terms of Fourier transform techniques and singular integral equations and are solved using Chebyshev polynomials. The SIFs and strain energy release rate are used to show the dynamic fracture behaviour of FGM under different loading. The current model reveals

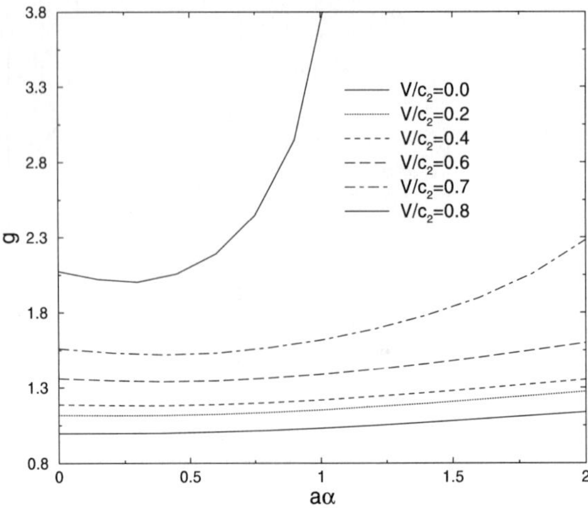

Fig. 9. Variation of the normalized strain energy release rate with αa under shear loading.

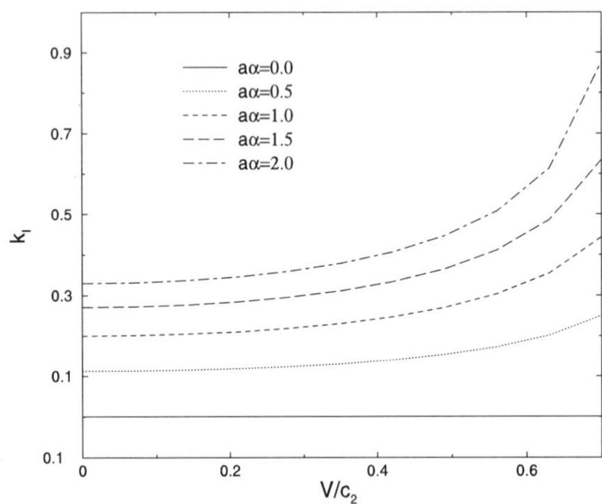

Fig. 10. Variation of the normalized SIF k_1 with v/c_2 under shear loading.

that crack closure exists under pure shear loading and that the gradient of the material properties and crack propagation speed influence the fracture behaviour considerably.

Acknowledgements

This work was supported by the Natural Sciences and Engineering Research Council of Canada.

Appendix A

The known functions e_1, e_2, e_3 and e_4 in Eq. (16) are,

$$e_1 = \frac{1}{\Delta_1}\left[\mathrm{i}s(\lambda_3 - \lambda_2)(v - b_2 b_3) + (b_2 - b_3)\left(vs^2 + \lambda_2\lambda_3\right)\right]$$

$$e_2 = \frac{1}{\Delta_1}\left[is(\lambda_3 - \lambda_4)(v - b_3 b_4) + (b_4 - b_3)(vs^2 + \lambda_3\lambda_4)\right]$$

$$e_3 = -\frac{1}{\Delta_1}\left[is(\lambda_1 - \lambda_2)(v - b_1 b_2) + (b_2 - b_1)(vs^2 + \lambda_1\lambda_2)\right]$$

$$e_4 = -\frac{1}{\Delta_1}\left[is(\lambda_1 - \lambda_4)(v - b_1 b_4) + (b_4 - b_1)(vs^2 + \lambda_1\lambda_4)\right]$$

$$\Delta_1 = (b_1 - b_3)(\lambda_1\lambda_3 + vs^2) + is(\lambda_1 - \lambda_3)(b_1 b_3 - v)$$

and $h_{ij}(i,j = 1,2)$ in (22) are given by

$$h_{11}(s,\xi_2) = \frac{i}{s\Delta}\left(\lambda_2 g_1 e^{\lambda_2\xi_2} - \lambda_4 g_3 e^{\lambda_4\xi_2}\right) + \frac{1}{\Delta}\left(b_4 g_3 e^{\lambda_4\xi_2} - b_2 g_1 e^{\lambda_2\xi_2}\right)$$

$$h_{12}(s,\xi_2) = \frac{i}{s\Delta}\left(\lambda_4 g_4 e^{\lambda_4\xi_2} - \lambda_2 g_2 e^{\lambda_2\xi_2}\right) + \frac{1}{\Delta}\left(b_2 g_2 e^{\lambda_2\xi_2} - b_4 g_4 e^{\lambda_4\xi_2}\right)$$

$$h_{21}(s,\xi_2) = \frac{v}{\Delta}\left(g_3 e^{\lambda_4\xi_2} - g_1 e^{\lambda_2\xi_2}\right) + \frac{i}{s\Delta}\left(b_2\lambda_2 g_1 e^{\lambda_2\xi_2} - b_4\lambda_4 g_3 e^{\lambda_4\xi_2}\right)$$

$$h_{22}(s,\xi_2) = \frac{v}{\Delta}\left(g_2 e^{\lambda_2\xi_2} - g_4 e^{\lambda_4\xi_2}\right) + \frac{i}{s\Delta}\left(b_4\lambda_4 g_4 e^{\lambda_4\xi_2} - b_2\lambda_2 g_2 e^{\lambda_2\xi_2}\right)$$

$g_{1k}^j,\ g_{2k}^j(j = 1,2)$ used in Eq. (29) are given by

$$g_{1k}^1 = ia\begin{cases}(-1)^n\int_0^\infty J_k(sa)\sin(s\xi_1)(h_{11}(0,s) - im_1)\,ds, & k = 2n \\ (-1)^{n+1}\int_0^\infty J_k(sa)\cos(s\xi_1)(h_{11}(0,s) - im_1)\,ds, & k = 2n+1\end{cases}$$

$$g_{1k}^2 = a\begin{cases}(-1)^n\int_0^\infty J_k(sa)\cos(s\xi_1)h_{12}(0,s)\,ds, & k = 2n \\ (-1)^n\int_0^\infty J_k(sa)\sin(s\xi_1)h_{12}(0,s)\,ds, & k = 2n+1\end{cases}$$

$$g_{2k}^1 = a\begin{cases}(-1)^n\int_0^\infty J_k(sa)\cos(s\xi_1)h_{21}(0,s)\,ds, & k = 2n \\ (-1)^n\int_0^\infty J_k(sa)\sin(s\xi_1)h_{21}(0,s)\,ds, & k = 2n+1\end{cases}$$

$$g_{2k}^2 = ia\begin{cases}(-1)^n\int_0^\infty J_k(sa)\sin(s\xi_1)(h_{22}(0,s) - im_2)\,ds, & k = 2n \\ (-1)^{n+1}\int_0^\infty J_k(sa)\cos(s\xi_1)(h_{22}(0,s) - im_2)\,ds, & k = 2n+1\end{cases}$$

References

[1] Delale F, Erdogan F. The crack problem for a nonhomogeneous plane. J Appl Mech 1983;50:609–14.
[2] Delale F, Erdogan F. On the mechanical modeling of the interfacial region in bonded half-planes. J Appl Mech 1988;55:317–24.
[3] Konda N, Erdogan F. The mixed mode crack problem in a nonhomogeneous elastic plane. Eng Fract Mech 1994;47:533–45.
[4] Eischen JW. Fracture of nonhomogeneous materials. Int J Fract 1987;34:3–22.
[5] Jin ZH, Noda N. Crack tip singular fields in nonhomogeneous materials. J Appl Mech 1994;61:738–40.
[6] Gu P, Asaro RJ. Cracks in functionally graded materials. Int J Solids Struct 1997;34(1):1–17.
[7] Yang W, Shih CF. Fracture along an interlayer. Int J Solids Struct 1994;31:985–1002.
[8] Erdogan F. Fracture mechanics of functionally graded materials. Compos Eng 1995;5(7):753–70.

[9] Jin ZH, Batra RC. Some basic fracture mechanics concepts in functionally graded materials. J Mech Phys Solids 1996;44(8): 1221–35.

[10] Ozturk M, Erdogan F. Mode I crack problem in an inhomogeneous orthotropic medium. Int J Eng Sci 1997;(35):869–83.

[11] Ozturk M, Erdogan F. The mixed mode crack problem in an inhomogeneous orthotropic medium. Int J Fract 1998;98:243–61.

[12] Marur PR, Tippur HV. Evaluation of mechanical properties of functionally graded materials. J Test Evaluat 1998;26:539–45.

[13] Butcher RJ, Rousseau CE, Tippur HV. A functionally graded particulate composite: preparation, measurements and failure analysis. Acta Mater 1999;47(1):259–68.

[14] Rousseau CE, Tippur HV. Compositionally graded materials with cracks normal to the elastic gradient. Acta Mater 2000; 48:4021–33.

[15] Gu P, Dao M, Asaro RJ. A simplified method for calculating the crack tip field of functionally graded materials using the domain integral. J Appl Mech 1999;66:101–8.

[16] Anlas G, Santare MH, Lambros J. Numerical calculation of stress intensity factors in functionally graded materials. Int J Fract 2000;104:131–43.

[17] Wang XD, Meguid SA. On the dynamic crack propagation in an interface with spatially varying elastic properties. Int J Fract 1994/1995;69:87–99.

[18] Parameswaran V, Shukla A. Crack-tip stress fields for dynamic fracture in functionally gradient materials. Mech Mater 1999;31:579–96.

[19] Parameswaran V, Shukla A. Dynamic fracture of a functionally gradient material having discrete property variation. J Mater Sci 1998;33:3303–11.

[20] Yoffe EH. The moving Griffith crack. Phil Mag 1951;7(42):739.

[21] Sih GC. Mechanics of Fracture 4: Elastodynamic Crack Problems. Leyden: Noordhoff International Publishing; 1977.

[22] Freund LB. Dynamic Fracture Mechanics. Cambridge, UK: Cambridge University Press; 1990.

PERGAMON

Engineering Fracture Mechanics 69 (2002) 1769–1790

**Engineering
Fracture
Mechanics**

www.elsevier.com/locate/engfracmech

A viscoelastic functionally graded strip containing a crack subjected to in-plane loading

Z.-H. Jin, Glaucio H. Paulino *

*Department of Civil and Environmental Engineering, Newmark Laboratory, University of Illinois at Urbana-Champaign,
205 North Mathews Avenue, Urbana, IL 61801, USA*

Received 20 October 2000; received in revised form 13 April 2001; accepted 17 April 2001

Abstract

In this paper, a crack in a viscoelastic strip of a functionally graded material (FGM) is studied under tensile loading conditions. The extensional relaxation modulus is assumed as $E = E_0 \exp(\beta y/h) f(t)$, where h is a scale length and $f(t)$ is a nondimensional function of time t either having the form $f(t) = E_\infty/E_0 + (1 - E_\infty/E_0) \exp(-t/t_0)$ for a linear standard solid or $f(t) = (t_0/t)^q$ for a power law material model, where E_0, E_∞, β, t_0 and q are material constants. An extensional relaxation function in the form $E = E_0 \exp(\beta y/h)[t_0 \exp(\delta y/h)/t]^q$ is also considered, in which the relaxation time depends on the Cartesian coordinate y exponentially with δ being a material constant describing the gradation of the relaxation time. The Poisson's ratio is assumed to have the form $v = v_0(1 + \gamma y/h) \exp(\beta y/h) g(t)$, where v_0 and γ are material constants, and $g(t)$ is a nondimensional function of time t. An elastic FGM crack problem is first solved and the "correspondence principle" is used to obtain both mode I and mode II stress intensity factors, and the crack opening/sliding displacements for the viscoelastic FGM considering various material models.
© 2002 Elsevier Science Ltd. All rights reserved.

Keywords: Crack; Stress intensity factor; Viscoelasticity; Standard linear solid; Power law material; Functionally graded material

1. Introduction

Materials exhibit creep and relaxation behavior at elevated temperatures. Upon such conditions, deformation creep occurs under constant stress state, while stress relaxation takes place under constant strain state. In the framework of linear continuum theory, such behavior can be studied by viscoelasticity. Generally speaking, the viscoelastic response may be obtained from the elastic solution via the *correspondence principle* [1], or the analogy between the Laplace transform of the viscoelastic solution and the corresponding elastic solution. Fracture behavior of homogeneous materials has been investigated by Broberg [2] who has given some examples of stress intensity factors (SIFs) for stationary cracks in viscoelastic solids. Atkinson and Chen [3,4] have investigated cracks in layered viscoelastic materials. Crack growth problems have been studied, for example, by Knauss [5] and Shapery [6–8].

---

* Corresponding author. Fax: +1-217-265-8041.
E-mail address: paulino@uiuc.edu (G.H. Paulino).

0013-7944/02/$ - see front matter © 2002 Elsevier Science Ltd. All rights reserved.
PII: S0013-7944(02)00049-8

Functionally graded materials (FGMs) are an alternative to homogeneous materials or layered composites, and are promising candidates for future advanced technological applications [9,10]. The material properties of FGMs are continuously graded with gradual change in *microstructural details* over predetermined geometrical orientations and distances, such as composition, morphology, and crystal structure [11,12]. In applications involving severe thermal gradients (e.g. thermal protection systems), FGM systems take advantage of heat and corrosion resistance typical of ceramics, and mechanical strength and toughness typical of metals. Several aspects of fracture mechanics of FGMs have been studied, for example, basic theory and review [13], crack tip fields [14,15], crack growth resistance curve [16], crack deflection [17], conservation laws [18], strain gradient theory [19], fracture testing [20], and statistical models [21].

Under elevated temperature conditions, FGMs also exhibit creep and relaxation behavior. For polymer-based FGMs (see, for example, [22–24]), their creep and relaxation behavior may be studied by viscoelasticity. However, in general, the correspondence principle, does not hold for FGMs. To avoid this problem, Paulino and Jin [25] have shown that the correspondence principle can still be used to obtain the viscoelastic solution *for a class of FGMs exhibiting relaxation (or creep) functions with separable kernels in space and time*. By using such *revisited correspondence principle* for FGMs, they have subsequently studied crack problems of FGM strips subjected to antiplane shear conditions [26,27]. Other studies on crack problems of nonhomogeneous viscoelastic materials directly solve the viscoelastic governing equations. For example, Schovanec and co-workers have considered stationary cracks [28], quasi-static crack propagation [29] and dynamic crack propagation [30] in nonhomogeneous viscoelastic media under antiplane shear conditions. Schovanec and Walton also considered quasi-static propagation of a plane strain mode I crack in a power-law inhomogeneous linearly viscoelastic body [31] and calculated the corresponding energy release rate [32]. Although a separable form for the relaxation functions was employed in Refs. [28–32], no use of the correspondence principle was made. Recently, Yang [33] performed stress analysis in FGM cylinders where steady-state creep conditions are considered only for the homogeneous material.

In the present study, a stationary crack in a viscoelastic FGM strip is investigated under tensile loading conditions. The extensional relaxation function of the material is assumed as

$$E = E_0 \exp(\beta y/h) f(t),$$

where h is a scale length and $f(t)$ is a nondimensional function of time t either having the form

$$f(t) = E_\infty/E_0 + (1 - E_\infty/E_0)\exp(-t/t_0): \quad \text{linear standard solid}$$

or

$$f(t) = (t_0/t)^q: \quad \text{power law material.}$$

We also consider the following variant form of the power law material model

$$E = E_0 \exp(\beta y/h)[t_0 \exp(\delta y/h)/t]^q,$$

in which *the relaxation time depends on y exponentially.* In the above expressions, the parameters E_0, E_∞, β, t_0; δ, q are material constants. The Poisson's ratio is assumed to take the form (also separable in space and time)

$$v = v_0(1 + \gamma y/h)\exp(\beta y/h)g(t).$$

where v_0 and γ are material constants, and $g(t)$ is a nondimensional function of time t. The material models considered above may be suitable, for example, for two phase polymeric/polymeric FGMs with their basic constitutents having different Young's moduli and Poisson's ratios but having approximately the same viscoelastic relaxation behavior. Since an FGM is a special composite of its constituents, the viscoelastic relaxation behavior may remain unchanged if its constitutents have the same relaxation behavior. Thus, the relaxation moduli of the FGM would have separable forms in space and time. Further, the independent material constants β and γ describe the spatial gradation in Young's modulus and Poisson's ratio.

According to the correspondence principle, we first consider a crack in an elastic strip of an FGM with the following properties:

$$E = E_0 \exp(\beta y/h), \quad v = v_0(1 + \gamma y/h)\exp(\beta y/h).$$

The Laplace transform of the viscoelastic solution is directly obtained from the elastic solution. For the traction boundary value problem, the stress intensity factors are the same as those for the nonhomogeneous elastic strip. The crack opening/sliding displacements, however, are functions of time.

The remainder of this paper is organized as follows. The basic equations of viscoelasticity are provided in the next section and some viscoelastic constitutive models for FGMs are discussed in Section 3. The boundary value problem of a crack in a viscoelastic FGM strip is presented in Section 4. SIFs and crack opening/sliding displacements are discussed in Sections 5 and 6, respectively. Relevant numerical aspects for solving the governing system of integral equations are given in Section 7. Numerical results of SIFs and crack opening/sliding displacements are given in Section 8. Finally, some concluding remarks and extensions are given in Section 9. Appendix A, containing the explicit form of the Fredholm kernels in the integral equations, supplements the paper.

2. Basic equations

The basic equations of quasi-static viscoelasticity of FGMs are the equilibrium equation (in the absence of body forces)

$$\sigma_{ij,j} = 0, \tag{1}$$

the strain–displacement relationship

$$\varepsilon_{ij} = \tfrac{1}{2}(u_{i,j} + u_{j,i}), \tag{2}$$

and the viscoelastic constitutive law

$$e_{ij} = \int_0^t J_1(\boldsymbol{x}, t - \tau)\frac{\mathrm{d}s_{ij}}{\mathrm{d}\tau}\,\mathrm{d}\tau, \quad \varepsilon_{kk} = \int_0^t J_2(\boldsymbol{x}, t - \tau)\frac{\mathrm{d}\sigma_{kk}}{\mathrm{d}\tau}\,\mathrm{d}\tau \tag{3}$$

with

$$s_{ij} = \sigma_{ij} - \frac{1}{3}\sigma_{kk}\delta_{ij}, \quad e_{ij} = \varepsilon_{ij} - \frac{1}{3}\varepsilon_{kk}\delta_{ij}, \tag{4}$$

where σ_{ij} are stresses, ε_{ij} are strains, s_{ij} and e_{ij} are deviatoric components of the stress and strain tensors, respectively, u_i are displacements, δ_{ij} is the Kronecker delta, $\boldsymbol{x} = (x_1, x_2, x_3)$, $J_1(\boldsymbol{x}, t)$ and $J_2(\boldsymbol{x}, t)$ are the creep functions, t denotes time, and the Latin indices have the range 1–3 with repeated indices implying the summation convention. Note that *for FGMs the creep functions also depend on spatial positions*, whereas in homogeneous viscoelasticity, they are only functions of time, i.e. $J_1 \equiv J_1(t)$ and $J_2 \equiv J_2(t)$ [1].

The creep functions $J_1(\boldsymbol{x}, t)$ and $J_2(\boldsymbol{x}, t)$ are related to the relaxation function in extension, $E(\boldsymbol{x}, t)$, and the relaxation function in Poisson's ratio, $v(\boldsymbol{x}, t)$, by the following equations [1]:

$$\bar{J}_1 = \frac{1 + p\bar{v}}{p^2\bar{E}}, \quad \bar{J}_2 = \frac{1 - 2p\bar{v}}{p^2\bar{E}}, \tag{5}$$

where a bar over a variable means the Laplace transform, and p is the Laplace transform variable.

Under plane stress conditions, the Laplace transforms of the basic equations (1)–(3) are reduced to

$$\frac{\partial \bar{\sigma}_{xx}}{\partial x} + \frac{\partial \bar{\sigma}_{xy}}{\partial y} = 0, \quad \frac{\partial \bar{\sigma}_{xy}}{\partial x} + \frac{\partial \bar{\sigma}_{yy}}{\partial y} = 0, \tag{6}$$

$$\bar{\epsilon}_{xx} = \frac{\partial \bar{u}}{\partial x}, \quad \bar{\epsilon}_{yy} = \frac{\partial \bar{v}}{\partial y}, \quad \bar{\epsilon}_{xy} = \frac{1}{2}\left(\frac{\partial \bar{u}}{\partial y} + \frac{\partial \bar{v}}{\partial x}\right), \tag{7}$$

$$\bar{\epsilon}_{xx} = \frac{1}{p\bar{E}}\left(\bar{\sigma}_{xx} - p\bar{v}\bar{\sigma}_{yy}\right), \quad \bar{\epsilon}_{yy} = \frac{1}{p\bar{E}}\left(\bar{\sigma}_{yy} - p\bar{v}\bar{\sigma}_{xx}\right), \quad \bar{\epsilon}_{xy} = \frac{2(1 + p\bar{v})}{p\bar{E}}\bar{\sigma}_{xy}, \tag{8}$$

where $x \equiv x_1$, $y \equiv x_2$, and u and v are displacements in the x and y directions, respectively.

3. Viscoelastic models for FGMs

This section describes three viscoelastic material models considered in this investigation. The first is a *standard linear solid* with constant relaxation time

$$E = E_0 \exp(\beta y/h)\left[\frac{E_\infty}{E_0} + \left(1 - \frac{E_\infty}{E_0}\right)\exp\left(-\frac{t}{t_0}\right)\right], \tag{9}$$

where β, E_0, E_∞ and t_0 are material constants and h is a scale length (e.g., the strip thickness). The second model is a *power law material* with *constant relaxation time*

$$E = E_0 \exp(\beta y/h)\left(\frac{t_0}{t}\right)^q, \quad 0 < q < 1, \tag{10}$$

where q is a material constant. The third model is still a power law material, but with *position-dependent relaxation time*

$$E = E_0 \exp(\beta y/h)\left[\frac{t_0 \exp(\delta y/h)}{t}\right]^q = E_0 \exp[(\beta + \delta q)y/h]\left(\frac{t_0}{t}\right)^q, \tag{11}$$

where δ is a material constant. For all three models, the Poisson's ratio is assumed as

$$v = v_0(1 + \gamma y/h)\exp(\beta y/h)g(t), \tag{12}$$

where v_0 and γ are material constants, and $g(t)$ is a nondimensional function of time t. The spatial position dependent part of the Possion ratio (12) was first proposed by Delale and Erdogan [34] to study non-homogeneous elastic crack problems. Noda and Jin [35] later used it to investigate thermal crack problems in nonhomogeneous solids. The Poisson's ratio given by Eq. (12) is subjected to the condition that $-1 \leq v \leq 0.5$ [36].

With the assumptions (9)–(12) on the relaxation modulus and the Poisson's ratio, the correspondence principle for viscoelastic FGMs [25] can be used to study crack problems, i.e. *the Laplace transformed viscoelastic FGM solution can be obtained directly from the solution of the corresponding elastic FGM solution by replacing E_0 and v_0 with $pE_0\bar{f}(p)$ and $pv_0\bar{g}(p)$, respectively. The final solution is realized upon inverting the transformed solution, where $\bar{f}(p)$ and $\bar{g}(p)$ are the Laplace transforms of $f(t)$ and $g(t)$, respectively.* For the standard linear solid (9), $f(t)$ is given by

$$f(t) = \left[\frac{E_\infty}{E_0} + \left(1 - \frac{E_\infty}{E_0}\right)\exp\left(-\frac{t}{t_0}\right)\right]. \tag{13}$$

For the *power law material* (10) or (11), $f(t)$ is

$$f(t) = \left(\frac{t_0}{t}\right)^q. \tag{14}$$

Since the function $g(t)$ does not affect the final governing equations and boundary conditions considered (see Section 4), no specific functional form is required.

It can be clearly seen from (9) to (12) that the relaxation moduli and the Poisson's ratio are separable functions in space and time. This is a necessary condition for applying the *revisited correspondence principle* [25]. This kind of relaxation functions may be appropriate for an FGM with its constituent materials having different Young's moduli and Poisson's ratios but having approximately the same viscoelastic relaxation behavior. Since the FGM is a special composite of its constitutents, the viscoelastic relaxation behavior may remain unchanged if its constitutents have the same relaxation behavior. Thus, the relaxation moduli of the FGM would have separable forms in space and time. Further, the material constants β and γ describe the spatial gradation in Young's modulus and Poisson's ratio. For model (9), the separable form of the extensional relaxation modulus means that the constituents should have the same ratio E_∞/E_0 and relaxation time t_0. For model (10), this implies that the constituents should have the same relaxation time t_0 and parameter q. For model (11), however, it is only required that the constituents have the same parameter q. The constituents may have different relaxation times. Potentially, this kind of FGMs may include some polymeric/polymeric materials such as propylene-homopolymer/Acetal-copolymer. The relaxation behavior of propylene homopolymer and Acetal copolymer are found to be similar—see Figs. 7.5 and 10.3, respectively, of Ogorkiewicz [37].

4. A crack in a viscoelastic FGM strip

Consider an infinite nonhomogeneous viscoelastic strip containing a crack of length $2a$, as shown in Fig. 1. The strip is subjected to a uniform tension $\sigma_0\Sigma(t)$ in the y-direction along both the lower boundary ($y = -h_1$) and the upper boundary ($y = h_2$), where σ_0 is a constant and $\Sigma(t)$ is a nondimensional function of time t. It is assumed that the crack lies on the x-axis, from $-a$ to a. The crack surfaces remain traction free. The boundary conditions of the crack problem, therefore, are

$$\sigma_{xy} = 0, \quad \sigma_{yy} = \sigma_0\Sigma(t), \quad y = -h_1, \quad |x| < \infty, \tag{15}$$

$$\sigma_{xy} = 0, \quad \sigma_{yy} = \sigma_0\Sigma(t), \quad y = h_2, \quad |x| < \infty, \tag{16}$$

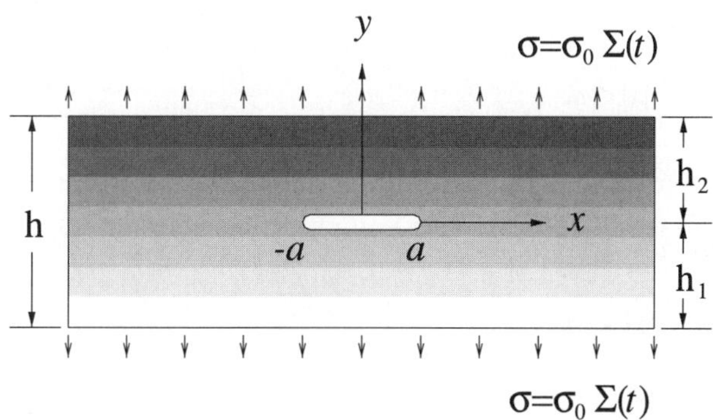

Fig. 1. A viscoelastic FGM strip occupying the region $|x| < \infty$ and $-h_1 \leqslant y \leqslant h_2$ with a crack at $|x| \leqslant a$ and $y = 0$. The boundaries of the strip ($y = -h_1, h_2$) are subjected to uniform traction $\sigma_0\Sigma(t)$.

$$\sigma_{xy} = \sigma_{yy} = 0, \quad y = 0, \quad |x| \le a, \tag{17}$$

$$\sigma_{xy}(x, 0^+) = \sigma_{xy}(x, 0^-), \quad a < |x| < \infty, \tag{18}$$

$$\sigma_{yy}(x, 0^+) = \sigma_{yy}(x, 0^-), \quad a < |x| < \infty, \tag{19}$$

$$u(x, 0^+) = u(x, 0^-), \quad a < |x| < \infty, \tag{20}$$

$$v(x, 0^+) = v(x, 0^-), \quad a < |x| < \infty. \tag{21}$$

According to the correspondence principle described in Ref. [25] and the previous section, one can first consider a nonhomogeneous elastic material with Young's modulus and Poisson's ratio

$$E = E_0 \exp(\beta y/h), \quad v = v_0(1 + \gamma y/h)\exp(\beta y/h), \tag{22}$$

respectively, and the viscoelastic solutions for models (9) and (10) may be obtained by the correspondence principle. Note that for material model (11) the viscoelastic solution can still be obtained by the correspondence principle provided that the corresponding elastic material has the following Young's modulus and Poisson's ratio:

$$E = E_0 \exp[(\beta + q\delta)y/h], \quad v = v_0(1 + \gamma y/h)\exp[(\beta + q\delta)y/h]. \tag{23}$$

For the elastic crack problem, the solution consists of a regular solution (for an uncracked strip) and a purturbed solution by the crack. The stresses for the regular solution are

$$\sigma_{xx} = v(y)\sigma_0 \Sigma(t) + E(y)(C_1 y + C_2), \tag{24}$$

$$\sigma_{yy} = \sigma_0 \Sigma(t), \tag{25}$$

$$\sigma_{xy} = 0, \tag{26}$$

where C_1 and C_2 are constants which can be determined by the displacement constraint conditions. These constants do not influence the final stress intensity factor and crack opening/sliding displacement results.

For the perturbed problem by the crack, the following boundary conditions need to be satisfied

$$\sigma_{xy} = 0, \quad \sigma_{yy} = 0, \quad y = -h_1, \quad |x| < \infty, \tag{27}$$

$$\sigma_{xy} = 0, \quad \sigma_{yy} = 0, \quad y = h_2, \quad |x| < \infty, \tag{28}$$

$$\sigma_{xy} = 0, \quad y = 0, \quad |x| \le a, \tag{29}$$

$$\sigma_{yy} = -\sigma_0 \Sigma(t), \quad y = 0, \quad |x| \le a, \tag{30}$$

$$\sigma_{xy}(x, 0^+) = \sigma_{xy}(x, 0^-), \quad a < |x| < \infty, \tag{31}$$

$$\sigma_{yy}(x, 0^+) = \sigma_{yy}(x, 0^-), \quad a < |x| < \infty, \tag{32}$$

$$u(x, 0^+) = u(x, 0^-), \quad a < |x| < \infty, \tag{33}$$

$$v(x, 0^+) = v(x, 0^-), \quad a < |x| < \infty. \tag{34}$$

The governing differential equation in terms of the Airy stress function $\Phi(x, y)$ for the nonhomogeneous elastic material (22) is [34]

$$\nabla^2\nabla^2\Phi - 2\left(\frac{\beta}{h}\right)\frac{\partial(\nabla^2\Phi)}{\partial y} + \left(\frac{\beta}{h}\right)^2\frac{\partial^2\Phi}{\partial y^2} = 0. \tag{35}$$

The stresses can be expressed in terms of Φ as

$$\sigma_{xx} = \frac{\partial^2\Phi}{\partial y^2}, \quad \sigma_{yy} = \frac{\partial^2\Phi}{\partial x^2}, \quad \sigma_{xy} = -\frac{\partial^2\Phi}{\partial x\partial y}. \tag{36}$$

The displacements are related to Φ by the constitutive relationship

$$\frac{\partial u}{\partial x} = \frac{1}{E_0}\left[\exp(-\beta y/h)\sigma_{xx} - v_0(1 + \gamma y/h)\sigma_{yy}\right], \tag{37}$$

$$\frac{\partial v}{\partial y} = \frac{1}{E_0}\left[\exp(-\beta y/h)\sigma_{yy} - v_0(1 + \gamma y/h)\sigma_{xx}\right], \tag{38}$$

$$\frac{\partial u}{\partial y} + \frac{\partial v}{\partial x} = \frac{2[\exp(-\beta y/h) + v_0(1 + \gamma y/h)]}{E_0}\sigma_{xy}. \tag{39}$$

By using the Fourier transform method (see, for example, Ref. [38]), the boundary value problem described by Eqs. (27)–(39) can be reduced to the following system of singular integral equations:

$$\int_{-1}^{1}\left[\frac{\varphi_1(s)}{s - r} + k_{11}(r,s,\beta)\varphi_1(s) + k_{12}(r,s,\beta)\varphi_2(s)\right]ds = 0, \quad |r| \leqslant 1,$$

$$\int_{-1}^{1}\left[\frac{\varphi_2(s)}{s - r} + k_{21}(r,s,\beta)\varphi_1(s) + k_{22}(r,s,\beta)\varphi_2(s)\right]ds = -2\pi\frac{\sigma_0\Sigma(t)}{E_0}, \quad |r| \leqslant 1, \tag{40}$$

where the unknown density functions $\varphi_1(r)$ and $\varphi_2(r)$ are defined by

$$\varphi_1(x) = \frac{\partial}{\partial x}[u(x,0^+) - u(x,0^-)],$$

$$\varphi_2(x) = \frac{\partial}{\partial x}[v(x,0^+) - v(x,0^-)], \tag{41}$$

the nondimensional coordinate r is

$$r = x/a, \tag{42}$$

and the Fredholm kernels $k_{ij}(r,s,\beta)$ $(i,j = 1, 2)$ can be found in Ref. [35] and are compiled in the Appendix A for completeness.

The functions $\varphi_1(r)$ and $\varphi_2(r)$ satisfy the following uniqueness condition

$$\int_{-1}^{1}\varphi_i(r)dr = 0, \quad (i = 1, 2), \tag{43}$$

They can be further expressed as

$$\varphi_i(r) = \psi_i(r)/\sqrt{1 - r^2}, \quad (i = 1, 2), \tag{44}$$

where $\psi_i(r)$ $(i = 1, 2)$ are continuous for $r \in [-1, 1]$. When $\varphi_i(r)$ are normalized by $\sigma_0\Sigma(t)/E_0$, the elastic Mode I and mode II SIFs, K_I^e and K_{II}^e, are obtained as

$$K_I^e = -\tfrac{1}{2}\psi_2(1,\beta)\sigma_0\Sigma(t)\sqrt{\pi a},$$

$$K_{II}^e = -\tfrac{1}{2}\psi_1(1,\beta)\sigma_0\Sigma(t)\sqrt{\pi a}. \tag{45}$$

Here, the notation $\psi_i(1,\beta)$ $(i=1,2)$ is adopted to emphasize the dependence of $\psi_1(1)$ and $\psi_2(1)$ on the material parameter β.

5. Stress intensity factors

The SIFs for the viscoelastic FGMs (9)–(11) with the Poisson ratio (12) can be obtained by using the correspondence principle between the elastic and the Laplace transformed viscoelastic equations. Since no other material parameters appear in the SIFs (45), except for the nondimensional parameter β, the SIFs for the viscoelastic FGMs are also given by Eq. (45) by means of the correspondence principle, i.e.

$$K_{\mathrm{I}} = -\tfrac{1}{2}\psi_2(1,\beta)\sigma_0\Sigma(t)\sqrt{\pi a} = K_{\mathrm{I}}^e,$$
$$K_{\mathrm{II}} = -\tfrac{1}{2}\psi_1(1,\beta)\sigma_0\Sigma(t)\sqrt{\pi a} = K_{\mathrm{II}}^e.$$

For the *power law material with position-dependent relaxation time* (11), the SIFs are

$$K_{\mathrm{I}} = -\tfrac{1}{2}\psi_2(1,\beta+q\delta)\sigma_0\Sigma(t)\sqrt{\pi a},$$
$$K_{\mathrm{II}} = -\tfrac{1}{2}\psi_1(1,\beta+q\delta)\sigma_0\Sigma(t)\sqrt{\pi a}. \tag{46}$$

It is seen that q and δ (parameters describing the position dependence of the relaxation time) affect the SIFs only through the combined parameter $(\beta+q\delta)$.

6. Crack opening/sliding displacements

Crack opening/sliding displacements are important parameters in assessing fracture behavior. For the crack problem considered here, those displacements will also evolve with time. Thus, the crack opening/sliding displacements are first given for general time-dependent loading by using the correspondence principle. Two special loading conditions, i.e. Heaviside step loading and exponential loading are subsequently considered.

It follows from Eqs. (41) and (44), and the correspondence principle, that the Laplace transforms of the crack opening/sliding displacements under time-dependent loading, $\sigma_0\Sigma(t)$, can be expressed by the density functions $\varphi_1(x)$ and $\varphi_2(x)$, or $\psi_1(r)$ and $\psi_2(r)$ (normalized by $\sigma_0\Sigma(t)/E_0$), as follows:

$$\overline{[u]} = \bar{u}(x,0^+) - \bar{u}(x,0^-) = \frac{\sigma_0\bar{\Sigma}(p)}{pE_0\bar{f}(p)}\int_{-a}^{x}\varphi_1(x')\mathrm{d}x' = \frac{a\sigma_0\bar{\Sigma}(p)}{pE_0\bar{f}(p)}\int_{-1}^{r}\frac{\psi_1(s)}{\sqrt{1-s^2}}\,\mathrm{d}s,$$
$$\overline{[v]} = \bar{v}(x,0^+) - \bar{v}(x,0^-) = \frac{\sigma_0\bar{\Sigma}(p)}{pE_0\bar{f}(p)}\int_{-a}^{x}\varphi_2(x')\mathrm{d}x' = \frac{a\sigma_0\bar{\Sigma}(p)}{pE_0\bar{f}(p)}\int_{-1}^{r}\frac{\psi_2(s)}{\sqrt{1-s^2}}\,\mathrm{d}s, \tag{47}$$

where $\bar{\Sigma}(p)$ is the Laplace transform of $\Sigma(t)$ and $\bar{f}(p)$ is the Laplace transform of $f(t)$ which is given in Eqs. (13) and (14).

The crack opening/sliding displacements are then obtained as follows:

$$[u] = u(x,0^+) - u(x,0^-) = \left(\frac{a\sigma_0}{E_0}\right)F(t)\int_{-1}^{r}\frac{\psi_1(s)}{\sqrt{1-s^2}}\,\mathrm{d}s,$$
$$[v] = v(x,0^+) - v(x,0^-) = \left(\frac{a\sigma_0}{E_0}\right)F(t)\int_{-1}^{r}\frac{\psi_2(s)}{\sqrt{1-s^2}}\,\mathrm{d}s, \tag{48}$$

where

$$F(t) = \mathscr{L}^{-1}\left[\frac{\overline{\Sigma}(p)}{p\overline{f}(p)}\right],$$
(49)

in which $\mathscr{L}^{-1}$ represents the inverse Laplace transform.

6.1. Heaviside step loading

For the Heaviside step loading,

$$\Sigma(t) = H(t) \rightarrow \overline{\Sigma}(p) = 1/p,$$
(50)

where $H(t)$ is the Heaviside step function. Thus the function $F(t)$ in Eqs. (48) and (49) becomes

$$F(t) = \frac{E_0}{E_\infty} - \left(\frac{E_0}{E_\infty} - 1\right)\exp\left(-\frac{E_\infty t}{E_0 t_0}\right),$$
(51)

for the *standard linear solid*, and

$$F(t) = \frac{(t/t_0)^q}{\Gamma(1-q)\Gamma(1+q)},$$
(52)

for the *power law material*, where $\Gamma(\cdot)$ is the Gamma function.

6.2. Exponential loading

Now consider the exponential loading [2] as another example

$$\Sigma(t) = \exp(-t/t_{\mathrm{L}}) \rightarrow \overline{\Sigma}(p) = 1/(p+1/t_{\mathrm{L}}),$$
(53)

where t_{L} is a positive constant (load relaxation time). The function $F(t)$ is then given by

$$F(t) = \frac{1}{E_\infty/E_0 - t_0/t_{\mathrm{L}}}\left[\left(1-\frac{t_0}{t_{\mathrm{L}}}\right)\exp\left(-\frac{t}{t_{\mathrm{L}}}\right) - \left(1-\frac{E_\infty}{E_0}\right)\exp\left(-\frac{E_\infty t}{E_0 t_0}\right)\right]$$
(54)

for the *standard linear solid*, and

$$F(t) = \frac{1}{\Gamma(1-q)\Gamma(q)t_0}\int_0^t \left(\frac{\tau}{t_0}\right)^{q-1}\exp\left(-\frac{t-\tau}{t_{\mathrm{L}}}\right)\mathrm{d}\tau,$$
(55)

for the *power law material*.

7. Numerical solution of the governing system of integral equations

To obtain the numerical solution of the system of governing integral equations (40), the density functions, $\psi_i(r)$ ($i = 1, 2$), are expanded into series of Chebyshev polynomials of the first kind. By noting the symmetric properties of the density functions (41), $\varphi_i(r)$ ($i = 1, 2$), they are expressed as follows:

$$\varphi_1(r) = \frac{1}{\sqrt{1-r^2}}\sum_{n=1}^{\infty} b_n T_{2n}(r), \quad |r| \leqslant 1,$$

$$\varphi_2(r) = \frac{1}{\sqrt{1-r^2}}\sum_{n=1}^{\infty} a_n T_{2n-1}(r), \quad |r| \leqslant 1,$$
(56)

where $T_n(r)$ are Chebyshev polynomials of the first kind, and a_n and b_n are unknown constants. It is noted that $\varphi_i(r)$ given by Eq. (56) already satisfy the condition (43).

By substituting the above equations into integral equation (40), we have

$$\sum_{n=1}^{\infty} \left\{ \pi U_{2n-1}(r)a_n + H_n^{11}(r)a_n + H_n^{12}(r)b_n \right\} = 0, \quad |r| \leqslant 1,$$

$$\sum_{n=1}^{\infty} \left\{ \pi U_{2n-2}(r)b_n + H_n^{21}(r)a_n + H_n^{22}(r)b_n \right\} = -2\pi \frac{\sigma_0 \Sigma(t)}{E_0}, \quad |r| \leqslant 1,$$

(57)

where $U_n(r)$ are Chebyshev polynomials of the second kind and $H_n^{ij}(r)$ ($i,j = 1, 2$) are given by

$$H_n^{11}(r) = \int_{-1}^{1} k_{11}(r,s,\beta) \frac{T_{2n}(s)}{\sqrt{1-s^2}} \, ds,$$

$$H_n^{12}(r) = \int_{-1}^{1} k_{12}(r,s,\beta) \frac{T_{2n-1}(s)}{\sqrt{1-s^2}} \, ds,$$

$$H_n^{21}(r) = \int_{-1}^{1} k_{21}(r,s,\beta) \frac{T_{2n}(s)}{\sqrt{1-s^2}} \, ds,$$

$$H_n^{22}(r) = \int_{-1}^{1} k_{22}(r,s,\beta) \frac{T_{2n-1}(s)}{\sqrt{1-s^2}} \, ds.$$

(58)

To solve the functional equations (57), the series on the left hand side are first truncated at the Nth term. A collocation technique [38] is then used and the collocation points, r_i, are chosen as the roots of the Chebyshev polynomials of the first kind

$$r_i = \cos \frac{(2i-1)\pi}{2N}, \quad i = 1, 2, \ldots, N.$$

(59)

The functional equations (57) are then reduced to a linear algebraic equation system

$$\sum_{n=1}^{N} \left\{ \pi U_{2n-1}(r_i)a_n + H_n^{11}(r_i)a_n + H_n^{12}(r_i)b_n \right\} = 0,$$

$$\sum_{n=1}^{N} \left\{ \pi U_{2n-2}(r_i)b_n + H_n^{21}(r_i)a_n + H_n^{22}(r_i)b_n \right\} = -2\pi \frac{\sigma_0 \Sigma(t)}{E_0}, \quad i = 1, 2, \ldots, N.$$

(60)

After a_n and b_n ($n = 1, 2, \ldots, N$) are determined, the nondimensional SIFs, K_I^* and K_{II}^*, are computed as follows:

$$K_I^* = -\frac{1}{2}\psi_2(1,\beta) = -\frac{1}{2}\sum_{n=1}^{N} b_n,$$

(61)

$$K_{II}^* = -\frac{1}{2}\psi_1(1,\beta) = -\frac{1}{2}\sum_{n=1}^{N} a_n.$$

(62)

In the following numerical calculations, it is found that 20 collocation points lead to a convergent SIF result.

8. Numerical results

Numerical results for SIFs are first calculated for a homogeneous elastic strip with a central crack (see Fig. 1). Table 1 shows the SIFs of the present calculations and those reported in the handbook by Tada et al. [39]. It is seen from the table that the SIFs are in good agreement. Another example considered is a semi-infinite homogeneous plane with a sub-surface crack, which was studied by Erdogan et al. [38]. The geometry is modeled by a cracked strip with the ratio $h_1/h = 0.01$ in this study (see Fig. 1). The SIFs from the present calculations and Ref. [38] are listed in Table 2. Again, it is observed that the SIFs are in good agreement. It should be noted that the existing results in Refs. [38,39] are also approximate solutions.

Fig. 2 shows normalized SIFs (see (61) and (62)), K_I^* and K_{II}^*, versus the nondimensional crack length $2a/h$ considering various nonhomogeneous parameter β for the linear standard solid and the power law model with constant relaxation time (see Eqs. (9) and (10)). As noted in Section 5, the SIFs are identical for both models. The crack is located in the middle of the strip, i.e. $h_2 = 0.5h$. Fig. 2(a) is the result for the mode I SIF. It is seen that the mode I SIF increases with increasing ratio $2a/h$ for all β values considered here. The SIF is higher than that of the corresponding homogeneous material ($\beta = 0$). The mode I SIF is an even function of β. However, this symmetry is valid only for the crack located in the center of the strip. Fig. 2(b) shows the result of mode II SIF. The mode II SIF is asymmetric about $\beta = 0$ (again, this anti-symmetry is valid only for the crack located in the center of the strip). The absolute value of the mode II SIF increases with increasing $2a/h$. As expected, the mode II SIF is zero for this crack geometry for a homogeneous strip ($\beta = 0$).

For the crack problem considered here, the SIFs for the FGM are, in general, higher than those for the corresponding homogeneous material (see Fig. 2 and the following figures). The higher SIFs do not necessarily mean that the FGM is inferior to the homogeneous materials. In fact, ceramic/metal FGMs usually have higher fracture toughness than monolithic ceramics. Therefore they can withstand higher SIFs. For example, Jin and Batra [16] have theoretically predicted that an alumina/nickel FGM may have a fracture toughness up to 30 times higher than that of alumina.

Fig. 3(a) and (b) shows the SIF results for a crack located at a distance of $h_2 = 0.1h$ from the upper edge of the strip (see Fig. 1). It is seen from Fig. 3(a) that in general, the mode I SIF is not symmetric about $\beta = 0$. This is anticipated since the geometry is not symmetrical about the crack line. The mode I SIF for negative β is slightly larger than that for the corresponding positive β with the same absolute value. For

Table 1
Mode I SIFs for a homogeneous strip with a central crack

$2a/h$	SIF (this study)	SIF (Ref. [39])
0.1	1.01	1.01
0.67	1.42	1.42
1.0	1.82	1.88
2.0	3.29	3.42

Table 2
Modes I and II SIFs for a semi-infinite homogeneous plane with a sub-surface crack

$2a/h_1$	Mode I SIF (this study)	Model I SIF (Ref. [38])	Mode II SIF (this study)	Model II SIF (Ref. [38])
0.67	1.0779	1.0778	0.0124	0.0123
1.0	1.1633	1.1621	0.0368	0.0367
2.0	1.5111	1.4860	0.1849	0.1796

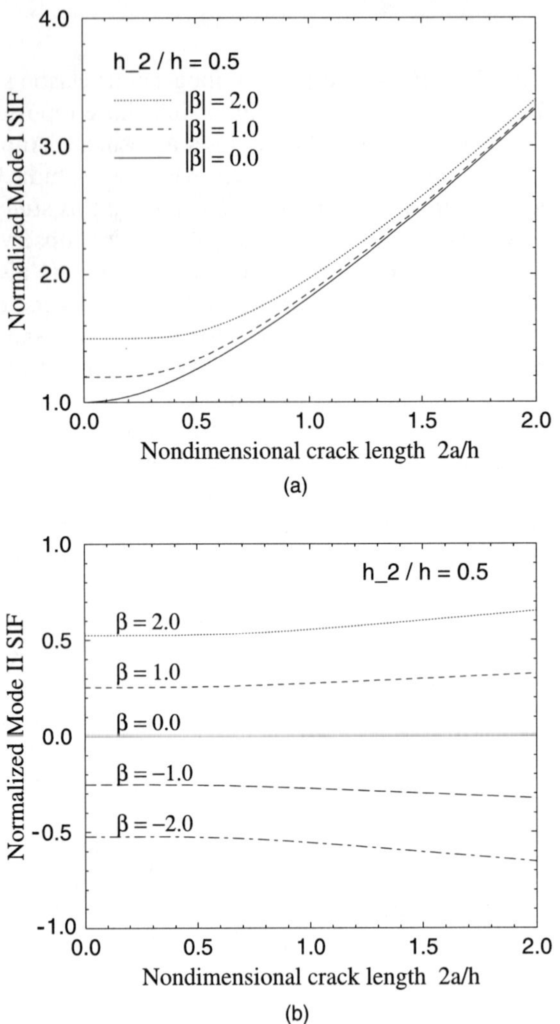

Fig. 2. Normalized SIFs versus nondimensional crack length, $2a/h$, for various material nonhomogeneous parameter β considering the linear standard solid and the power law material with constant relaxation time ($h_2 = 0.5h$), (a) mode I SIF; (b) mode II SIF.

very small nondimensional crack length $2a/h_2$, however, the mode I SIF is still approximately symmetric about $\beta = 0$. This is because the crack seems located in an infinite plate. It is observed from Fig. 3(b) that the magnitude of the mode II SIF increases with increasing $2a/h_2$ for $\beta < 0$. For $\beta > 0$, however, the magnitude of the mode II SIF may decrease or increase with increasing $2a/h_2$ depending on the value of β and the range of $2a/h_2$.

Fig. 4 shows normalized SIFs versus the nondimensional crack length $2a/h$ for $\beta = 2$ and various values of δ for the power law material with position-dependent relaxation time (see Eq. (11)). The crack is located in the middle of the strip. The effect of spatial position dependence of the relaxation time on the SIFs is reflected through the parameter δ. The parameter q is taken as 0.4 in all calculations. Thus the curves for $\delta = \pm 1$ may be obtained from the curve $\delta = 0$ by shifting this curve by $\beta = \mp 0.4$. It is clear from Fig. 4 that with respect to the corresponding model with constant relaxation time (i.e. $\delta = 0$), a positive δ increases SIFs when $\beta > 0$.

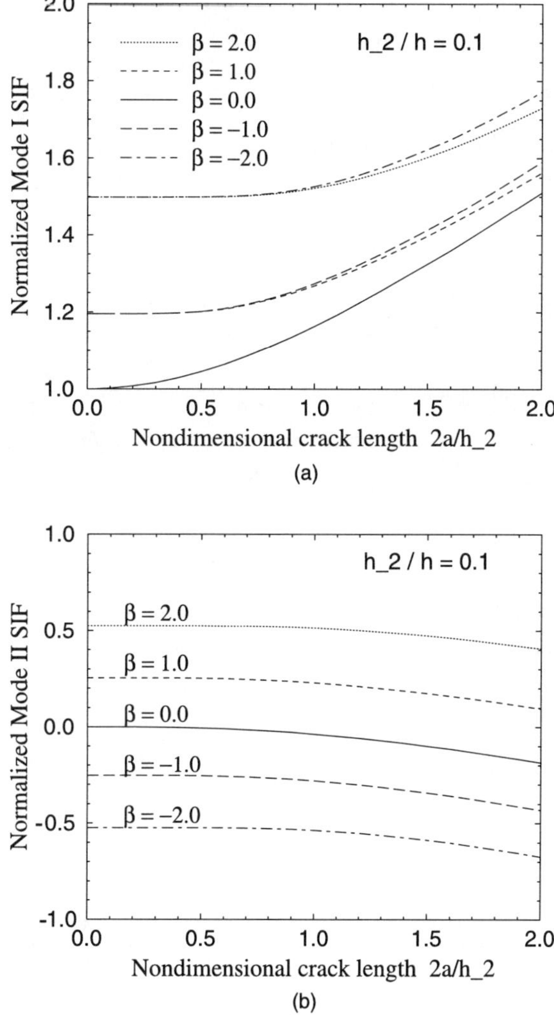

Fig. 3. Normalized SIFs versus nondimensional crack length, $2a/h_2$, for various material nonhomogeneous parameter β considering the linear standard solid and the power law material with constant relaxation time ($h_2 = 0.1h$), (a) mode I SIF; (b) mode II SIF.

Fig. 5 shows normalized SIFs versus the nondimensional crack length $2a/h$ for $\beta = -2$ and various values of δ for the power law material with position-dependent relaxation time (see Eq. (11)). The crack is located in the middle of the strip. In contrast with the result in Fig. 4, a negative δ increases the magnitudes of SIFs when $\beta < 0$.

Special attention needs to be paid when $|\beta|$ is relatively small compared with $|q\delta|$. In this case, the variation of SIFs with δ follows different paths depending on the value of $\beta + q\delta$. For example, Fig. 6 shows normalized SIFs versus the nondimensional crack length $2a/h$ for $\beta = 0.1$ and various values of δ for the power law material with position-dependent relaxation time (see Eq. (11)). The crack is again located in the middle of the strip. It is observed that the mode I SIF increases with an increase in the absolute value of δ (Fig. 6(a)). The magnitude of the mode II SIF also increases with an increase in the absolute value of δ (Fig. 6(b)).

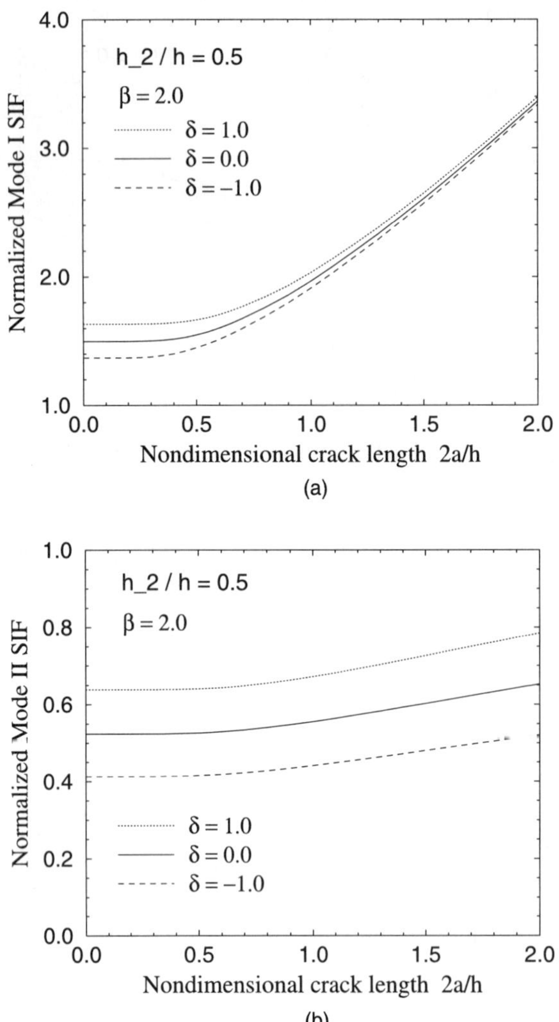

Fig. 4. Normalized SIFs versus nondimensional crack length, $2a/h$, for various material nonhomogeneous parameter δ considering the power law material with position-dependent relaxation time ($\beta = 2.0$, $q = 0.4$, $h_2 = 0.5h$), (a) mode I SIF; (b) mode II SIF.

Fig. 7 shows the SIF results for $\beta = 2$ and various values of δ for the power law material with position-dependent relaxation time (see Eq. (11)). The crack is located at a distance of $h_2 = 0.1h$ from the upper edge of the strip (see Fig. 1). The SIFs are found to follow a similar trend to that for a central crack except that the magnitude of mode II SIF decreases with increasing $2a/h_2$.

Fig. 8 shows the SIF results for $\beta = -2$ and various values of δ for the power law material with position-dependent relaxation time (see Eq. (11)). The crack is again located at a distance of $h_2 = 0.1h$ from the upper edge of the strip (see Fig. 1). The mixed mode SIFs follow a similar trend to that for a central crack (cf. Fig. 5).

Fig. 9(a) shows the crack opening displacement for the standard linear solid and the power law material with constant relaxation time (see Eq. (48)). The crack is located in the middle of the strip with $2a/h = 1$.

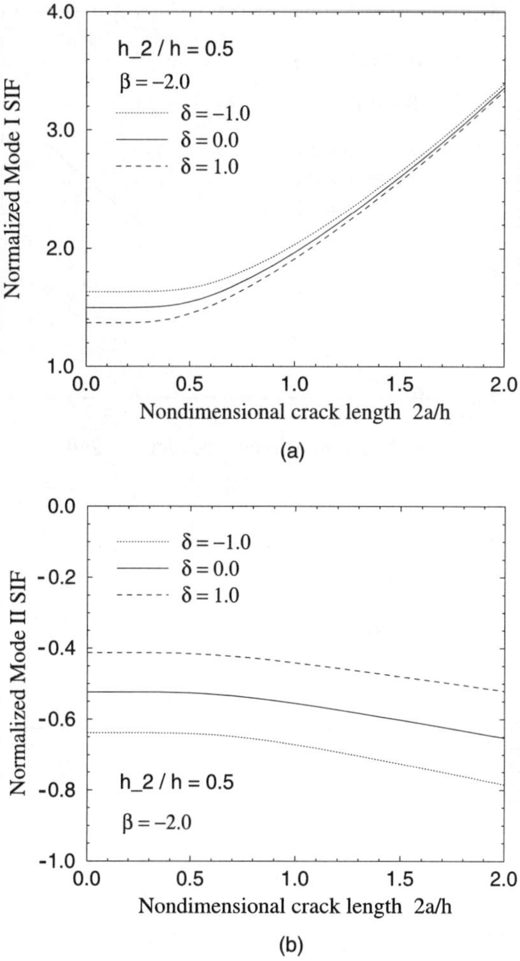

Fig. 5. Normalized SIFs versus nondimensional crack length, $2a/h$, for various material nonhomogeneous parameter δ considering the power law material with position-dependent relaxation time ($\beta = -2.0$, $q = 0.4$, $h_2 = 0.5h$), (a) mode I SIF; (b) mode II SIF.

For this special configuration, the crack opening displacement is symmetric about $\beta = 0$ (i.e. the same crack opening displacement is obtained for $\pm\beta$). It is seen from the figure that the crack opening displacement increases with increasing β. Fig. 9(b) depicts the crack sliding displacement for the same viscoelastic models. It is clear that the absolute value of the crack sliding displacement increases with increasing β. The crack sliding displacement vanishes for $\beta = 0$. This is anticipated because when the crack is located in the middle of a homogeneous strip, there is no mode II deformation.

Fig. 10 shows the crack opening/sliding displacements for a crack located in the middle of the strip of power law material with graded relaxation time. Three δ values are considered and β is taken as 2.0. As in the case of SIFs, with respect to the corresponding power law model with constant relaxation time (i.e. $\delta = 0$), a positive δ increases the crack opening displacement when $\beta > 0$. A negative δ would increase the crack opening displacement when $\beta < 0$, and increase the absolute value of the crack sliding displacement. This observation is consistent with the SIF investigation (see Fig. 4).

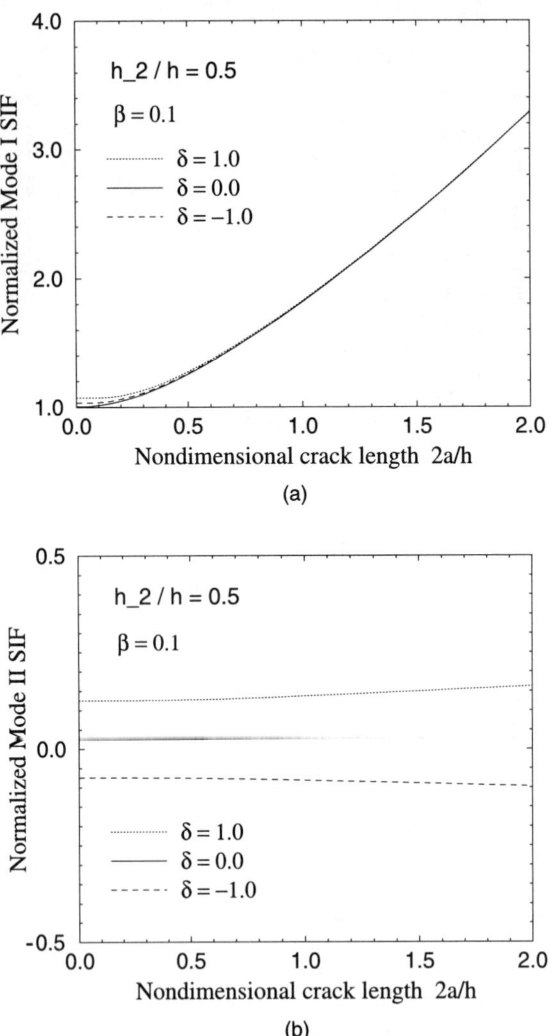

Fig. 6. Normalized SIFs versus nondimensional crack length, $2a/h$, for various material nonhomogeneous parameter δ considering the power law material with position-dependent relaxation time ($\beta = 0.1$, $q = 0.4$, $h_2 = 0.5h$), (a) mode I SIF; (b) mode II SIF.

9. Concluding remarks and extensions

The correspondence principle is used to study the SIFs and crack opening/sliding displacements for a crack in a viscoelastic FGM strip with relaxation functions having separable forms in space and time. Three viscoelastic models are considered, i.e. standard linear solid, power law material with constant relaxation time, and power law material with position-dependent relaxation time. Under traction boundary conditions, the SIFs for the models with constant relaxation times are the same as those for the corresponding nonhomogeneous elastic materials, while for the power law material with graded relaxation time, the SIFs are influenced by the gradation of the relaxation time. The crack opening/sliding displacements evolve with time reflecting the creep behavior of the material under traction boundary conditions.

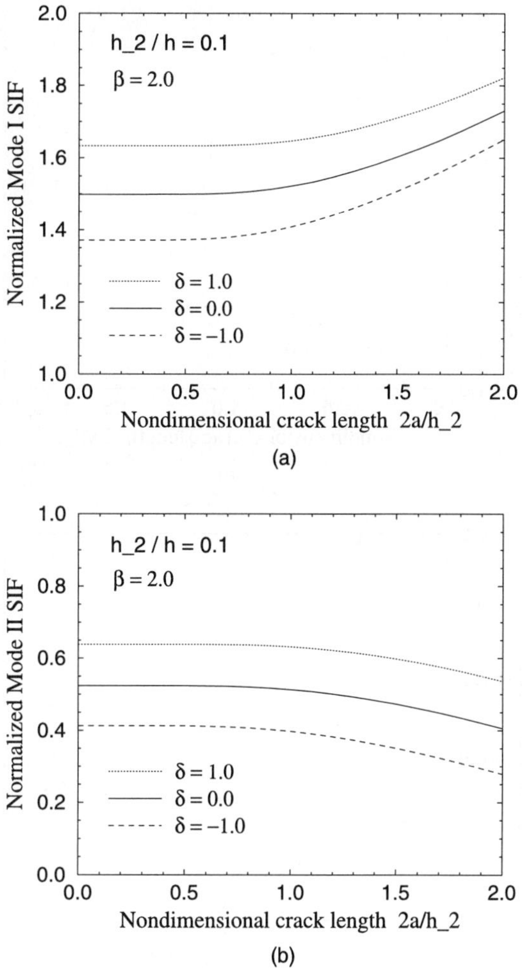

Fig. 7. Normalized SIFs versus nondimensional crack length, $2a/h_2$, for various material nonhomogeneous parameter δ considering the power law material with position-dependent relaxation time ($\beta = 2.0$, $h_2 = 0.1h$), (a) mode I SIF; (b) mode II SIF.

Natural extensions of this work include investigation of *displacement* (rather than traction) boundary conditions, *crack propagation*, and *experimental verification*. The solution of crack boundary value problems with displacement boundary conditions on the outer boundaries of the FGM strip may follow the methodology presented in this paper. The crack propagation modeling in viscoelastic FGMs can be thought as an extension of the techniques proposed by Wnuk [40], Knauss [5], and Shapery [6–8]. The experimental verification can be done by adapting the basic setup for fabricating large scale polymeric FGMs by Lambros et al. [23] to creep/relaxation and/or viscoelastic fracture testing. These topics are presently under consideration by the authors.

Acknowledgements

We would like to acknowledge the support from the National Science Foundation (NSF) under grant no. CMS-9996378 (Mechanics and Materials Program).

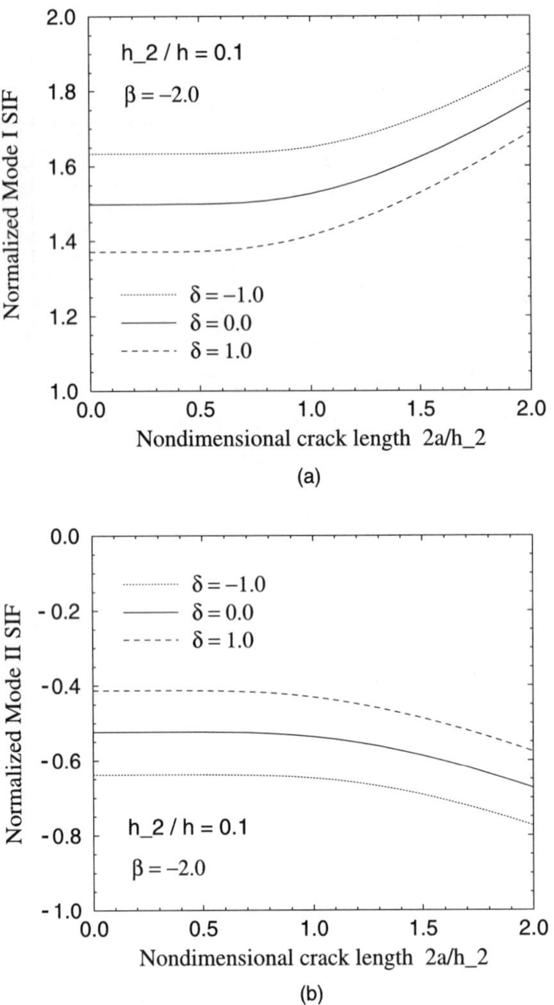

Fig. 8. Normalized SIFs versus nondimensional crack length, $2a/h_2$, for various material nonhomogeneous parameter δ considering the power law material with position-dependent relaxation time ($\beta = -2.0$, $h_2 = 0.1h$), (a) mode I SIF; (b) mode II SIF.

Appendix A

The Fredholm kernels $k_{ij}(r, s, \beta)$ $(i, j = 1, 2)$ in the system of integral equations (40) are given below [35]:

$$k_{11}(r, s, \beta) = \int_0^\infty [1 + 4\xi f_{11}(\xi)] \sin(r - s)\xi \, d\xi,$$

$$k_{12}(r, s, \beta) = \int_0^\infty 4\xi f_{12}(\xi) \cos(r - s)\xi \, d\xi,$$

$$k_{21}(r, s, \beta) = -\int_0^\infty 4\xi^2 f_{21}(\xi) \cos(r - s)\xi \, d\xi,$$

$$k_{22}(r, s, \beta) = \int_0^\infty [1 + 4\xi^2 f_{22}(\xi)] \sin(r - s)\xi \, d\xi.$$

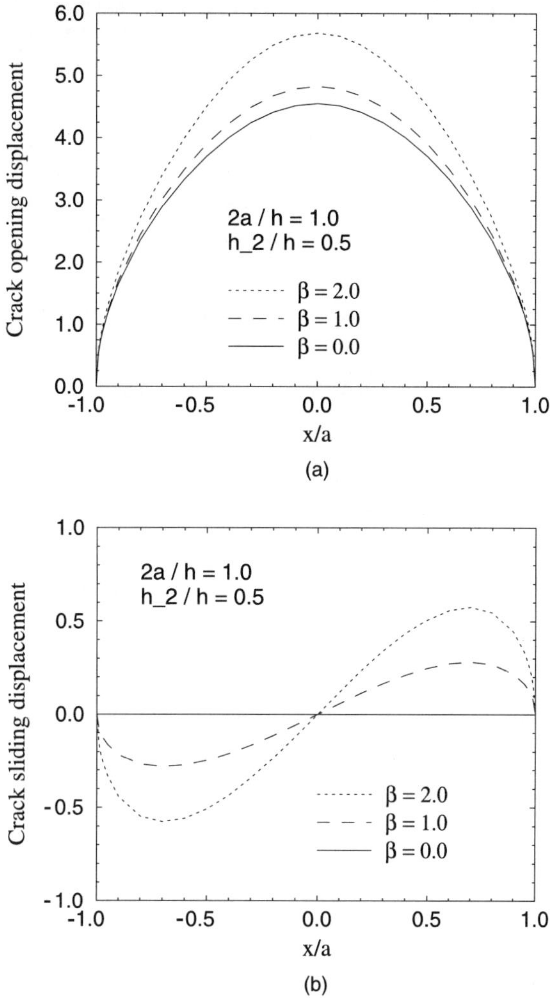

Fig. 9. Crack opening/sliding displacements for three β values, standard linear solid and power law material with constant relaxation time ($h_2 = 0.5h$, $2a/h = 1.0$), (a) mode I case; (b) mode II case.

The functions $f_{ij}(\xi)$ $(i, j = 1, 2)$ in the above equations are given by

$$f_{11}(\xi) = [-\beta H_{11} + s_2(s_1 - s_2)H_{12}]/(s_1 - s_2)^3,$$
$$f_{12}(\xi) = [-2\xi H_{11} - \xi(s_1 - s_2)H_{12}]/(s_1 - s_2)^3,$$
$$f_{21}(\xi) = [-\beta H_{21} + s_2(s_1 - s_2)H_{22}]/(s_1 - s_2)^3,$$
$$f_{22}(\xi) = [-2\xi H_{21} - \xi(s_1 - s_2)H_{22}]/(s_1 - s_2)^3,$$

in which

$$s_1 = -s - \beta/2, \quad s_2 = s - \beta/2, \quad s = \sqrt{\xi^2 + \beta^2/4},$$

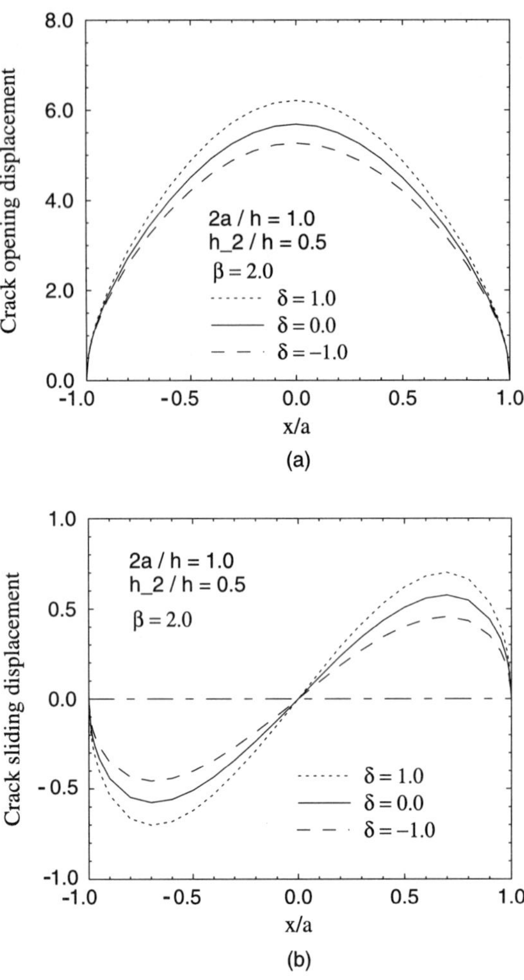

Fig. 10. Crack opening/sliding displacements for three δ values, power law material with position dependent relaxation time ($\beta = 2.0$, $h_2 = 0.5h$, $2a/h = 1.0$), (a) mode I case; (b) mode II case.

and

$$H_{11}(\xi) = (h_{11}d_{11} + h_{12}d_{21})/D_A,$$
$$H_{12}(\xi) = (h_{11}d_{12} + h_{12}d_{22})/D_A,$$
$$H_{21}(\xi) = (h_{21}d_{11} + h_{22}d_{21})/D_A,$$
$$H_{12}(\xi) = (h_{21}d_{12} + h_{22}d_{22})/D_A.$$

In the above expressions, the functions $h_{ij}(\xi)$ $(i,j = 1, 2)$ are given by

$$h_{11}(\xi) = -s_1 + \exp(-2sh_1)(s_1 + 2h_1ss_2),$$
$$h_{12}(\xi) = 1 - \exp(-2sh_1)[1 - 2h_1s(1 - h_1s_2)],$$
$$h_{21}(\xi) = 1 - \exp(-2sh_1)(1 + 2sh_1),$$
$$h_{22}(\xi) = 2h_1^2 s \exp(-2sh_1),$$

the functions $d_{ij}(\xi)$ $(i, j = 1, 2)$ are

$$d_{11}(\xi) = \exp(-2hs)[1 - 2sh(1 - 2h_1 s) - (1 - 2h_1 s + 4h_1^2 s^2) + \exp(-2sh_2)]$$
$$+ \exp(-2sh_2)(1 + 2h_2 s + 4h_2^2 s^2) - 1,$$
$$d_{12}(\xi) = 2s\exp(-2hs)[h(h_1 - h_2 - 2h_1 h_2 s) - h_1^2 \exp(-2sh_2)] + 2h_2^2 s\exp(-2sh_2),$$
$$d_{21}(\xi) = 4s^2 \exp(-2hs)[h - h_1 \exp(-2sh_2)] - 4h_2^2 s\exp(-2sh_2),$$
$$d_{22}(\xi) = \exp(-2hs)[(1 + 2h_1 s)(1 + 2h_2 s - \exp(-2sh_2) + 4h_2^2 s^2)] + (1 - 2h_2 s)\exp(-2sh_2) - 1,$$

and $D_A(\xi)$ is

$$D_A(\xi) = 1 - 2(1 + 2h^2 s^2)\exp(-2hs) + \exp(-4hs).$$

References

[1] Christensen RM. Theory of Viscoelasticity. New York: Academic Press; 1971.

[2] Broberg KB. Cracks and Fracture. London: Academic Press; 1999.

[3] Atkinson C, Chen CY. The influence of layer thickness on the stress intensity factor of a crack lying in an elastic (viscoelastic) layer embedded in a different elastic (viscoelastic) medium (mode III analysis). Int J Eng Sci 1996;34:639–58.

[4] Atkinson C, Chen CY. The influence of layer thickness on the stress intensity factor of a crack lying in an (visco)elastic layer embedded in a different (visco)elastic medium (plane strain, mode I analysis). Proc Royal Soc London: A 1997;453:1445–71.

[5] Knauss WG. The mechanics of polymer fracture. Appl Mech Rev 1973;26:1–17.

[6] Shapery RA. A theory of crack initiation and growth in viscoelastic media, I. Theoretical development. Int J Fract 1975;11:141–58.

[7] Shapery RA. A theory of crack initiation and growth in viscoelastic media, II. Approximate methods of analysis. Int J Fract 1975;11:549–62.

[8] Shapery RA. A theory of crack initiation and growth in viscoelastic media, III. Analysis of continuous growth. Int J Fract 1975;11:141–58.

[9] Hirai T. Functionally gradient materials. In: Brook RJ, editor. Materials Science and Technology, Processing of Ceramics, Part 2, vol. 17B. Weinheim, Germany: VCH Verlagsgesellschaft mbH; 1996. p. 292–341.

[10] Suresh S, Mortensen A. Functionally Graded Materials. London: The Institute of Materials, IOM Communications Ltd; 1998.

[11] Koizumi M. The concept of FGMs. In: Holt JB, Koizumi M, Hirai T, Munir Z, editors. Functionally Graded Materials, 34. Ohio: American Ceramic Society; 1993. p. 3–10.

[12] Aboudi J, Pindera MJ, Arnold SM. Higher-order theory for functionally graded materials. Compos Part B: Eng 1999;30B: 777–832.

[13] Erdogan F. Fracture mechanics of functionally graded materials. Compos Eng 1995;5:753–70.

[14] Eischen JW. Fracture of nonhomogeneous materials. Int J Fract 1987;34:3–22.

[15] Jin Z-H, Noda N. Crack-tip singular fields in nonhomogeneous materials. ASME J Appl Mech 1994;61:738–40.

[16] Jin Z-H, Batra RC. Some basic fracture mechanics concepts in functionally graded materials. J Mech Phys Solids 1996;44: 1221–35.

[17] Gu P, Asaro RJ. Crack deflection in functionally graded materials. Int J Solids Struct 1996;34:3085–98.

[18] Honein T, Herrmann G. Conservation laws in nonhomogeneous plane elastostatics. J Mech Phys Solids 1997;45:789–805.

[19] Paulino GH, Fannjiang AC, Chan YS. Gradient elasticity theory for a mode III crack in a functionally graded material. Mater Sci Forum 1999;308–311:971–6.

[20] Carpenter RD, Paulino GH, Munir ZA, Gibeling JC. A novel technique to generate sharp cracks in metallic/ceramic functionally graded materials by reverse 4-point bending. Scripta Materialia 2000;43:547–52.

[21] Becker Jr. TL, Cannon RM, Ritchie RO. A statistical RKR fracture model for the brittle fracture of functionally graded materials. Mater Sci Forum 1999;308–311:957–62.

[22] Parameswaran V, Shukla A. Dynamic fracture of functionally gradient material having discrete property variation. J Mater Sci 1998;33:3303–11.

[23] Lambros J, Santare MH, Li H, Sapna III GH. A novel technique for the fabrication of laboratory scale model functionally graded materials. Exp Mech 1999;39:184–90.

[24] Marur PR, Tippur HV. Dynamic response of bimaterial and graded interface cracks under impact loading. Int J Fract 2000;103:95–109.

[25] Paulino GH, Jin Z-H. Correspondence principle in viscoelastic functionally graded materials. ASME J Appl Mech 2001;68: 129–32.

[26] Paulino GH, Jin Z-H. Viscoelastic functionally graded materials subjected to antiplane shear fracture. ASME J Appl Mech 2001;68:284–93.

[27] Paulino GH, Jin Z-H. A crack in a viscoelastic functionally graded material layer embedded between two dissimilar homogeneous viscoelastic layers—antiplane shear analysis. Int J Fract 2001;111:283–303.

[28] Alex R, Schovanec L. An anti-plane crack in a nonhomogeneous viscoelastic body. Eng Fract Mech 1996;55:727–35.

[29] Herrmann JM, Schovanec L. Quasi-static mode III fracture in a nonhomogeneous viscoelastic body. Acta Mechanica 1990;85:235–49.

[30] Herrmann JM, Schovanec L. Dynamic steady-state mode III fracture in a nonhomogeneous viscoelastic body. Acta Mechanica 1994;106:41–54.

[31] Schovanec L, Walton JR. The quasi-static propagation of a plane strain crack in a power-law inhomogeneous linearly viscoelastic body. Acta Mechanica 1987;67:61–77.

[32] Schovanec L, Walton JR. The energy release rate for a quasi-static mode I crack in a nonhomogeneous linearly viscoelastic body. Eng Fract Mech 1987;28:445–54.

[33] Yang YY. Time-dependent stress analysis in functionally graded materials. Int J Solids Struct 2000;37:7593–608.

[34] Delale F, Erdogan F. On the mechanical modeling of the interfacial region in bonded half planes. ASME J Appl Mech 1988;55:317–24.

[35] Noda N, Jin Z-H. Thermal stress intensity factors for a crack in a strip of a functionally gradient material. Int J Solids Struct 1993;30:1039–56.

[36] Fung YC. Foundations of Solid Mechanics. Englewood Cliffs, NJ: Prentice Hall; 1965.

[37] Ogorkiewicz RM. Engineering Properties of Thermoplastics. London: Wiley-Interscience; 1970.

[38] Erdogan F, Gupta GD, Cook TS. Numerical solution of singular integral equations. In: Sih GC, editor. Mechanics of Fracture, vol. 1. Leyden: Noordhoff; 1973. p. 368–425.

[39] Tada H, Paris P, Irwin G. The Stress Analysis of Cracks Handbook. Hellertown, PA, USA: Del Research Corporation; 1973.

[40] Wnuk MP. Subcritical growth of fracture (inelastic fatigue). Int J Fract Mech 1971;7:383–407.

PERGAMON

Engineering Fracture Mechanics 69 (2002) 1791–1809

Engineering Fracture Mechanics

www.elsevier.com/locate/engfracmech

Transient thermoelastic responses of functionally graded materials containing collinear cracks

N. Noda [a,*], B.L. Wang [b]

[a] *Department of Mechanical Engineering, Shizuoka University, Hamamatsu 432-8561, Japan*
[b] *Center for Composite Materials, Harbin Institute of Technology, Harbin 150001, China*

Received 26 December 2000; received in revised form 31 July 2001; accepted 6 August 2001

Abstract

A generalized method to treat the collinear cracks in functionally graded materials (FGMs) subjected to thermal loads is established. The integral transform technique is used to reduce the problem to a set of singular integral equations which is solved numerically. A metal/ceramic FGM plate and a metal layer/ceramic layer bonded plate are analyzed as examples. Numerical results are obtained to show the influence of crack spacing on field intensity factors ahead of the crack-tips. It is observed that thermal stress intensity factors can increase or decrease with the crack spacing. Through the introduction of the FGM, the thermal stress intensity factors can be reduced.
© 2002 Published by Elsevier Science Ltd.

1. Introduction

Functionally graded materials (FGMs) have attracted much attention for high-temperature applications since it is possible with these materials to obtain a combination of properties that cannot be achieved in conventional monolithic materials. This is obtained by varying the composition from one side of the material to the other either continuously or stepwise as in multiplayer materials. FGMs are inhomogeneous materials, which are tailored in such a way as to derive beneficial behavior from their inhomogeneity.

Under high-temperature loading, various types of cracks may occur in FGMs. Naturally, an important aspect that needs to be addressed in engineering applications of FGMs is the fracture related failure. Plane elasticity problems involving crack in FGM are solved by specifically assuming a functional form, usually a linear or an exponential function. Assuming an exponential spatial variation of the elastic modulus, Atkinson and List [1], Dhaliwal and Singh [2], and Delale and Erdogan [3] solved crack problems for non-homogeneous materials subjected to mechanical loads. By further assuming the exponential variation of thermal properties of the material, Jin and Noda [4] and Erdogan and Wu [5] computed thermal stress intensity factor (TSIF) for non-homogeneous solids. Gu and Asaro [6,7] considered a semi-infinite crack in

---

* Corresponding author.
E-mail address: tmnnoda@ipc.shizuoka.ac.jp (N. Noda).

0013-7944/02/$ - see front matter © 2002 Published by Elsevier Science Ltd.
PII: S0013-7944(02)00055-3

a strip of FGM under edge loading and obtained stress intensity factor relations for many commonly used fracture specimen configurations. Recently, Erdogan [8] has reviewed the elementary concepts of fracture mechanics of FGM and identified a number of typical problems relating to fracture of FGM. Jin and Batra [9] summarized the crack-tip fields in a general non-homogeneous material [9]. Jin and Batra studied the thermal stress relaxation problem at the tip of an edge crack in FGMs under thermal shock [10]. Jin and Batra also considered some aspects of thermal fracture mechanics of ceramics with temperature-dependent properties [11]. More recently, Kokini and Takeuchi investigated the mechanism of multiple crack formation at the surface of graded thermal barrier coatings subjected to transient heating and cooling loads [12]. Ueda and Shindo studied small crack kinking in FGMs subjected a constant initial strain resulting from stress relaxation [13]. Li, Lambros, Cheeseman and Santare described the fracture testing of FGMs [14]. The nature of the singular field around the crack in FGM was analyzed parametrically using finite element method [15].

FGMs are special types of non-homogeneous materials with properties varying continuously. The analysis of these specific types of inhomogeneous materials requires new analytic and numerical techniques. The problem has been considered extensively for special cases where the material properties assume certain specific forms. Most of the inhomogeneous materials exhibit one-dimensional inhomogeneity. Analytical models for the fracture of such inhomogeneous materials with arbitrarily varied material properties have already been developed [16–21].

On the other hand, the geometry of collinear cracks is of practical importance in the fracture of FGMs since experimental studies have shown that collinear cracks are developed prior to forming a main crack in FGM. Accordingly, the goal of this paper is to discuss the collinear cracks in FGMs under thermal load. The loads applied to the materials can be transient or steady state. Using the integral transform method, the problem is reduced to the solution of a set of singular integral equation. This paper is completed by including graphical plots of the TSIFs around crack-tips. For collinear cracks in homogeneous elastic solids, we refer the reader to [22–26].

2. Structure description

Consider a functionally graded material of height h with properties that vary as a function of coordinate y. The geometry of the problem to be formulated is shown in Fig. 1. The medium is infinite in the x-direction. The medium is treated as a laminate containing many thin layers (say N layers). For each layer, the material properties in the center of the layer are assigned as the layer constant properties. For the Jth layer, the thickness is h_J, throughout the paper the subscript J is associated with Jth layer, counted up from the bottom of the laminate, the subscript j denotes the interface number between the Jth layer and the $(J+1)$th layer. We construct a local coordinate y_J with origin at the point where y_J axis intersects the bottom surface of the Jth layer. The adjacent two layers are perfectly bonded or partly separated by a pair of equal collinear Griffith cracks which are located symmetrically with reference to $x = 0$ plane. The left and the right tips of the jth crack are located at $x = c_j - a_j$ and $x = c_j + a_j$, respectively, where a_j is the half crack length.

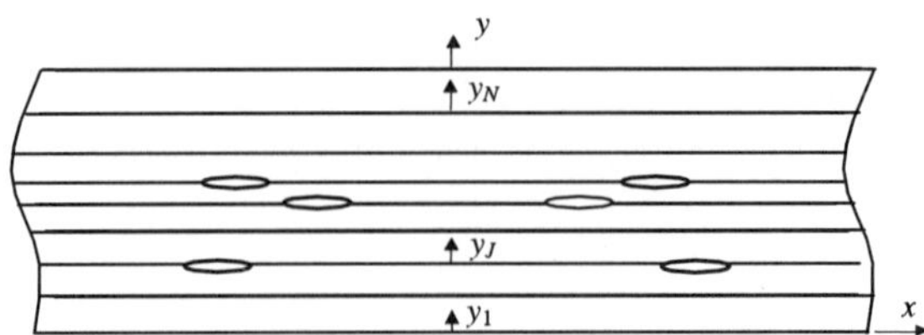

Fig. 1. Geometry and coordinates of a functionally graded material plate with many collinear Griffith cracks.

Introduce the generalized vectors

$$\{V(x,y)\} = (u \quad v \quad T)^{\mathrm{T}} \tag{1}$$

and

$$\{S(x,y)\} = (\sigma_{xy} \quad \sigma_{yy} \quad q_y)^{\mathrm{T}} \tag{2}$$

where u and v are the x- and y-displacement components, respectively, T represents the temperature variation, σ and q stand, respectively, for stress and thermal flow. Without loss in generality, the plate is taken to be free from mechanical traction since the solution due to mechanical loading can be added onto that of thermal loading for the total effect. The boundary conditions on the bottom surface and the top surface of the plate are specified as follows:

$$\{S_1(x,0)\} = \{0 \quad 0 \quad -h_a(T_1(x,0) - T_a(x))\} \tag{3a}$$

$$\{S_N(x,h_N)\} = \{0 \quad 0 \quad h_b(T_N(x,h_N) - T_b(x))\} \tag{3b}$$

where h_a and h_b are coefficients of relative surface heat transfer, T_a and T_b are environment temperature. We assume that T_a and T_b can be Fourier transformed.

The cracks are taken to be thermally insulated and free from mechanical traction, i.e.,

$$\{t_j(x)\} = 0, \quad c_j - a_j < x < c_j + a_j, \quad j = 1, 2, \ldots, N - 1 \tag{4}$$

The continuity conditions on bonded interfaces are

$$\{V_{(J+1)}(x,0)\} = \{V_J(x,h_J)\}, \quad x < c_j - a_j, \quad j = 1, 2, \ldots, N - 1 \tag{5a}$$

$$\{V_{(J+1)}(x,0)\} = \{V_J(x,h_J)\}, \quad x > c_j + a_j, \quad j = 1, 2, \ldots, N - 1 \tag{5b}$$

$$\{S_J(x,y_J = h_J)\} = \{S_{(J+1)}(x,y_{J+1} = 0)\}, \quad 0 < x < \infty, \quad j = 1, 2, \ldots, N - 1 \tag{5c}$$

3. Governing equations

The elasticity and thermal flow fields for each homogeneous layer can be written as

$$\left. \begin{array}{l} \sigma_x = c_{11}\dfrac{\partial u}{\partial x} + c_{12}\dfrac{\partial v}{\partial y} - \lambda T \\[2mm] \sigma_y = c_{12}\dfrac{\partial u}{\partial x} + c_{11}\dfrac{\partial v}{\partial y} - \lambda T \\[2mm] \sigma_{xy} = c_{66}\left(\dfrac{\partial u}{\partial y} + \dfrac{\partial v}{\partial x}\right) \end{array} \right\} \tag{6a}$$

$$\left\{ \begin{array}{c} q_x \\ q_y \end{array} \right\} = - \begin{bmatrix} k & 0 \\ 0 & k \end{bmatrix} \left\{ \begin{array}{c} \partial T/\partial x \\ \partial T/\partial y \end{array} \right\} \tag{6b}$$

where the symbols λ are stress–temperature coefficients, c_{ij} elastic constants, and k thermal conductive coefficients. The strains ε_{ij} are related to the mechanical displacements u_i by $\varepsilon_{ij} = (u_{i,j} + u_{j,i})/2$, where a comma indicates partial derivative. For an isotropic material, the stiffness coefficients are related to Young's module E and Poisson's ratio v by $c_{11} = E(1 - v)/(1 + v)(1 - 2v)$, $c_{12} = vE/(1 + v)(1 - 2v)$ and $c_{66} = E/2(1 + v)$, the stress–temperature coefficients are related to thermal expansion coefficient α as $\lambda = E\alpha/(1 - 2v)$.

The equilibrium equation for the thermal flows and the stresses are

$$\frac{\partial q_x}{\partial x} + \frac{\partial q_y}{\partial y} = -\rho c_v \frac{\partial T}{\partial t} \tag{7a}$$

$$\frac{\partial \sigma_{xx}}{\partial x} + \frac{\partial \sigma_{xy}}{\partial y} = 0 \tag{7b}$$

$$\frac{\partial \sigma_{xy}}{\partial x} + \frac{\partial \sigma_{yy}}{\partial y} = 0 \tag{7c}$$

where c_v and ρ being specific heat and mass density, respectively.

4. Method of solution

We now replace the mixed boundary conditions (5a)–(5c) by the conditions

$$\partial(\{V_{J+1}(x,0)\} - \{V_J(x,h_J)\})/\partial x = \{d_j(x)\} \tag{8a}$$

$$\{d_j(x)\} = 0, \quad 0 < x < c_j - a_j, \quad x > c_j + a_j \tag{8b}$$

$$\int_{c_j-a_j}^{c_j+a_j} \{d_j(x)\}\, \mathrm{d}x = 0 \tag{8c}$$

Let $\{t_j(x)\}$ represent $\{S_J(x, y_J = h_J)\}$ and $\{S_{(J+1)}(x, y_{J+1} = 0)\}$. Combine Eqs. (A.12) and (8a), we have

$$\{d_j(r)\} = \frac{2}{\pi} \int_0^\infty [G_1(sr)]\{\Omega_j(s)\}\, \mathrm{d}s, \quad j = 1, 2, \ldots, N-1 \tag{9}$$

where

$$[G_1(sx)] = \mathrm{diag}(-\cos sx \quad \sin sx \quad \sin sx) \tag{10}$$

$$\{\Omega_j(s)\} = [L_j]\{\bar{t}_{j-1}(s)\} + [M_j]\{\bar{t}_j(s)\} + [N_j]\{\bar{t}_{j+1}(s)\} \tag{11}$$

and where

$$[L_j(s)] = [D_{bJ}(h_J)], \quad [M_j(s)] = [D_{aJ}(h_J)] - [D_{b(J+1)}(0)], \quad [N_j(s)] = -[D_{a(J+1)}(0)] \tag{12}$$

The Fourier transform of (9) is

$$[D(s)]\{\Sigma(s)\} = \{\delta(s)\} - \{\delta_b(s)\} \tag{13}$$

where

$$\{\delta(s)\} = \left\{ \int_{c_1-a_1}^{c_1+a_1} [G_1(sr)]\{d_1(r)\}\, \mathrm{d}r, \ldots, \quad \int_{c_j-a_j}^{c_j+a_j} [G_1(sr)]\{d_j(r)\}\, \mathrm{d}r, \ldots, \quad \int_{c_{N-1}-a_{N-1}}^{c_{N-1}+a_{N-1}} [G_1(sr)]\{d_{(N-1)}(r)\}\, \mathrm{d}r \right\}^{\mathrm{T}} \tag{14}$$

$$\{\Sigma(s)\} = \left(\{\bar{t}_1(s)\}, \ldots, \quad \{\bar{t}_j(s)\}, \ldots, \quad \{\bar{t}_{N-1}(s)\} \right)^{\mathrm{T}} \tag{15}$$

$$[D(s)] = \begin{bmatrix} M_1 & N_1 & & & & \\ L_2 & M_2 & N_2 & & & \\ & & \ddots & & & \\ & & & L_{N-2} & M_{N-2} & N_{N-2} \\ & & & & L_{N-1} & M_{N-1} \end{bmatrix} \tag{16}$$

and $\{\delta_b(s)\}$ is a vector of $3(N-1)$ rows, with the first three and the last three elements the only non-zero elements, these elements are related to the boundary conditions by

$$\{\delta_{b1}(s)\} = [L_1(s)]\{\bar{t}_0(s)\} \tag{17a}$$

$$\{\delta_{b(N-1)}(s)\} = [N_{N-1}(s)]\{\bar{t}_N(s)\} \tag{17b}$$

Refer to matrix $[K_{mn}]$, which denotes the $(3m-2)$th to $(3m)$th rows and the $(3n-2)$th to $(3n)$th columns sub-matrix in matrix $[D(s)]^{-1}$, and make use of inverse Fourier transform to Eq. (13), we have

$$\{t_j(x)\} + \{t_{bj}(x)\} = \sum_{k=1}^{N-1} \int_{c_k-a_k}^{c_k+a_k} [R_{jk}]\{d_k(r)\}\,dr \tag{18}$$

where

$$[R_{jk}(x,r)] = \frac{2}{\pi} \int_0^\infty [G(sx)][K_{jk}(s)][G_1(sr)]\,ds \tag{19}$$

$$\{t_{bj}(x)\} = \frac{2}{\pi} \int_0^\infty [G(sx)]([K_{j1}(s)]\{\delta_{b1}(s)\} + [K_{j(N-1)}(s)]\{\delta_{b(N-1)}(s)\})\,ds \tag{20}$$

where $[G(sx)] = \mathrm{diag}(\sin sx \ \cos sx \ \cos sx)$. The singular behavior of the kernel $[R_{jk}(x,r)]$ may be obtained from the asymptotic analysis of the integral in (19). For large values of s, only $[K_{jj}(s)] \to [M_j(\infty)]^{-1}$ are non-zero elements in matrix $[D(s)]^{-1}$. It can be deduced from (19) that

$$[R_{jk}(x,r)] = \frac{1}{\pi}\frac{[M_j(\infty)]^{-1}}{r-x} + \frac{\begin{bmatrix} -1 & & \\ & 1 & \\ & & 1 \end{bmatrix}[M_j(\infty)]^{-1}}{\pi}\frac{1}{r+x} + \frac{2}{\pi}[A_{jk}(x,r)] \tag{21}$$

where

$$[A_{jk}(x,r)] = \int_0^\infty [G(sx)]([K_{jk}(s)] - [M_j(\infty)]^{-1})[G_1(sr)]\,ds \tag{22}$$

So far $\{d_j\}$ is the only unknown function in the problem which may be determined from the crack face boundary conditions. We observe that (18) is valid outside and inside the crack. For the later, $\{t_j(x)\}$ is zero, Eq. (18) reads

$$\{t_{bj}(x)\} = \frac{[M_j(\infty)]^{-1}}{\pi} \int_{c_j-a_j}^{c_j+a_j} \frac{1}{r-x}\{d_j(r)\}\,dr + \frac{\begin{bmatrix} -1 & & \\ & 1 & \\ & & 1 \end{bmatrix}[M_j(\infty)]^{-1}}{\pi} \int_{c_j-a_j}^{c_j+a_j} \frac{1}{r+x}\{d_j(r)\}\,dr$$

$$+ \frac{2}{\pi} \sum_{k=1}^{N-1} \int_{c_k-a_k}^{c_k+a_k} [A_{jk}]\{d_k(r)\}\,dr \tag{23}$$

Introducing the following two non-dimensional variables $\bar{r}$ and $\bar{x}$

$$\bar{r} = (r - c_j)/a_j, \quad \bar{x} = (x - c_j)/a_j \tag{24}$$

Thus, the integral equation (23) can be reduced to the standard form

$$\{t_{bj}(x)\} = \frac{[M_j(\infty)]^{-1}}{\pi} \int_{-1}^{1} \frac{1}{\bar{r} - \bar{x}} \{d_j(r)\}\, \mathrm{d}\bar{r} + \frac{\begin{bmatrix} -1 & & \\ & 1 & \\ & & 1 \end{bmatrix} [M_j(\infty)]^{-1}}{\pi} \int_{-1}^{1} \frac{1}{\bar{r} + \bar{x} + 2c_j/a_j} \{d_j(r)\}\, \mathrm{d}\bar{r}$$
$$+ \frac{2}{\pi} \sum_{k=1}^{N-1} \int_{-1}^{1} [\overline{\Lambda}_{jk}]\{d_k(r)\}\, \mathrm{d}\bar{r} \tag{25}$$

where $[\overline{\Lambda}_{jk}] = a_k[\Lambda_{jk}]$.

The numerical solution of the above system of equations has been obtained by employing a weighted residual technique to reduce the singular integral equations to a set of algebraic equations in unknown coefficients of Chebychev polynomials of the first kind. Note that the solutions for $\{d_j(r)\}$ has the following form:

$$\{d_j(r)\} = \{g(r)\}/\sqrt{1 - \bar{r}^2} \tag{26}$$

where $\{g_j(r)\}$ is a continuous and bounded vector. The stress intensity factors for mode I and II (Tables 1 and 2), and the thermal flow intensity factor (denoted by mode 0) can be calculated as

$$(K_{\mathrm{II}} \quad K_{\mathrm{I}} \quad K_0)_j = \sqrt{a_j}[M_j(\infty)]\{g_j(-a_j)\} \tag{27a}$$

Table 1
Influence of layer thickness on steady mode I TSIF $K_{\mathrm{I}}/\alpha_c E_c T_0 \sqrt{a}$ for FGM

c/a	Inner crack-tips		Outer crack-tips	
	$N = 10$	$N \geqslant 30$	$N = 10$	$N \geqslant 30$
1			0.0288	0.0259
1.1	0.0195	0.0166	0.00872	0.00849
1.2	0.0148	0.0121	0.00709	0.00688
1.3	0.0112	0.00987	0.00612	0.00586
1.5	0.00732	0.00709	0.00480	0.00463
2	0.00384	0.00396	0.00389	0.00373
3	0.00423	0.00408	0.00506	0.00485
5	0.00644	0.00620	0.00649	0.00635
∞	0.00658	0.00650	0.00658	0.00650

Table 2
Influence of layer thickness on steady mode II TSIF $K_{\mathrm{II}}/\alpha_c E_c T_0 \sqrt{a}$ for FGM

c/a	Inner crack-tips		Outer crack-tips	
	$N = 10$	$N \geqslant 30$	$N = 10$	$N \geqslant 30$
1			0.118	0.114
1.1	−0.0222	−0.0231	0.0752	0.0728
1.2	−0.0263	−0.0273	0.0691	0.0675
1.3	−0.0305	−0.0314	0.0670	0.0648
1.5	−0.0349	−0.0362	0.0621	0.0608
2	−0.0404	−0.0427	0.0579	0.0561
3	−0.0453	−0.0472	0.0539	0.0523
5	−0.0483	−0.0499	0.0490	0.0503
∞	−0.0495	−0.0502	0.0495	0.0502

for an inner crack-tip, and

$$(K_{II} \quad K_I \quad K_0)_j = -\sqrt{a_j}[M_j(\infty)]\{g_j(a_j)\} \tag{27b}$$

for an outer crack-tip.

By now, we have arrived at the solutions in Laplace transform plane. It remains to invert the resulting Laplace transform expression. This is achieved by adopting the numerical technique outlined in [27] and [28]. This method is based on approximating the inverse Laplace transform $f(t)$ by a finite series of the form

$$f(t) = \frac{e^{dt}}{T_d}\left[\frac{f^*(d)}{2} + \sum_{k=1}^{M}\{\mathrm{Re}\{f^*(d + k\pi i/T_d)\}\cos(k\pi t/T_d) - \mathrm{Im}\{f^*(d + k\pi i/T_d)\}\sin(k\pi t/T_d)\}\right] \tag{28}$$

where M is a sufficiently large positive integer, d and T_d are parameters chosen such that the following accuracy criterion is achieved:

$$e^{dt}[\mathrm{Re}\{f^*(d + k\pi i/T_d)\}\cos(M\pi t/T_d) - \mathrm{Im}\{f^*(d + k\pi i/T_d)\}\sin(M\pi t/T_d)] < \varepsilon \tag{29}$$

where ε is the pre-selected degree of accuracy, and T_d is related to the particular time t at which the function is to be evaluated by the relation $0 < t < T_d$. Special care must be taken during the numerical inversion since the Laplace inversion is not straightforward. We can choice several values for M and d to see the convergence of the solution. For the present problem, the variation of the temperature with time is quite slow, the method used above give a very good convergence for a M value larger than 100.

Suppose that $f^*(p)$ is the Laplace transform of function $f(t)$. Then

$$\lim_{p \to 0} p f^*(p) = \lim_{t \to \infty} f(t) \tag{30}$$

Therefore, the static solutions of the problem are easily obtained using Eq. (30).

5. Numerical example and discussion

Considered are a FGM plate and a Ni/TiC bonded (non-FGM) plate, as shown in Fig. 2. The FGM is made of metal Ni and ceramic TiC. The elastic modulus E, Poisson's ratio v, coefficient of thermal expansion α, density ρ, thermal conductivity k, and specific heat c, are, respectively, $E_m = 200$ GPa, $v_m = 0.31$, $\alpha_m = 18 \times 10^{-6}$/K, $\rho_m = 6825$ kg/m^3, $k_m = 54$ W/m K, $c_m = 595$ J/kg K, $E_c = 460$ GPa, $v_c = 0.34$, $\alpha_c = 7.4 \times 10^{-6}$/K, $\rho_c = 4127$ kg/m^3, $k_c = 27$ W/m K, $c_c = 682$ J/kg K, where the subscript m and c stand for metal and ceramics, respectively. The FGM is pure metal at the bottom and pure ceramic at the top. At any position y in the FGM, the local volume fraction of metal is assumed to be $V_m(y)$ which obeys a power-law type relation $1 - (y/h)^g$, where g is known as gradient exponent or in-homogeneity parameter. In our numerical analysis, the volume fraction of metal V_m is a linear function of the position y, i.e., $g = 1$. The three-phase model [29] is utilized to determine the effective shear modulus and thermal conductivity of the FGM. Composite spheres model [29] is utilized to determine effective bulk modulus and coefficient of thermal expansion. The effective density and the specific heat of FGM are evaluated by rule-of-mixture. In the analysis, the graded region is treated as a number of thin layers, with each layer being assigned slightly different material properties.

The geometric parameters are $a = 0.125h$, where h is the total thickness of the plates. The metal layer and the ceramics layer of the non-FGM plate have the same thickness. Initially at $t = 0$, the elastic plate at an initial temperature zero is assumed to correspond to zero stress state. The temperature on the top surfaces of the two plates is suddenly changed to T_0, while the bottom surfaces are kept at the initial temperature, thereby producing a through-thickness temperature gradient at all times. The respective field intensity

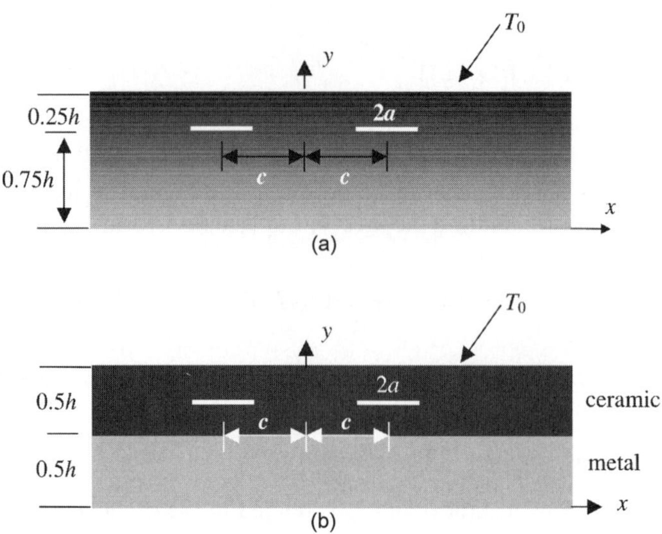

Fig. 2. A cracked FGM specimen and a non-FGM specimen.

factors (thermal stress intensity factors K_I, K_{II} and the thermal flow intensity factor K_0) are calculated at the tips $x = c - a$ (inner tip) and $x = c + a$ (outer tip) of the cracks.

We at first discuss an accuracy of the numerical results and convergence of the numerical results. In the foregoing analysis, we have treated the functionally graded material as N thin layers with each layer being assigned with slightly different material properties. The thickness of each thin layer is h_g/N, where h_g is the total thickness of the FGM material. To validate the numerical procedure, the steady TSIFs for different N values are tabulated in Tables 1 and 2. It is clear that as N increase the result converge to some steady values.

Figs. 3 and 4 show the temperature distribution as functions of y at a non-dimensional time 0.1 and ∞, respectively. The cross-sections are at $x = 0$, crack-tip, and crack center, respectively. It is seen that the

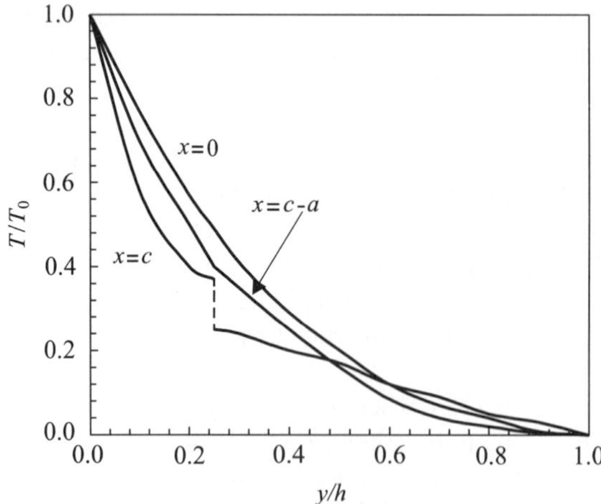

Fig. 3. Temperature distributions for different cross-sections at $t/(\rho_c c_c h^2/k_c) = 0.1$ ($c = 5a$).

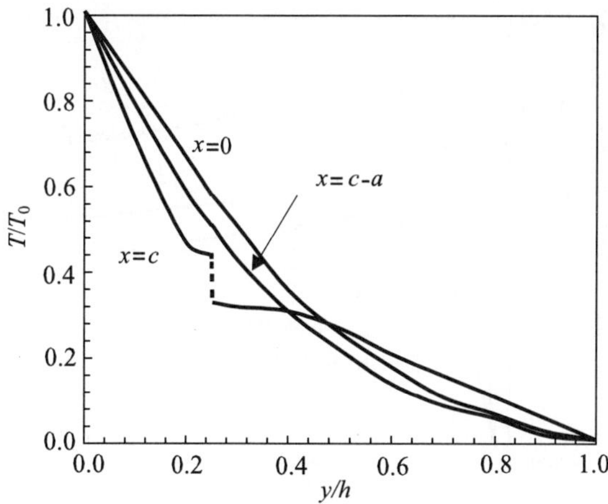

Fig. 4. Steady temperature distributions for different cross-sections ($c = 5a$).

temperature gradient along the through-thickness direction increases as the cross-section approaches the crack-tip. For a cross-section across the crack, there is a temperature jump, due to the fact that the cracks are thermal insulated.

Figs. 5–7 show, respectively, the thermal flow intensity factors and the TSIFs with time. The figures show that at time increases there are general peak values of the field intensity factor. The peak values reach at some time after thermal shock. The times at which the peak values are reached are slightly different for FGM and non-FGM.

Tables 3–8 show the peak and the steady values of the field intensity factor for various crack space c/a values. The results are also graphical depicted in Figs. 8–13. Note that $c/a = 1$ means that the inner tips of the two cracks have approached together. It is observed that as the space of the cracks increases the

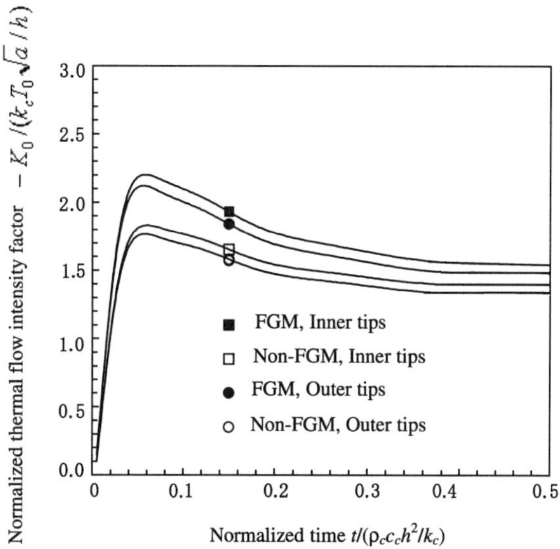

Fig. 5. Variation of thermal flow intensity factors with time ($c/a = 1.5$).

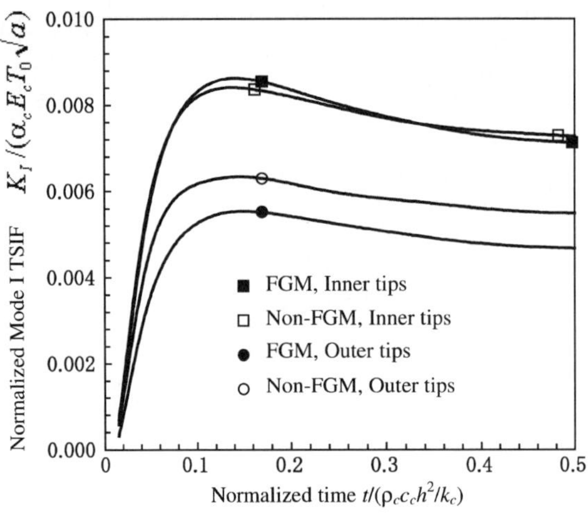

Fig. 6. Variation of mode I TSIFs with time ($c/a = 1.5$).

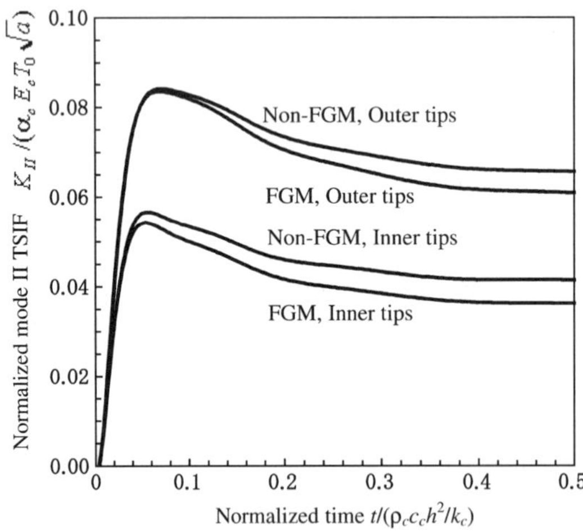

Fig. 7. Variation of mode II TSIFs with time ($c/a = 1.5$).

Table 3
Peak values of the thermal flow intensity factor $-K_0/(k_c T_0 \sqrt{a}/h)$ for FGM and non-FGM

c/a	Inner crack-tips		Outer crack-tips	
	FGM	Non-FGM	FGM	Non-FGM
1			2.43	2.02
1.1	2.83	2.35	2.18	1.81
1.2	2.47	2.05	2.15	1.79
1.3	2.33	1.93	2.13	1.78
1.5	2.20	1.83	2.12	1.77
2	2.12	1.77	2.11	1.76
3	2.11	1.76	2.10	1.76
5	2.10	1.76	2.10	1.76
∞	2.10	1.76	2.10	1.76

Table 4
Steady values of the thermal flow intensity factor $-K_0/(k_c T_0 \sqrt{a}/h)$ for FGM and non-FGM

c/a	Inner crack-tips		Outer crack-tips	
	FGM	Non-FGM	FGM	Non-FGM
1			1.82	1.64
1.1	2.02	1.82	1.54	1.39
1.2	1.75	1.58	1.51	1.37
1.3	1.64	1.48	1.49	1.35
1.5	1.54	1.39	1.47	1.33
2	1.47	1.33	1.45	1.32
3	1.44	1.31	1.44	1.31
5	1.43	1.30	1.43	1.30
∞	1.43	1.30	1.43	1.30

Table 5
Peak values of the mode I TSIF $K_I/\alpha_c E_c T_0 \sqrt{a}$ for FGM and non-FGM

c/a	Inner crack-tips		Outer crack-tips	
	FGM	Non-FGM	FGM	Non-FGM
1			0.0304	0.0308
1.1	0.0204	0.0190	0.0103	0.0104
1.2	0.0149	0.0141	0.00831	0.00857
1.3	0.0121	0.0116	0.00706	0.00752
1.5	0.00855	0.00846	0.00554	0.00634
2	0.00473	0.00554	0.00449	0.00569
3	0.00515	0.00635	0.00622	0.00723
5	0.00808	0.00877	0.00821	0.00888
∞	0.00825	0.00895	0.00825	0.00895

Table 6
Steady values of the mode I TSIF $K_I/\alpha_c E_c T_0 \sqrt{a}$ for FGM and non-FGM

c/a	Inner crack-tips		Outer crack-tips	
	FGM	Non-FGM	FGM	Non-FGM
1			0.0259	0.0270
1.1	0.0166	0.0164	0.00849	0.00897
1.2	0.0121	0.0120	0.00688	0.00747
1.3	0.00987	0.00979	0.00586	0.00656
1.5	0.00709	0.00730	0.00463	0.00553
2	0.00396	0.00483	0.00373	0.00488
3	0.00408	0.00523	0.00485	0.00593
5	0.00620	0.00713	0.00635	0.00727
∞	0.00650	0.00743	0.00650	0.00743

magnitudes of the mode I TSIFs steadily decrease, and then increase and ultimately tend to the values for the single crack problem. For small crack spaces, the mode I TSIFs have smaller values at outer tips than at inner tips. The opposite trends are found for mode large crack space. In that case the mode I TSIFs have larger values at outer tips than at inner tips.

Table 7
Peak values of the mode II TSIF $K_{II}/\alpha_c E_c T_0 \sqrt{a}$ for FGM and non-FGM

c/a	Inner crack-tips		Outer crack-tips	
	FGM	Non-FGM	FGM	Non-FGM
1			0.140	0.148
1.1	−0.0336	−0.0359	0.0984	0.0993
1.2	−0.0406	−0.0438	0.0928	0.0936
1.3	−0.0469	−0.0502	0.0889	0.0901
1.5	−0.0551	−0.0579	0.0835	0.0845
2	−0.0648	−0.0671	0.0784	0.0793
3	−0.0702	−0.0716	0.0740	0.0757
5	−0.0725	−0.0739	0.0728	0.0744
∞	−0.0730	−0.0744	0.0730	0.0744

Table 8
Steady values of the mode II TSIF $K_{II}/\alpha_c E_c T_0 \sqrt{a}$ for FGM and non-FGM

c/a	Inner crack-tips		Outer crack-tips	
	FGM	Non-FGM	FGM	Non-FGM
1			0.114	0.122
1.1	−0.0231	−0.0264	0.0728	0.0787
1.2	−0.0273	−0.0314	0.0675	0.0731
1.3	−0.0314	−0.0361	0.0648	0.0703
1.5	−0.0362	−0.0416	0.0608	0.0660
2	−0.0427	−0.0487	0.0561	0.0614
3	−0.0472	−0.0529	0.0523	0.0579
5	−0.0499	−0.0554	0.0503	0.0559
∞	−0.0502	−0.0558	0.0502	0.0558

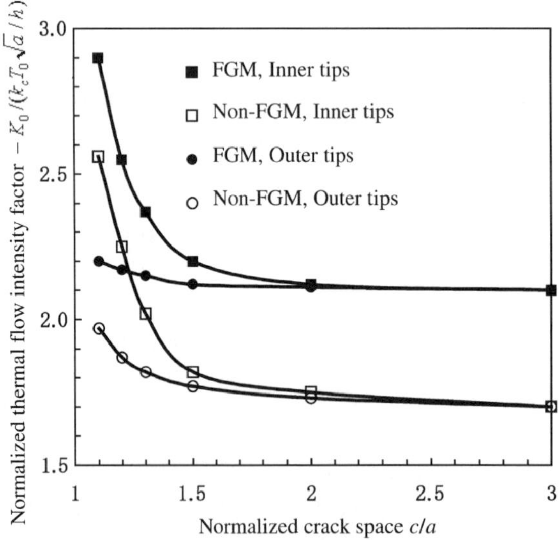

Fig. 8. Variation of peak thermal flow intensity factors with crack space.

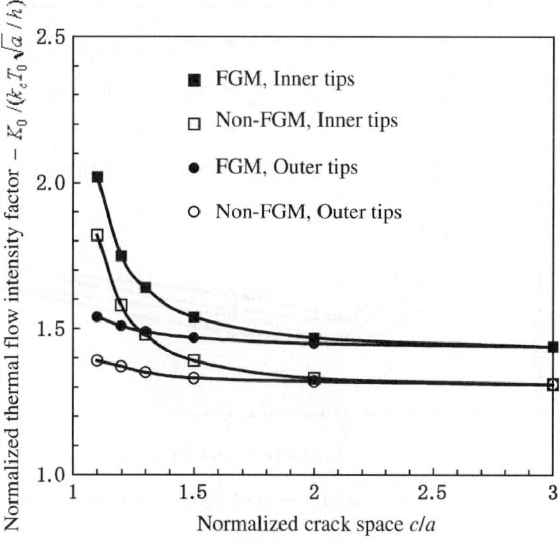

Fig. 9. Variation of steady thermal flow intensity factors with crack space.

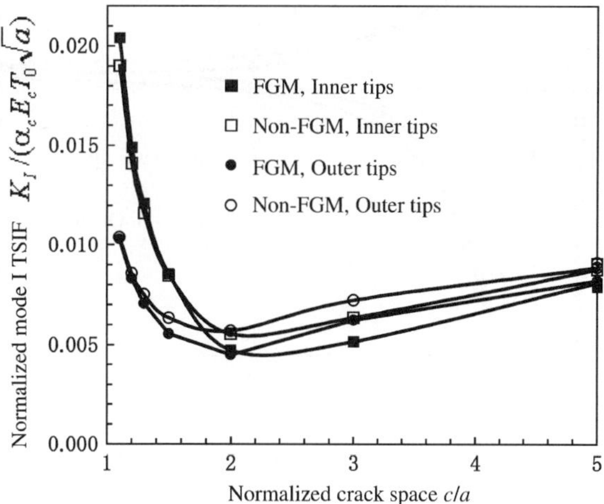

Fig. 10. Variation of peak mode I TSIFs with crack space.

Tabulated in Tables 7 and 8 are peak and steady mode II (shear) TSIFs for different crack space c/a values. The TSIFs at inner tips and outer tips show different c/a dependence. The results are also plotted in Figs. 12 and 13. The values of the TSIF increase for inner tips but decrease for outer tips with increasing crack space. As was expected, they approach the same value when c/a value is sufficiently large.

In considered problem the most pronounced values of TSIF are mode II TSIFs. It is observed that the TSIFs can be decreased considerable by using functionally graded materials. If one further considers the fact that the fracture toughness of a metal/ceramics mixture is larger than that of a pure ceramics, the fracture strength of a FGM is predicated to be much higher than that of a pure ceramics.

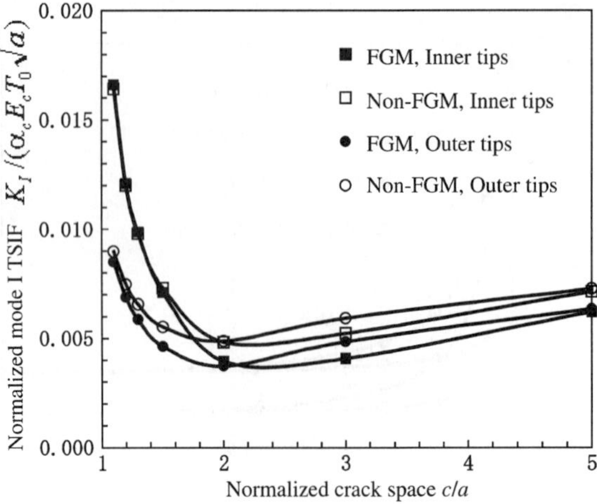

Fig. 11. Variation of steady mode I TSIFs with crack space.

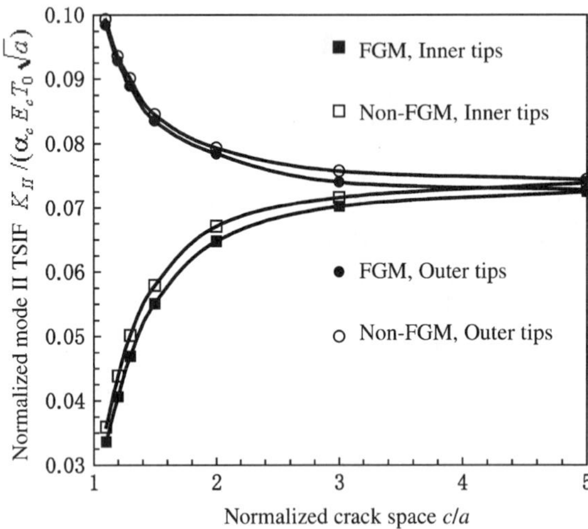

Fig. 12. Variation of peak mode II TSIFs with crack space (negative values for inner tips).

6. Conclusions

In this study, we considered collinear crack problems in functionally graded materials whose properties are described by arbitrary functions of coordinate y. The collinear cracks are placed to normal to the gradient direction. We used laminate model and the integral transform technique to reduce the problem to a set of singular integral equation. Numerical analysis was performed to evaluate the influence of crack space on the time-related thermoelastic singular stress fields for a functionally graded material plate and a non-FGM plate.

It has been experimentally observed that surface cracking in the material gradient direction at the ceramic side is the most common failure mode of a metal–ceramic FGM when it is subjected to a thermal shock. Another major failure mode is surface cracking that corresponds to crack parallel to the material

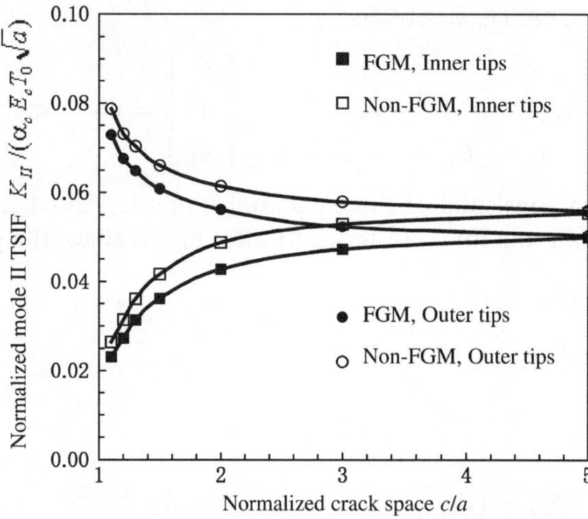

Fig. 13. Variation of steady mode II TSIFs with crack space (negative values for inner tips).

gradient direction. The crack may deflect from the original direction and then grow in a different direction. The problems of thermally induced surface cracking of FGMs with arbitrarily varied material properties need to be analyzed.

Appendix A

Make use of Eq. (1) and apply Laplace transform over the time variable t, Eqs. (7a)–(7c) can be reduced to:

$$\left.\begin{aligned}
k\frac{\partial^2 T^*}{\partial x^2} + k\frac{\partial^2 T^*}{\partial y^2} &= \rho c_v p T^* \\
c_{11}\frac{\partial^2 u^*}{\partial x^2} + c_{66}\frac{\partial^2 u^*}{\partial y^2} + (c_{12}+c_{66})\frac{\partial^2 v^*}{\partial x \partial y} &= \lambda\frac{\partial T^*}{\partial x} \\
c_{66}\frac{\partial^2 v^*}{\partial x^2} + c_{22}\frac{\partial^2 v^*}{\partial y^2} + (c_{12}+c_{66})\frac{\partial^2 u^*}{\partial x \partial y} &= \lambda\frac{\partial T^*}{\partial y}
\end{aligned}\right\}, \quad \text{for the } J\text{th layer} \tag{A.1}$$

where "p" is the Laplace transform parameter. A quantity with a superscript "$*$" denotes the Laplace transform. If the problem is steady state, p is zero.

In the following, the superscript "$*$" will be omitted for simplicity. Using Fourier transform to the space variable x, Eq. (A.1) may be solved to give the displacement and temperature in each layer of the functionally graded material:

$$\{V_J(x, y_J)\} = \frac{2}{\pi}\int_0^\infty [G(sx)](A_1 \quad A_2 \quad A_3)^{\mathrm{T}} F_J e^{s\xi y_J}\, \mathrm{d}s \tag{A.2}$$

where $[G(sx)] = \mathrm{diag}(\sin sx \ \cos sx \ \cos sx)$, F_J is an unknown function to be determined. Here, and in the following, the subscript J refers to the quantities associated with the materials occupying the Jth layer. The subscript j refers to the quantities associated with the jth interface.

Substituting from (A.2) into (A.1), we obtain

$$
\begin{bmatrix}
c_{66}\xi^2 - c_{11} & -(c_{12}+c_{66})\xi & \lambda/s \\
(c_{12}+c_{66})\xi & c_{22}\xi^2 - c_{66} & -\xi\lambda/s \\
0 & 0 & -k\xi^2 + k + \rho c_v p/s^2
\end{bmatrix}
\begin{Bmatrix} A_1 \\ A_2 \\ A_3 \end{Bmatrix} = 0, \quad \text{for the } J\text{th layer}
\tag{A.3}
$$

This is an eigenvalue problem consisting of three equations; non-trivial (A_1, A_2, A_3) exists if λ is a root of the determinant. There are six roots for λ. In terms of these eigenvalues, the general solution to Eq. (A.1) is

$$
\{V_J(x, y_J)\} = \frac{2}{\pi} \int_0^\infty [G(sx)][A_J]\{F_J\}\, \mathrm{d}s
\tag{A.4}
$$

where

$$
[A_J(y_J)] = \lfloor A_{k\alpha} e^{s\xi_\alpha y_J} \rfloor, \quad \{F_J\} = \{F_\alpha\}^{\mathrm{T}}, \quad k = 1, 2, 3, \quad \alpha = 1, \ldots, 6
\tag{A.5}
$$

The generalized vector $\{S_J\}$ defined in Eq. (2) can be obtained by using (1) and (A.4)

$$
\{S_J(x, y_J)\} = \frac{2}{\pi} \int_0^\infty [G(sx)][C_J]\{F_J\} s\, \mathrm{d}s
\tag{A.6}
$$

where $[C(y_J)]$ is a 3×6 matrix with the αth column being

$$
\begin{bmatrix}
c_{66}\xi_\alpha A_{1\alpha} - c_{66}A_{2\alpha} \\
c_{12}A_{1\alpha} + c_{22}\xi_\alpha A_{2\alpha} - \lambda A_{3\alpha}/s \\
-k\xi_\alpha A_{3\alpha}
\end{bmatrix}, \quad \alpha = 1, \ldots, 6
\tag{A.7}
$$

For $y_J = 0$ and $y_J = h_J$, Eq. (A.6) reads

$$
\{t_{j-1}(x)\} = \frac{2}{\pi} \int_0^\infty [G(sx)][C_J(0)]\{F_J\} s\, \mathrm{d}s
\tag{A.8a}
$$

$$
\{t_j(x)\} = \frac{2}{\pi} \int_0^\infty [G(sx)][C_J(h_J)]\{F_J\} s\, \mathrm{d}s
\tag{A.8b}
$$

where we have used the notations $\{t_{j-1}(x)\} = \{S_J(x, y_J = 0)\}$ and $\{t_j(x)\} = \{S_J(x, y_J = h_J)\}$. By applying inverse Fourier transform to Eqs. (A.8a) and (A.8b) one can express $\{F_J\}$ in terms of $\{t_j\}$ and $\{t_{j-1}\}$

$$
\{F_J\} = \frac{1}{s}[B_J]^{-1}\begin{Bmatrix} \bar{t}_j(s) \\ \bar{t}_{j-1}(s) \end{Bmatrix}
\tag{A.9}
$$

where

$$
[B_J] = \begin{bmatrix} C_J(h_J) \\ C_J(0) \end{bmatrix}
\tag{A.10}
$$

$$
\{\bar{t}_j(s)\} \int_0^\infty [G(sx)]\{t_j(x)\}\, \mathrm{d}x
\tag{A.11}
$$

Substituting (A.9) into (A.4), we obtain

$$
\{V_J(x, y_J)\} = \frac{2}{\pi} \int_0^\infty \frac{1}{s}[G(sx)][D_{aJ}\ D_{bJ}]\begin{Bmatrix} \bar{t}_j(s) \\ \bar{t}_{j-1}(s) \end{Bmatrix}\, \mathrm{d}s
\tag{A.12}
$$

where $[D_{aJ}]$ and $[D_{bJ}]$ are 4×4 matrices formed from

$$[D_{aJ}(y_J)\ D_{bJ}(y_J)] = [A_J(y_J)][B_J]^{-1} \tag{A.13}$$

It should be mentioned that since the generalized vectors $\{S_1(x,0)\} = \{t_0(x)\}$ for the bottom surface, and $\{S_N(x,h_N)\} = \{t_N(x)\}$ for the top surface of the plate are unknown, Eq. (A.12) may not be applied directly to the first layer and the last layer. Eqs. (A.10) and (A.11) should be modified according to Appendices B and C.

Appendix B

For lower surface and upper surface of the first layer, Eqs. (A.4) and (A.7) become

$$\{V_1(x,0)\} = \frac{2}{\pi} \int_0^\infty [G(sx)][A_1(0)]\{F_1\}\,\mathrm{d}s \tag{B.1}$$

$$\{t_0(x)\} = \frac{2}{\pi} \int_0^\infty [G(sx)][C_1(0)]\{F_1\}s\,\mathrm{d}s \tag{B.2}$$

$$\{t_1(x)\} = \frac{2}{\pi} \int_0^\infty [G(sx)][C_1(h_1)]\{F_1\}s\,\mathrm{d}s \tag{B.3}$$

Since the generalized vector $\{S_1(x,0)\} = \{t_0(x)\}$ for the bottom surface of the plate is unknown, we replace the boundary condition (3a) with

$$\{t_0(x)\} = \{S_1(x,0)\} = \boldsymbol{h}_a(\{V_1(x,0)\} - \{V_a(x)\}) \tag{B.4}$$

where

$$\boldsymbol{h}_a = \mathrm{diag}(0\quad 0\quad -h_a) \tag{B.5}$$

$$V_a(x) = (0\quad 0\quad T_a)^{\mathrm{T}} \tag{B.6}$$

Eqs. (B.1), (B.2) and (B.4) can be combined to show that

$$\boldsymbol{h}_a\{V_a(x)\} = \frac{2}{\pi} \int_0^\infty [G(sx)]\overline{C}_1\{F_1\}s\,\mathrm{d}s \tag{B.7}$$

in which

$$\overline{C}_1 = [C_1(0)] - \boldsymbol{h}_a[A_1(0)]/s \tag{B.8}$$

where $[C_1(0)]$ is calculated from Eq. (A.7) with $J = 1$.

Solve $\{F_1\}$ from Eqs. (B.3) and (B.7), we have

$$\{F_1\} = \frac{1}{s}[B_1]^{-1}\begin{Bmatrix} \bar{t}_1(s) \\ \bar{t}_0(s) \end{Bmatrix} \tag{B.9}$$

where

$$[B_1] = \begin{bmatrix} C_1(h_1) \\ \overline{C}_1 \end{bmatrix} \tag{B.10}$$

$$\{\bar{t}_0(s)\} = -\int_0^\infty [G(sx)]\boldsymbol{h}_a\{V_a(x)\}\,\mathrm{d}x = \left(0\quad 0\quad \int_0^\infty h_a \cos sx T_a(x)\,\mathrm{d}x\right)^{\mathrm{T}} \tag{B.11}$$

Appendix C

For the top layer we replace the boundary condition (3b) with

$$\{t_N(x)\} = \{S_N(x, h_N)\} = \boldsymbol{h}_b(\{V_N(x, h_N)\} - \{V_b(x)\}) \tag{C.1}$$

where

$$\boldsymbol{h}_b = \text{diag}(0 \quad 0 \quad h_b) \tag{C.2}$$

$$V_b(x) = (0 \quad 0 \quad T_b)^{\mathrm{T}} \tag{C.3}$$

Following the same procedure as described in Appendix B, $\{F_N\}$ can be found to be

$$\{F_N\} = \frac{1}{s}[B_N]^{-1}\left\{ \begin{array}{c} \bar{t}_N(s) \\ \bar{t}_{N-1}(s) \end{array} \right\} \tag{C.4}$$

where

$$[B_N] = \left[\begin{array}{c} \overline{C}_N \\ C_N(0) \end{array} \right] \tag{C.5}$$

$$\{\bar{t}_N(s)\} = -\int_0^\infty [G(sx)]\boldsymbol{h}_b\{V_b(x)\}\,\mathrm{d}x = \left(0 \quad 0 \quad -\int_0^\infty h_b \cos sx T_b(x)\,\mathrm{d}x \right)^{\mathrm{T}} \tag{C.6}$$

and where

$$\overline{C}_N = [C_N(h_N)] - \boldsymbol{h}_b[A_N(h_N)]/s \tag{C.7}$$

References

[1] Atkinson C, List RD. Steady state crack propagation into media with spatially varying elastic properties. Int J Eng Sci 1978;16:717–30.
[2] Dhaliwal RS, Singh BM. On the theory of elasticity of a non-homogeneous medium. J Elasticity 1978;8:3–22.
[3] Delale F, Erdogan F. The crack problem for a non-homogeneous plane. J Appl Mech 1983;50:609–14.
[4] Jin Z-H, Noda N. An internal crack parallel to the boundary of a non-homogeneous half plane under thermal loading. Int J Eng Sci 1993;31:793–806.
[5] Erdogan F, Wu BH. Analysis of FGM specimens for fracture toughness testing. In: Ceramic Transactions: Functionally Gradient Materials, vol. 34. Westerville, Ohio: American Ceramic Society; 1993. p. 39–46.
[6] Gu P, Asaro RJ. Cracks in functionally graded materials. Int J Solid Struct 1997;34:1–17.
[7] Gu P, Asaro RJ. Crack deflection in functionally graded materials. Int J Solid Struct 1997;34:3085–98.
[8] Erdogan F. Fracture of functionally graded materials. Composite Eng 1995;5:753–70.
[9] Jin Z-H, Batra RC. Some basic fracture mechanics concepts in functionally graded materials. J Mech Phys Solids 1996;44:1221–35.
[10] Jin Z-H, Batra RC. Stress intensity relaxation at the tip of an edge crack in a functionally graded materials subjected to a thermal shock. J Therm Stress 1996;19:317–39.
[11] Jin Z-H, Batra RC. Thermal fracture of ceramics with temperature-dependent properties. J Therm Stress 1998;21:157–76.
[12] Kokini K, Takeuchi YR. Multiple surface fracture of graded ceramic coatings. J Therm Stress 1998;21:715–25.
[13] Ueda S, Shindo Y. Cracking kinking in functionally graded materials due to an initial strain resulting from stress relaxation. J Therm Stress 2000;23:285–90.
[14] Li H, Lambros J, Cheeseman BA, Santare MH. Experimental investigation of the quasi-static fracture of functionally graded materials. Int J Solid Struct 2000;37:3715–32.
[15] Marur PR, Tippur HV. Numerically analysis of crack-tip fields in functionally graded materials with a crack normal to the elastic gradient. Int J Solid Struct 2000;37:5353–70.

[16] Wang BL, Han JC, Du SY. Crack problems in inhomogeneous composites subjected to dynamic loading. Int J Solid Struct 2000;37:1251–74.

[17] Wang BL, Han JC, Du SY. Multi-crack problems for inhomogeneous composites subjected to dynamic anti-plane loading. Int J Fract 1999;100:343–53.

[18] Wang BL, Han JC, Du SY. Crack problem in graded composite materials under transient thermal loading. J Therm Stress 2000;23:143–68.

[19] Wang BL, Han JC, Du SY. Fracture mechanics for multilayers with penny-shaped cracks under dynamic torsional loading. Int J Eng Sci 2000;38:893–901.

[20] Wang BL, Han JC, Du SY. Functionally graded penny-shaped cracks under dynamic loading. Theor Appl Fract Mech 1999;32:165–75.

[21] Wang BL, Han JC, Du SY. Thermoelastic fracture mechanics for nonhomogeneous material subjected to unsteady thermal load. J Appl Mech 2000;67:87–95.

[22] De J, Patra B. Propagation of two collinear Griffith cracks in an orthotropic strip. Eng Fract Mech 46:835–42.

[23] Zhou ZG, Bai YY, Zhang XW. Two collinear Griffith cracks subjected to uniform tension in infinitely long strip. Int J Solid Struct 1999;36:5597–609.

[24] Das S, Patra B, Debnath L. Stress intensity factor around two co-planar Griffith cracks in an orthotropic layer sandwiched between two identical orthotropic half planes. Int J Eng Sci 2000;38:121–33.

[25] Parihar KS, Sowdamini S. Three collinear cracks in an infinite elastic medium. Int J Eng Sci 1985;23:151–62.

[26] Shouetsu, Qian R. Thermal stresses around two collinear Griffith cracks in an adhesive layer between two dissimilar elastic half-planes. J Therm Stress 1995;18:185–96.

[27] Crump KS. Numerical Inversion of Laplace transforms using a Fourier series Approximation. J Assoc Comput Machine 1976;23(1):89–96.

[28] Honig G, Hirdes U. A method for the numerical inversion of the Laplace transform. J Comput Appl Math 1984;10:113–32.

[29] Christensen RM. Mechanics of Composite Materials. New York: Wiley; 1979.